*The Statistical Analysis of Experimental Data*

# The Statistical Analysis of Experimental Data

JOHN MANDEL

National Bureau of Standards
Washington, D.C.

DOVER PUBLICATIONS, INC., *New York*

Published in Canada by General Publishing Company, Ltd., 30 Lesmill Road, Don Mills, Toronto, Ontario.
Published in the United Kingdom by Constable and Company, Ltd.

This Dover edition, first published in 1984, is an unabridged and corrected republication of the work first published by Interscience Publishers, a division of John Wiley & Sons, N.Y., in 1964.

Manufactured in the United States of America
Dover Publications, Inc., 31 East 2nd Street, Mineola, N.Y. 11501

**Library of Congress Cataloging in Publication Data**

Mandel, John, 1914–
    The statistical analysis of experimental data.

    Reprint. Originally published: New York : Interscience, 1964.
    Bibliography: p.
    Includes index.
    1. Science—Statistical methods.   2. Mathematical statistics.   I. Title.
Q175.M343   1984          507.2          83-20599
ISBN 0-486-64666-1

# *Preface*

The aim of this book is to offer to experimental scientists an appreciation of the statistical approach to data analysis. I believe that this can be done without subjecting the reader to a complete course in mathematical statistics. However, a thorough understanding of the *ideas* underlying the modern theory of statistics is indispensable for a meaningful use of statistical methodology. Therefore, I have attempted to provide the reader with such an understanding, approaching the subject from the viewpoint of the physical scientist.

Applied statistics is essentially a tool in the study of other subjects. The physical scientist interested in statistical inference will look for an exposition written in terms of his own problems, rather than for a comprehensive treatment, even on an elementary level, of the entire field of mathematical statistics. It is not sufficient for such an exposition merely to adopt the *language* of the physical sciences in the illustrative examples. The very *nature* of the statistical problem is likely to be different in physics or chemistry than it is in biology, agronomy, or economics. I have attempted, in this book, to follow a pattern that flows directly from the problems that confront the physical scientist in the analysis of experimental data.

The first four chapters, which present fundamental mathematical definitions, concepts, and facts will make the book largely self-contained.

The remainder of the book, starting with Chapter 5, deals with statistics primarily as an interpretative tool. The reader will learn that some of the most popular methods of statistical design and analysis have a more limited scope of applicability than is commonly realized. He is urged, throughout this book, to analyze and re-analyze his *problem* before analyzing his *data*, to be certain that the data analysis fits the objectives and nature of the experiment.

An important area of application of statistics is the emerging field of "materials science." Organizations such as the National Aeronautics and Space Administration, the National Bureau of Standards, and the American Society for Testing and Materials are actively engaged in studies of the properties of materials. Such studies require, in the long run, the aid of fully standardized test methods, and standardization of this type requires, in turn, the use of sound statistical evaluation procedures.

I have devoted a lengthy chapter to the statistical study of test methods and another chapter to the comparison of two or more alternative methods of test. While these chapters are primarily of practical interest, for example, in standardization work, they should be of value also to the laboratory research worker.

The large number of worked-out examples in this book are, with very few exceptions, based on genuine data obtained in the study of real laboratory problems. They serve a dual purpose. Primarily they are intended to establish the indispensable link between the physical or chemical problem and the statistical technique proposed for its solution. Secondly, they illustrate the numerical computations involved in the statistical treatment of the data. I deliberately discuss some problems to which the statistical analysis has provided no conclusive solution. Such problems are excellent illustrative material, because they show how a statistical analysis reveals the *inadequacy* of some data, i.e., their inability to provide the sought-for answers. Examples of this type reveal the role of statistics as a diagnostic tool. A careful reading of the examples should aid the reader in achieving both a better insight into the statistical procedures and greater familiarity with their technical details.

It is a pleasant task to express my appreciation to a number of my colleagues who were kind enough to read substantial parts of the manuscript. In particular I wish to thank Charles E. Weir, Frank L. McCrackin, and Mary G. Natrella, whose many suggestions were invaluable.

Thanks are also due to Samuel G. Weissberg, Theodore W. Lashof, Grant Wernimont, Max Tryon, Frederic J. Linnig, and Mary Nan Steel, all of whom helped with useful discussions, suggestions and encouragement. Robert F. Benjamin did a valuable job in editing.

In expressing my appreciation to all of these individuals, I wish to emphasize that any shortcomings of the book are my own. I shall welcome comments and suggestions from the readers.

JOHN MANDEL

*June, 1964*
*Washington, D.C.*

# Contents

# INTRODUCTION

To say that measurement is at the heart of modern science is to utter a commonplace. But it is no commonplace to assert that the ever increasing importance of measurements of the utmost precision in science and technology has created, or rather reaffirmed, the need for a systematic science of data analysis. I am sorry to say that such a science does not, as yet, exist. It is true that many scientists are superb data analysts. They acquire this skill by combining a thorough understanding of their particular field with a knowledge, based on long experience, of how data behave and, alas, too often also misbehave. But just as the fact that some people can calculate exceedingly well with an abacus is no reason for condemning electrical and electronic computers, so the existence of some excellent data analysts among scientists should not deter us from trying to develop a true science of data analysis. What has been, so far, mostly a combination of intuition and experience should be transformed into a systematic body of knowledge with its own principles and working rules. Statistical principles of inference appear· to constitute a good starting point for such an enterprise. The concept of a frequency distribution, which embodies the behavior of chance fluctuations, is a felicitous one for the description of many pertinent aspects of measurement. If this concept is combined with the principle of least squares, by which the inconsistencies of measurements are compensated, and with the modern ideas underlying "inverse probability," which allow us to make quantitative statements about the causes of observed chance events, we obtain an impressive body of useful knowledge.   Nevertheless, it is by no means certain that a systematic science of data analysis, if and when it finally will be developed, will be based *exclusively* on probabilistic concepts.   Undoubtedly probability will always play an important role in data analysis but it is rather likely that principles of a different nature will also be invoked in the final formation of such a science.   In the meantime we must make use of whatever methods are available to us for a meaningful approach to the analysis of experimental data.   This book is an attempt to present, not only some of these methods, but also, and perhaps with even greater emphasis, the type of reasoning that underlies them.

The expression "design and analysis of experiments," which couples two major phases of experimentation, is often encountered in statistics.   In a general sense, the design of an experiment really involves the entire reasoning process by which the experimenter hopes to link the facts he wishes to learn from the experiment with those of which he is already reasonably certain.   Experimentation without design seldom occurs in science and the analysis of the data necessarily reflects the design of the experiment. In this book we shall be concerned implicitly with questions of experimental design, in the general sense just described, and also in a more specific sense: that of design as the structural framework of the experiment.   The structure given by the design is generally also used to exhibit the data, and in this sense too design and analysis are indissolubly linked.

The expression "statistical design of experiments" has an even more specific meaning.   Here, *design* refers to completely defined arrangements, prescribing exactly the manner in which samples for test or experimentation shall be selected, or the order in which the measurements are to be made, or special spatial arrangements of the objects of experimentation, or a combination of these things.   The object of these schemes is to compensate, by an appropriate arrangement of the experiment, for the many known or suspected sources of bias that can vitiate the results.   A good example of the pertinence of the statistical design of experiments is provided by the road testing of automobile tires to compare their rates of tread wear.   Differences in road and climatic conditions from one test period to another, effects of the test vehicle and of the wheel position in which the tire is mounted—these and many other variables constitute systematic factors that affect the results.   Through a carefully planned system of rotation of the tires among the various wheels in successive test periods it is possible to compensate for these disturbances.   Thus, the statistical design of experiments is particularly important where limitations in space or time impose restrictions in the manner in which the experiment can be carried out.

The analysis of data resulting from experiments designed in accordance with specific statistical designs is mostly predetermined: it aims mainly at compensation for various sources of bias. Data analysis as dealt with in this book is taken in a wider sense, and concerns the manner in which diagnostic inferences about the basic objectives of the experiment can be drawn from the data. We will attempt an exposition in some depth of the underlying rationale of data analysis in general, and of the basic techniques available for that purpose.

To most experimenters the object of primary interest is the scientific interpretation of their findings; they do not, in general, consider data as a subject of intrinsic value, but only as a means to an end: to measure properties of interest and to test scientific hypotheses. But experimental data have, in some sense, intrinsic characteristics. Statistics is concerned with the behavior of data under varying conditions. The basic idea underlying the application of statistics to scientific problems is that a thorough knowledge of the behavior of data is a prerequisite to their scientific interpretation. I believe that an increasing number of scientists are gradually adopting this view. I also believe that the field of applied statistics, and even of theoretical statistics, can be appreciably enriched by practicing scientists who recognize the statistical nature of some of their problems. In the end, the successful use of statistical methods of data analysis will be determined both by the interest of scientists in this area and by the skill of statisticians in solving the problems proposed by the scientists.

While mathematics plays an important part in statistics, it is fortunately not necessary to be an accomplished mathematician to make effective use of statistical methods. A knowledge of ordinary algebra, the more elementary aspects of calculus and analytical geometry, combined with a willingness to think systematically, should suffice to acquire an excellent understanding of statistical methods. I use the word "understanding" deliberately for *without it*, the application of statistical techniques of data analysis may well become a treacherous trap; *with it*, the use of statistics becomes not only a powerful tool for the interpretation of experiments but also a task of real intellectual gratification. It is hoped that this book will, in some small measure at least, contribute to a more widespread appreciation of the usefulness of statistics in the analysis of experimental results.

# chapter 1

# THE NATURE OF
# MEASUREMENT

## 1.1 TYPES OF MEASUREMENT

The term measurement, as commonly used in our language, covers many fields of activities. We speak of measuring the diameter of the sun, the mass of an electron, the intelligence of a child, and the popularity of a television show. In a very general sense all of these concepts may be fitted under the broad definition, given by Campbell (1), of measurement as the "assignment of numerals to represent properties." But a definition of such degree of generality is seldom useful for practical purposes.

In this book the term measurement will be used in a more restricted sense: we will be concerned with measurement in the physical sciences only, including in this category, however, the technological applications of physics and chemistry and the various fields of engineering. Furthermore, it will be useful to distinguish between three types of measurements.

1. Basic to the physical sciences is the determination of fundamental constants, such as the velocity of light or the charge of the electron. Much thought and experimental work have gone into this very important but rather specialized field of measurement. We will see that statistical methods of data analysis play an important role in this area.

2. The purpose behind most physical and chemical measurements is to characterize a particular material or physical system with respect to a given property. The material might be an ore, of which it is required to determine the sulfur content. The physical system could be a microscope,

of which we wish to determine the magnification factor. Materials subjected to chemical analysis are generally homogeneous gases, liquids or solids, or finely ground and well-mixed powders of known origin or identity. Physical systems subjected to measurement consist mostly of specified component parts assembled in accordance with explicit specifications. A careful and precise description of the material or system subjected to measurement as well as the property that is to be measured is a necessary requirement in all physical science measurements. In this respect, the measurements in the second category do not differ from those of category 1. The real distinction between the two types is this: a method of type 1 is in most cases a specific procedure, applicable only to the determination of a single fundamental constant and aiming at a unique number for this constant, whereas a method of type 2 is a technique applicable to a large number of objects and susceptible of giving any value within a certain range. Thus, a method for the measurement of the velocity of light *in vacuo* need not be applicable to measuring other velocities, whereas a method for determining the sulfur content of an ore should retain its validity for ores with varying sulfur contents.

3. Finally, there are methods of *control* that could be classified as measurements, even though the underlying purpose for this type of measurement is quite different from that of the two previous types. Thus, it may be necessary to make periodic determinations of the pH of a reacting mixture in the production of a chemical or pharmaceutical product. The purpose here is not to establish a value of intrinsic interest but rather to insure that the fluctuations in the pH remain within specified limits. In many instances of this type, one need not even know the value of the measurement since an automatic mechanism may serve to control the desired property.

We will not be concerned, in this book, with measurements of type 3. Our greatest emphasis by far will be on measurements belonging to the second type. Such measurements involve three basic elements: a material or a physical system, a physical or chemical property, and a procedure for determining the value of such a property for the system considered. Underlying this type of measurement is the assumption that the measuring procedure must be applicable for a range of values of the property under consideration.

## 1.2  MEASUREMENT AS A PROCESS

The process of assigning numerals to properties, according to Campbell's definition, is of course not an arbitrary one. What is actually involved is a set of rules to be followed by the experimenter. In this respect, the measurement procedure is rather similar to a manufacturing process. But whereas a manufacturing process leads to a physical object, the measuring process has as its end result a mere number (or an ordered set of numbers). The analogy can be carried further. Just as in a manufacturing process, environmental conditions (such as the temperature of a furnace, or the duration of a treatment) will in general affect the quality of the product, so, in the measuring process, environmental conditions will also cause noticeable variations in the numbers resulting from the operation. These variations have been referred to as *experimental error*. To the statistician, experimental error is distinctly different from mistakes or blunders. The latter result from *departures* from the prescribed procedure. Experimental error, on the other hand, occurs even when the rules of the measuring process are strictly observed, and it is due to whatever looseness is inherent in these rules. For example, in the precipitation step of a gravimetric analysis, slight differences in the rate of addition of the reagent or in the speed of agitation are unavoidable, and may well affect the final result. Similarly, slight differences in the calibration of spectrophotometers, even of the same type and brand, may cause differences in the measured value.

## 1.3  MEASUREMENT AS A RELATION

Limiting our present discussion to measurements of the second of the three types mentioned in Section 1.1, we note an additional aspect of measurement that is of fundamental importance. Measurements of this type involve a *relationship*, similar to the relationship expressed by a mathematical function. Consider for example a chemical analysis made by a spectrophotometric method. The property to be measured is the concentration, $c$, of a substance in solution. The measurement, $T$, is the ratio of the transmitted intensity, $I$, to the incident intensity, $I_0$. If Beer's law (3) applies, the following relation holds:

$$T = \frac{I}{I_0} = e^{-kc} \tag{1.1}$$

Thus, the measured quantity, $T$, is expressible as a mathematical function of the property to be measured, $c$. Obviously, the two quantities, $T$ and $c$, are entirely distinct. It is only because of a relationship such as Eq. 1.1 that we can also claim to have measured the concentration $c$ by this process.

Many examples can be cited to show the existence of a relationship in measuring processes. Thus, the amount of bound styrene in synthetic rubber can be measured by the refractive index of the rubber. The

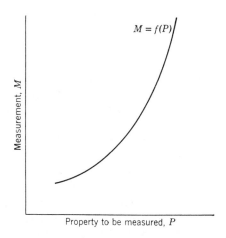

Fig. 1.1   A monotonic relationship associated with a measuring process.

measurement of forces of the order of magnitude required for rocket propulsion is accomplished by determining changes in the electrical properties of proving rings subjected to these forces.   In all these cases, three elements are present: a *property* to be determined ($P$), a *measured quantity* ($M$), and a *relationship* between these two quantities:

$$M = f(P) \tag{1.2}$$

Figure 1.1 is a graphical representation of the relationship associated with a measuring process.

## 1.4   THE ELEMENTS OF A MEASURING PROCESS

The description of a measuring process raises a number of questions. In the first place, the quantity $P$ requires a definition.   In many cases $P$ cannot be defined in any way other than as the result of the measuring process itself; for this particular process, the relationship between measurement and property then becomes the identity $M \equiv P$; and the study of any new process, $M'$, for the determination of $P$ is then essentially the study of the relationship of two measuring processes, $M$ and $M'$.

In some technological problems, $P$ may occasionally remain in the form of a more or less vague concept, such as the degree of vulcanization of rubber, or the surface smoothness of paper.   In such cases, the relation Eq. 1.2 can, of course, never be known.   Nevertheless, this relation remains useful as a conceptual model even in these cases, as we will see in greater detail in a subsequent chapter.

Cases exist in which the property of interest, $P$, is but one of the parameters of a *statistical distribution function*, a concept which will be defined in Chapter 3.   An example of such a property is furnished by the number average molecular weight of a polymer.   The weights of the molecules of the polymer are not all identical and follow in fact a statistical distribution function.   The number average molecular weight is the average of the weights of all molecules.   But the existence of this distribution function makes it possible to define other parameters of the distribution that are susceptible of measurement, for example, the weight average molecular weight.   Many technological measuring processes fall in this category.   Thus, the elongation of a sheet of rubber is generally determined by measuring the elongation of a number of dumbbell specimens cut from the sheet.   But these individual measurements vary from specimen to specimen because of the heterogeneity of the material, and the elongation of the entire sheet is best defined as a central parameter of the statistical distribution of these individual elongations.   This central parameter is not necessarily the arithmetic average.   The *median** is an equally valid parameter and may in some cases be more meaningful than the average.

A second point raised by the relationship aspect of measuring processes concerns the nature of Eq. 1.2.   Referring to Fig. 1.1, we see that the function, in order to be of practical usefulness, must be *monotonic*, i.e., $M$ must either consistently increase or consistently decrease, when $P$ increases. Figure 1.2 represents a non-monotonic function; two different values, $P_1$ and $P_2$ of the property give rise to the same value, $M$, of the measurement. Such a situation is intolerable unless the process is limited to a range of $P$ values, such as $PP'$, within which the curve is indeed monotonic.

The relation between $M$ and $P$ is specific for any particular measuring process.   It is generally different for two different processes, even when the property $P$ is the same in both instances.   As an example we may consider two different analytical methods for the determination of per cent chromium in steel, the one gravimetric and the other spectrophotometric. The property $P$, per cent chromium, is the same in both cases; yet the curve relating measurement and property is different in each case.   It is important to realize that this curve varies also with the type of material or the nature of the physical system.   The determination of sulfur in an ore is an entirely different process from the determination of sulfur in vulcanized rubber, even though the property measured is per cent sulfur in both cases.   An instrument that measures the smoothness of paper may react differently for porous than for non-porous types of paper.   In order that the relationship between a property and a measured quantity be

---

* The median is a value which has half of all measurements below it, and half of them above it.

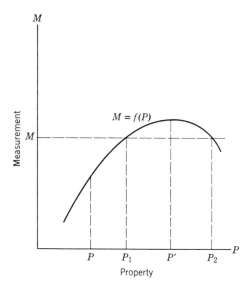

Fig. 1.2   A non-monotonic function—two different values, $P_1$ and $P_2$, of the property give rise to the same value, $M$, of the measurement.

sharply determined, it is necessary to properly identify the types of materials to which the measuring technique is meant to apply.   Failure to understand this important point has led to many misunderstandings. A case in point is the problem that frequently arises in technological types of measurement, of the *correlation* between different tests.   For example, in testing textile materials for their resistance to abrasion one can use a silicon carbide abrasive paper or a blade abradant.   Are the results obtained by the two methods correlated?   In other words, do both methods rank different materials in the same order?   A study involving fabrics of different constructions (4) showed that there exists no unique relationship between the results given by the two procedures.   If the fabrics differ from each other only in terms of one variable, such as the number of yarns per inch in the filling direction, a satisfactory relationship appears.   But for fabrics that differ from each other in a number of respects, the correlation is poor or non-existent.   The reason is that the two methods differ in the kind of abrasion and the rate of abrasion. Fabrics of different types will therefore be affected differently by the two abradants.   For any one abradant, the relation between the property and the measurement, considered as a curve, depends on the fabrics included in the study.

Summarizing so far, we have found that a measuring process must deal with a properly identified property $P$; that it involves a properly specified procedure yielding a measurement $M$; that $M$ is a monotonic function of $P$ over the range of $P$ values to which the process applies; and that the systems or materials subjected to the process must belong to a properly circumscribed class.    We must now describe in greater detail the aspect of measurement known as experimental error.

## 1.5   SOURCES OF VARIABILITY IN MEASUREMENT

We have already stated that error arises as the result of fluctuations in the conditions surrounding the experiment.    Suppose that it were possible to "freeze" temporarily all environmental factors that might possibly affect the outcome of the measuring process, such as temperature, pressure, the concentration of reagents, the amount of friction in the measuring instrument, the response time of the operator, and others of a similar type.    Variation of the property $P$ would then result in a mathematically defined response in the measurement $M$, giving us the curve $M = f_1(P)$. Such a curve is shown in Fig. 1.3.    Now, we "unfreeze" the surrounding world for just a short time, allowing all factors enumerated above to change slightly and then "freeze" it again at this new state.    This time we will obtain a curve $M = f_2(P)$ which will be slightly different from the first curve, because of the change in environmental conditions.    To perceive the true nature of experimental error, we merely continue

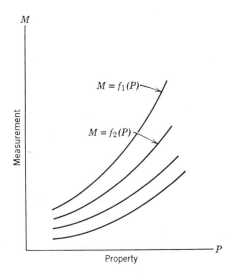

**Fig. 1.3**   Bundle of curves representing a measuring process.

indefinitely this conceptual experiment of "freezing" and "unfreezing" the environment for each set of measurements determining the curve. The process will result in a bundle of curves, each one corresponding to a well defined, though unknown state of environmental conditions. The entirety of all the curves in the bundle, corresponding to the infinite variety in which the environmental factors can vary for a given method of measurement, constitutes a mathematical representation of the measuring process defined by this method (2). We will return to these concepts when we examine in detail the problem of evaluating a method of measurement. At this time we merely mention that the view of error which we have adopted implies that the variations of the environmental conditions, though partly unpredictable, are nevertheless subject to some limitations. For example, we cannot tolerate that during measurements of the density of liquids the temperature be allowed to vary to a large extent. It is this type of limitation that is known as *control* of the conditions under which any given type of measurement is made. The width of our bundle of curves representing the measuring process is intimately related to the attainable degree of control of environmental conditions. This attainable degree of control is in turn determined by the *specification* of the measuring method, i.e., by the exact description of the different operations involved in the method, with prescribed *tolerances*, beyond which the pertinent environmental factors are not allowed to vary. Complete control is humanly impossible because of the impossibility of even being aware of all pertinent environmental factors. The development of a method of measurement is to a large extent the discovery of the most important environmental factors and the setting of tolerances for the variation of each one of them (6).

## 1.6  SCALES OF MEASUREMENT

The preceding discussion allows us to express Campbell's idea with greater precision. For a given measuring process, the assignment of numbers is not made in terms of "properties," but rather in terms of the different "levels" of a given property. For example, the different metal objects found in a box containing a standard set of "analytical weights" represent different "levels" of the single property "weight." Each of the objects is assigned a number, which is engraved on it and indicates its weight. Is this number unique? Evidently not, since a weight bearing the label "5 grams" could have been assigned, with equal justification, the numerically different label "5000 milligrams." Such a change of label is known as a *transformation of scale*. We can clarify our thoughts about this subject by visualizing the assignment of numbers to the levels of a property as a sort of "mapping" (5): the object of interest, analogous to the geographic territory to be mapped, is the property under study; the *representation* of this property is a scale of numbers, just as the map is a representation of the territory. But a map must have *fidelity* in the sense that relationships inferred from it, such as the relative positions of cities and roads, distances between them, etc., be at least approximately correct. Similarly, the relationships between the numbers on a scale representing a measuring process must be a faithful representation of the corresponding relations between the measured properties. Now, the relationships that exist between numbers depend upon the particular system of numbers that has been selected. For example, the system composed of positive integers admits the relations of equality, of directed inequality (greater than, or smaller than), of addition and of multiplication. In regard to subtraction and division the system imposes some limitations since we cannot subtract a number from a smaller number and still remain within the system of positive numbers. Nor can we divide a number by another number not its divisor and still remain within the system of integers. Thus, a property for which the operations of subtraction and division are meaningful even when the results are negative or fractional should not be "mapped" onto the system of positive integers. For each property, a system must be selected in which both the symbols (numbers) and the relations and operations defined for those symbols are meaningful counterparts of similar relations and operations in terms of the measured property. As an example, we may consider the temperature scale, say that known as "degrees Fahrenheit." If two bodies are said to be at temperatures respectively equal to 75 and 100 degrees Fahrenheit, it is a meaningful statement that the second is at a temperature 25 degrees higher than the first. It is *not* meaningful to state that the second is at a temperature 4/3 that of the first, even though the statement is true enough as an arithmetic fact about the numbers 75 and 100. The reason lies in the physical definition of the Fahrenheit scale, which involves the assignment of fixed numerals to only two well-defined physical states and a subdivision of the interval between these two numbers into equal parts. The operation of division for this scale is meaningful only when it is carried out on *differences* between temperatures, not on the temperatures themselves. Thus, if one insisted on computing the *ratio* of the temperature of boiling water to that of freezing water, he would obtain the value 212/32, or 6.625 using the Fahrenheit scale; whereas, in the Centigrade scale he would obtain 100/0, or infinity. Neither number expresses a physical reality.

The preceding discussion shows that for each property we must select an appropriate scale. We now return to our question of whether this scale is in any sense unique. In other words, could two or more essentially different scales be equally appropriate for the representation of the same property? We are all familiar with numerous examples of the existence of alternative scales for the representation of the same property: lengths can be expressed in inches, in feet, in centimeters, even in light years. Weights can be measured in ounces, in pounds, in kilograms. In these cases, the passage of one scale to another, the transformation of scales, is achieved by the simple device of multiplying by a fixed constant, known as the *conversion factor*. Slightly more complicated is the transformation of temperature scales into each other. Thus, the transformation of degrees Fahrenheit into degrees Centigrade is given by the relation

$$°C = \frac{5}{9} (°F) - \frac{160}{9} \tag{1.3}$$

Whereas the previous examples required merely a proportional relation, the temperature scales are related by a non-proportional, but still *linear* equation. This is shown by the fact that Eq. 1.3, when plotted on a graph, is represented by a straight line.

Are *non-linear* transformations of scale permissible? There is no reason to disallow such transformations, for example, of those relations involving powers, polynomials, logarithms, trigonometric functions, etc. But it is important to realize that not all the mathematical operations that can be carried out on numbers in any particular scale are necessarily meaningful in terms of the property represented by this scale. Transformations of scale can have drastic repercussions in regard to the pertinence of such operations. For example, when a scale $x$ is transformed into its logarithm, $\log x$, the operation of addition on $x$ has no simple counterpart in $\log x$, etc. Such changes also affect the mathematical form of relationships between different properties. Thus, the ideal gas law $pV = RT$ which is a multiplicative type of relation becomes an additive one, when logarithmic scales are used for the measurement of pressure, volume, and temperature:

$$\log p + \log V = \log R + \log T \tag{1.4}$$

Statistical analyses of data are sometimes considerably simplified through the choice of particular scales. It is evident, however, that a transformation of scale can no more change the intrinsic characteristics of a measured property or of a measuring process for the evaluation of this property than the adoption of a new map can change the topographic aspects of a terrain. We will have an opportunity to discuss this important matter further in dealing with the comparison of alternative methods of measurement for the same property.

## 1.7 SUMMARY

From a statistical viewpoint, measurement may be considered as a process operating on a physical system. The outcome of the measuring process is a number or an ordered set of numbers. The process is influenced by environmental conditions, the variations of which constitute the major cause for the uncertainty known as experimental error.

It is also useful to look upon measurement as a relationship between the magnitude $P$ of a particular property of physical systems and that of a quantity $M$ which can be obtained for each such system. The relationship between the measured quantity $M$ and the property value $P$ depends upon the environmental conditions that prevail at the time the measurement is made. By considering the infinity of ways in which these conditions can fluctuate, one arrives at the notion of a bundle of curves, each of which represents the relation between $M$ and $P$ for a fixed set of conditions.

The measured quantity $M$ can always be expressed in different numerical scales, related to each other in precise mathematical ways. Having adopted a particular scale for the expression of a measured quantity, one must always be mindful of the physical counterpart of the arithmetic operations that can be carried out on the numbers of the scale; while some of these operations may be physically meaningful, others may be devoid of physical meaning. Transformations of scale, i.e., changes from one scale to another, are often useful in the statistical analysis of data.

## REFERENCES

1. Campbell, N. R., *Foundations of Science*, Dover, New York, 1957.
2. Mandel, J., "The Measuring Process," *Technometrics*, **1**, 251–267 (Aug. 1959).
3. Meites, Louis, *Handbook of Analytical Chemistry*, McGraw-Hill, New York, 1963.
4. Schiefer, H. F., and C. W. Werntz, "Interpretation of Tests for Resistance to Abrasion of Textiles," *Textile Research Journal*, **22**, 1–12 (Jan. 1952).
5. Toulmin, S., *The Philosophy of Science, An Introduction*, Harper, New York, 1960.
6. Youden, W. J., "Experimental Design and the ASTM Committees," *Research and Standards*, 862–867 (Nov. 1961).

*chapter 2*

# STATISTICAL MODELS AND
# STATISTICAL ANALYSIS

## 2.1 EXPERIMENT AND INFERENCE

When statistics is applied to experimentation, the results are often stated in the language of mathematics, particularly in that of the theory of probability. This mathematical mode of expression has both advantages and disadvantages. Among its virtues are a large degree of objectivity, precision, and clarity. Its greatest disadvantage lies in its ability to hide some very inadequate experimentation behind a brilliant facade. Let us explain this point a little further. Most statistical procedures involve well described formal computations that can be carried out on *any* set of data satisfying certain formal *structural* requirements. For example, data consisting of two columns of numbers, $x$ and $y$, such that to each $x$ there corresponds a certain $y$, and vice-versa, can always be subjected to calculations known as *linear regression*, giving rise to at least two distinct straight lines, to correlation analysis, and to various tests of significance. Inferences drawn from the data by these methods may be not only incorrect but even thoroughly misleading, despite their mathematically precise nature. This can happen either because the assumptions underlying the statistical procedures are not fulfilled or because the problems connected with the data were of a completely different type from those for which the particular statistical methods provide useful information. In other cases the inferences may be pertinent and valid, so far as they go, but they may fail to call attention to the basic insufficiency of the experiment.

Indeed, most sets of data provide some useful information, and this information can often be expressed in mathematical or statistical language, but this is no guarantee that the information actually desired has been obtained.   The evaluation of methods of measurement is a case in point. We will discuss in a later chapter the requirements that an experiment designed to evaluate a test method must fulfill in order to obtain not only *necessary* but also *sufficient* information.

The methods of statistical analysis are intimately related to the problems of *inductive inference*: drawing inferences from the particular to the general.   R. A. Fisher, one of the founders of the modern science of statistics, has pointed to a basic and most important difference between the results of induction and deduction (2).   In the latter, conclusions based on partial information are always correct, despite the incompleteness of the premises, provided that this partial information is itself correct.   For example, the theorem that the sum of the angles of a plane triangle equals 180 degrees is based on certain postulates of geometry, but it does not necessitate information as to whether the triangle is drawn on paper or on cardboard, or whether it is isosceles or not.   If information of this type is subsequently added, it cannot possibly alter the fact expressed by the theorem.   On the other hand, inferences drawn by induction from incomplete information may be entirely wrong, even when the information given is unquestionably correct.   For example, if one were given the data of Table 2.1 on the pressure and volume of a fixed mass of gas, one might

**TABLE 2.1**  Volume–Pressure  Relation  for  Ethylene,  an  Apparently Proportional Relationship

| Molar volume (liters) | Pressure (atmospheres) |
|---|---|
| 0.182 | 54.5 |
| 0.201 | 60.0 |
| 0.216 | 64.5 |
| 0.232 | 68.5 |
| 0.243 | 72.5 |
| 0.258 | 77.0 |
| 0.280 | 83.0 |
| 0.298 | 89.0 |
| 0.314 | 94.0 |

infer, by induction, that the pressure of a gas is proportional to its volume, a completely erroneous statement.   The error is due, of course, to the fact that another important item of information was omitted, namely that each pair of measurements was obtained at a different temperature, as indicated in Table 2.2. Admittedly this example is artificial and extreme; it was introduced merely to emphasize the basic problem in inductive reasoning: the dependence of inductive inferences not only on the *correctness* of the data, by also on their *completeness*. Recognition of the danger of drawing false inferences from incomplete, though correct information

**TABLE 2.2**  Volume–Pressure–Temperature Relation for Ethylene

| Molar volume (liters) | Pressure (atmospheres) | Temperature (degrees Centigrade) |
|---|---|---|
| 0.182 | 54.5 | 15.5 |
| 0.201 | 60.0 | 25.0 |
| 0.216 | 64.5 | 37.7 |
| 0.232 | 68.5 | 50.0 |
| 0.243 | 72.5 | 60.0 |
| 0.258 | 77.0 | 75.0 |
| 0.280 | 83.0 | 100.0 |
| 0.298 | 89.0 | 125.0 |
| 0.314 | 94.0 | 150.0 |

has led scientists to a preference for designed experimentation above mere observation of natural phenomena. An important aspect of statistics is the help it can provide toward designing experiments that will provide reliable and sufficiently complete information on the pertinent problems. We will return to this point in Section 2.5.

## 2.2   THE NATURE OF STATISTICAL ANALYSIS

The data resulting from an experiment are always susceptible of a large number of possible manipulations and inferences. Without proper guidelines, "analysis" of the data would be a hopelessly indeterminate task. Fortunately, there always exist a number of natural limitations that narrow the field of analysis. One of these is the *structure* of the experiment. The structure is, in turn, determined by the basic *objectives* of the experiment. A few examples may be cited.

1. In testing a material for conformance with specifications, a number of specimens, say 3 or 5, are subjected to the same testing process, for example a tensile strength determination. The *objective* of the experiment is to obtain answers to questions of the following type: "Is the tensile strength of the material equal to at least the specified lower limit, say S pounds per square inch?" The *structure* of the data is the simplest

possible: it consists of a statistical sample from a larger collection of items. The statistical analysis in this case is not as elementary as might have been inferred from the simplicity of the data-structure. It involves an inference from a small sample (3 or 5 specimens) to a generally large quantity of material. Fortunately, in such cases one generally possesses information apart from the meager bit provided by the data of the experiment. For example, one may have reliable information on the repeatability of tensile measurements for the type of material under test. The relative difficulty in the analysis is in this case due not to the structure of the data but rather to matters of sampling and to the questions that arise when one attempts to give mathematically precise meaning to the basic problem. One such question is: what is meant by the tensile strength of the material? Is it the average tensile strength of all the test specimens that could theoretically be cut from the entire lot or shipment? Or is it the weakest spot in the lot? Or is it a tensile strength value that will be exceeded by 99 per cent of the lot? It is seen that an apparently innocent objective and a set of data of utter structural simplicity can give rise to fairly involved statistical formulations.

2. In studying the effect of temperature on the rate of a chemical reaction, the rate is obtained at various preassigned temperatures. The *objective* here is to determine the relationship represented by the curve of reaction rate against temperature. The *structure* of the data is that of two columns of paired values, temperature and reaction rate. The statistical analysis is a curve fitting process, a subject to be discussed in Chapter 11. But what curve are we to fit? Is it part of the statistical analysis to make this decision? And if it is not, then what is the real objective of the statistical analysis? Chemical considerations of a theoretical nature lead us, in this case, to plot the logarithm of the reaction rate against the reciprocal of the absolute temperature and to expect a reasonably straight line when these scales are used. The statistician is grateful for any such directives, for without them the statistical analysis would be mostly a groping in the dark. The purpose of the analysis is to confirm (or, if necessary, to deny) the presumed linearity of the relationship, to obtain the best values for the parameters characterizing the relationship, to study the magnitude and the effect of experimental error, to advise on future experiments for a better understanding of the underlying chemical phenomena or for a closer approximation to the desired relationship.

3. Suppose that a new type of paper has been developed for use in paper currency. An experiment is performed to compare the wear characteristics of the bills made with the new paper to that of bills made from the conventional type of paper (4). The *objective* is the comparison of the wear characteristics of two types of bills. The *structure* of the data

depends on how the experiment is set up. One way would consist in sampling at random from bills collected by banks and department stores, to determine the age of each bill by means of its serial number and to evaluate the condition of the bill in regard to wear. Each bill could be classified as either "fit" or "unfit" at the time of sampling. How many samples are required? How large shall each sample be? The *structure* of the data in this example would be a relatively complex classification scheme. How are such data to be analyzed? It seems clear that in this case, the analysis involves counts, rather than measurements on a continuous scale. But age is a continuous variable. Can we transform it into a finite set of categories? How shall the information derived from the various samples be pooled? Are there any criteria for detecting biased samples?

From the examples cited above, it is clear that the statistical analysis of data is not an isolated activity. Let us attempt to describe in a more systematic manner its role in scientific experimentation.

In each of the three examples there is a more or less precisely stated objective: (a) to determine the value of a particular property of a lot of merchandise; (b) to determine the applicability of a proposed physical relationship; (c) to compare two manufacturing processes from a particular viewpoint. Each example involves data of a particular structure, determined by the nature of the problem and the judgment of the experimenter in designing the experiment. In each case the function of the data is to provide answers to the questions stated as objectives. This involves inferences from the particular to the general or from a sample to a larger collection of items. Inferences of this type are inductive, and therefore uncertain. Statistics, as a science, deals with uncertain inferences, through the concept of probability. However, the concept of probability, as it pertains to inductive reasoning, is not often used by physical scientists. Physicists are not likely to "bet four to one against Newtonian mechanics" or to state that the "existence of the neutron has a probability of 0.997." Why, then, use statistics in questions of this type? The reason lies in the unavoidable fluctuations encountered both in ordinary phenomena and in technological and scientific research. No two dollar bills are identical in their original condition, nor in the history of their usage. No two analyses give identical results, though they may appear to do so as a result of rounding errors and our inability to detect differences below certain thresholds. No two repetitive experiments of reaction rates yield identical values. Finally, no sets of measurements met in practice are found to lie on an absolutely smooth curve. Sampling fluctuations and experimental errors of measurement are always present to vitiate our observations. Such fluctuations therefore introduce a certain lack of definiteness in our inferences.   And it is the role of a statistical analysis to determine the extent of this lack of definiteness and thereby to ascertain the limits of validity of the conclusions drawn from the experiment.

Rightfully, the scientist's object of primary interest is the *regularity* of scientific phenomena.   Equally rightfully, the statistician concentrates his attention on the *fluctuations* marring this regularity.   It has often been stated that statistical methods of data analysis are justified in the physical sciences wherever the errors are "large," but unnecessary in situations where the measurements are of great precision.   Such a viewpoint is based on a misunderstanding of the nature of physical science.   For, to a large extent, the activities of physical scientists are concerned with determining the boundaries of applicability of physical laws or principles. As the precision of the measurements increases, so does the accuracy with which these boundaries can be described, and along with it, the insight gained into the physical law in question.

Once a statistical analysis is understood to deal with the uncertainty introduced by errors of measurement as well as by other fluctuations, it follows that statistics should be of even greater value in situations of high precision than in those in which the data are affected by large errors.   By the very nature of science, the questions asked by the scientist are always somewhat ahead of his ability to answer them; the availability of data of high precision simply pushes the questions a little further into a domain where still greater precision is required.

## 2.3  STATISTICAL MODELS

We have made use of the analogy of a mapping process in discussing scales of measurement.   This analogy is a fruitful one for a clearer understanding of the nature of science in general (5).   We can use it to explain more fully the nature of statistical analyses.

Let us consider once more the example of the type of 2 above.   When Arrhenius' law holds, the rate of reaction, $k$, is related to the temperature at which the reaction takes place, $T$, by the equation:

$$\ln k = \left(-\frac{E}{R}\right)\frac{1}{T} \tag{2.1}$$

where $E$ is the activation energy and $R$ the gas constant.   Such an equation can be considered as a *mathematical model* of a certain class of physical phenomena.   It is not the function of a model to establish causal relationships, but rather to express the relations that exist between different physical entities, in the present case reaction rate, activation energy, and temperature.   Models, and especially mathematical ones, are our most powerful tool for the study of nature. They are, as it were, the "maps" from which we can read the interrelationships between natural phenomena. When is a model a statistical one? When it contains statistical concepts. Arrhenius' law, as shown in Eq. 2.1 is not a statistical model. The contribution that statistics can make to the study of this relation is to draw attention to the experimental errors affecting the measured values of physical quantities such as $k$ and $T$, and to show what allowance must be made for these errors and how they might affect the conclusions drawn from the data concerning the applicability of the law to the situations under study.

Situations exist in which the statistical elements of a model are far more predominant than in the case of Arrhenius' law, even after making allowances for experimental errors. Consider, for example, a typical problem in *acceptance sampling*: a lot of 10,000 items (for example, surgical gloves) is submitted for inspection to determine conformance with specifications requirements. The latter include a destructive physical test, such as the determination of tensile strength. How many gloves should be subjected to the test and how many of those tested should have to "pass" the minimum requirements in order that the lot be termed acceptable? The considerations involved in setting up a model for this problem are predominantly of a statistical nature: they involve such statistical concepts* as the *probability of acceptance* of a "bad" lot and the *probability of rejection* of a "good" lot, where "good" and "bad" lots must be properly defined. The probability distributions involved in such a problem are well known and their mathematical theory completely developed. Where errors of measurement are involved, rather than mere sampling fluctuations, the statistical aspects of the problem are generally less transparent; the model contains, in the latter cases, both purely mathematical and statistical elements. One of the objectives of this book is to describe the statistical concepts that are most likely to enter problems of measurement.

## 2.4  STATISTICAL MODELS IN THE STUDY OF MEASUREMENT

In this section we will describe two measurement problems, mainly for the purpose of raising a number of questions of a general nature. A more detailed treatment of problems of this type is given in later chapters.

1. Let us first consider the determination of fundamental physical constants, such as the velocity of light, Sommerfeld's fine-structure constant, Avogadro's number, or the charge of the electron. Some of these can be determined individually, by direct measurement. In the

* For a discussion of these concepts, see Chapter 10.

case of others, all one can measure is a number of functions involving two or more of the constants. For example, denoting the velocity of light by $c$, Sommerfeld's fine-structure constant by $\alpha$, Avogadro's number by $N$, and the charge of the electron by $e$, the following quantities can be measured (1): $c$; $Ne/c$; $Ne^2/\alpha^3c^2$; $\alpha^3c/e$; $\alpha^2c$.

Let us denote these five functions by the symbols $Y_1$, $Y_2$, $Y_3$, $Y_4$, and $Y_5$, respectively. Theoretically, four of these five quantities suffice for the determination of the four physical constants $c$, $\alpha$, $N$, and $e$. However, all measurements being subject to error, one prefers to have an abundance of measurements, in the hope that the values inferred from them will be more reliable than those obtained from a smaller number of measurements. Let us denote the experimental errors associated with the determination of the functions $Y_1$ to $Y_5$, by $\varepsilon_1$ to $\varepsilon_5$, respectively. Furthermore, let $y_1$, $y_2$, $y_3$, $y_4$, and $y_5$ represent the experimental values (i.e., the values actually obtained) for these functions. We then have:

$$y_1 = Y_1 + \varepsilon_1 = c + \varepsilon_1$$
$$y_2 = Y_2 + \varepsilon_2 = Ne/c + \varepsilon_2$$
$$y_3 = Y_3 + \varepsilon_3 = Ne^2/\alpha^3c^2 + \varepsilon_3$$
$$y_4 = Y_4 + \varepsilon_4 = \alpha^3c/e + \varepsilon_4$$
$$y_5 = Y_5 + \varepsilon_5 = \alpha^2c + \varepsilon_5 \qquad (2.2)$$

These equations are the core of the statistical model. The $y$'s are the results of measurement, i.e., numbers; the symbols $c$, $\alpha$, $N$, and $e$ represent the unknowns of the problem. What about the errors $\varepsilon$? Clearly the model is incomplete if it gives us no information in regard to the $\varepsilon$. In a subsequent chapter we will discuss in detail what type of information about the $\varepsilon$ is required for a valid statistical analysis of the data. The main purpose of the analysis is of course the evaluation of the constants $c$, $\alpha$, $N$, and $e$. A second objective, of almost equal importance, is the determination of the reliability of the values inferred by the statistical analysis.

2. Turning now to a different situation, consider the study of a method of chemical analysis, for example the titration of fatty acid in solutions of synthetic rubber, using an alcoholic sodium hydroxide solution as the reagent (3). Ten solutions of known fatty acid concentration are prepared. From each, a portion of known volume (aliquot) is titrated with the reagent. The number of milliliters of reagent is recorded for each titration. Broadly speaking, the objective of the experiment is to evaluate the method of chemical analysis. This involves the determination of the precision and accuracy of the method. What is the relationship between the measurements (milliliters of reagent) and the known concentrations of the prepared solutions? Let $m$ denote the number of milliliters of

reagent required for the titration of a solution containing one milligram of fatty acid. Furthermore, suppose that a "blank titration" of $b$ milliliters of reagent is required for the neutralization of substances, other than the fatty acid, that react with the sodium hydroxide solution. A solution containing $x_i$ mg. of fatty acid will therefore require a number of milliliters of reagent equal to

$$Y_i = mx_i + b \qquad (2.3)$$

Adding a possible experimental error, $\varepsilon_i$, to this theoretical amount, we finally obtain the *model equation* for the number, $y_i$, of milliliters of reagent:

$$y_i = Y_i + \varepsilon_i = mx_i + b + \varepsilon_i \qquad (2.4)$$

In this relation, $x_i$ is a known quantity of fatty acid, $y_i$ a measured value, and $b$ and $m$ are parameters that are to be evaluated from the experiment.

It is of interest to observe that in spite of very fundamental differences between the two examples described in this section, the models, when formulated as above, present a great deal of structural similarity.

The analogy is shown in Table 2.3. In the first example, we wish to determine the parameters $c$, $\alpha$, $N$, and $e$, and the reliability of the estimates we obtain for them. In the second example, we are also interested in parameters, namely $b$ and $m$, and in the reliability of their estimates, these quantities being pertinent in the determination of accuracy. In addition, we are interested in the errors $\varepsilon$, which are pertinent in the evaluation of

**TABLE 2.3** Comparison of Two Experiments

|  | Determination of physical constants | Study of analytical method |
|---|---|---|
| Parameters to be measured | $c$, $\alpha$, $N$, $e$ | $b$, $m$ |
| Functions measured | $Y_1 = c$ <br> $Y_2 = Ne/c$ <br> $Y_3 = Ne^2/\alpha^3 c^2$ <br> $Y_4 = \alpha^3 c/e$ <br> $Y_5 = \alpha^2 c$ | $Y_1 = b + mx_1$ <br> $Y_2 = b + mx_2$ <br> $\vdots$ <br> $Y_{10} = b + mx_{10}$ |
| Measurements | $y_1 = Y_1 + \varepsilon_1$ <br> $y_2 = Y_2 + \varepsilon_2$ <br> $y_3 = Y_3 + \varepsilon_3$ <br> $y_4 = Y_4 + \varepsilon_4$ <br> $y_5 = Y_5 + \varepsilon_5$ | $y_1 = Y_1 + \varepsilon_1$ <br> $y_2 = Y_2 + \varepsilon_2$ <br> $\vdots$ <br> $y_{10} = Y_{10} + \varepsilon_{10}$ |

the precision of the method.   The structural analogy is reflected in certain similarities in the statistical analyses, but beyond the analogies there are important differences in the objectives.   No statistical analysis is satisfactory unless it explores the model in terms of all the major objectives of the study; a good analysis goes well beyond mere structural resemblance with a conventional model.

## 2.5   THE DESIGN OF EXPERIMENTS

In example 2 of Section 2.2, we were concerned with a problem of curve fitting.   The temperature effect on reaction rate was assumed to obey Arrhenius' law expressed by Eq. 2.1.   This equation is the essential part of the model underlying the analysis.   One of the objectives of the experiment is undoubtedly the determination of the activation energy, $E$. This objective hinges, however, on another more fundamental aspect, viz., the extent to which the data conform to the theoretical Eq. 2.1. The question can be studied by plotting $\ln k$ versus $1/T$, and observing whether the experimental points fall close to a single straight line.   Thus, the statistical analysis of experimental data of this type comprises at least two parts: the verification of the assumed model, and the determination of the parameters and of their reliability.   In setting up the experiment, one must keep both these objectives in mind.   The first requires that enough points be available to determine curvature, if any should exist. The second involves the question of how precisely the slope of a straight line can be estimated.   The answer depends on the precision of the measuring process by which the reaction rates are determined, the degree of control of the temperature, the number of experimental points and their spacing along the temperature axis.

The example just discussed exemplifies the types of precautions that must be taken in order to make an experiment successful.   These precautions are an important phase of the design of the experiment.   In the example under discussion the basic model exists independently of the design of the experiment.   There are cases, on the other hand, in which the design of the experiment determines, to a large extent, the model underlying the analysis.

In testing the efficacy of a new drug, say for the relief of arthritis, a number of patients suffering from the disease are treated with the drug. A statistical analysis may consist in determining the proportion of patients whose condition has materially improved after a given number of treatments.   A parameter is thus estimated.   By repeating the experiment several times, using new groups of patients, one may even evaluate the precision of the estimated parameter.   The determination of this parameter and of its precision may have validity.  Yet from the viewpoint of determining the efficacy of the drug for the treatment of arthritis the experiment, standing by itself, is valueless.  Indeed, the parameter determined by the experiment, in order to be an index of the efficacy of the drug, must be compared with a similarly determined parameter for other drugs or for no drug at all.  Such information may be available from other sources and in that case a comparison of the values of the parameter for different drugs, or for a particular drug versus no drug, may shed considerable light on the problem.  It is generally more satisfactory to combine the two necessary pieces of information into a single experiment by taking a double-sized group of patients, administering the drug under test to one group, and a "placebo" to the other group.  Alternatively, the second group may be administered another, older drug.  By this relatively simple modification of the original experiment, the requirements for a meaningful answer to the stated problem are built into the experiment.  The parameters of interest can be compared and the results answer the basic question.  The modified experiment does not merely yield a parameter; the parameter obtained can be interpreted in terms of the problem of interest.

Note that in the last example, unlike the previous one, the model is determined by the design of the experiment.  Indeed, in the former example, the basic model is expressed by Arrhenius' law, Eq. 2.1, a mathematical relation derived from physical principles.  In the example of drug testing no such equation is available.  Therefore the statistical treatment itself provides the model.  It does this by postulating the existence of two statistical populations,* corresponding to the two groups of patients.  Thus the model in this example is in some real sense the creation of the experimenter.  We will not, at this point, go into further aspects of statistical design.  As we proceed with our subject, we will have many occasions to raise questions of experimental design and its statistical aspects.

## 2.6  STATISTICS AS A DIAGNOSTIC TOOL

It has often been stated that the activities of a research scientist in his laboratory cannot or should not be forced into a systematic plan of action, because such a plan would only result in inhibiting the creative process of good research.  There is a great deal of truth in this statement, and there are indeed many situations where the laying out of a fixed plan, or the prescription of a predetermined statistical design would do more harm than good.  Nevertheless, the examination of data, even from such

* This concept will be defined in Chapter 3.

incompletely planned experimentation, is generally more informative when it is carried out with the benefit of statistical insight. The main reason for this lies in the diagnostic character of a good statistical analysis. We have already indicated that one of the objectives of a statistical analysis is to provide clues for further experimentation. It could also serve as the basis for the formulation of tentative hypotheses, subject to further experimental verification. Examples that illustrate the diagnostic aspects of statistical analysis will be discussed primarily in Chapters 8 and 9. We will see that the characteristic feature of these examples is that rather than adopting a fixed model for the data, the model we actually use as a basis for the statistical analysis is one of considerable flexibility; this allows the experimenter to make his choice of the final model as a result of the statistical examination of the data rather than prior to it.

It must be admitted that the diagnostic aspect of the statistical analysis of experimental data has not received the attention and emphasis it deserves.

## 2.7  SUMMARY

The methods of statistical analysis are closely related to those of inductive inference, i.e., to the drawing of inferences from the particular to the general. A basic difference between deductive and inductive inference is that in deductive reasoning, conclusions drawn from correct information are always valid, even if this information is incomplete; whereas in inductive reasoning even correct information may lead to incorrect conclusions, namely when the available information, although correct, is incomplete. This fact dominates the nature of statistical analysis, which requires that the data to be analyzed be considered within the framework of a statistical model. The latter consists of a body of assumptions, generally in the form of mathematical relationships, that describe our understanding of the physical situation or of the experiment underlying the data. The assumptions may be incomplete in that they do not assign numerical values to all the quantities occurring in the relationships. An important function of the statistical analysis is to provide estimates for these unknown quantities. Assessing the reliability of these estimates is another important function of statistical analysis.

Statistical models, as distinguished from purely mathematical ones, contain provisions for the inclusion of experimental error. The design of experiments, understood in a broad sense, consists in prescribing the conditions for the efficient running of an experiment, in such a way that the data will be likely to provide the desired information within the framework of the adopted statistical model.

An important, though insufficiently explored aspect of the statistical analysis of data lies in its potential diagnostic power: by adopting a flexible model for the initial analysis of the data it is often possible to select a more specific model as the result of the statistical analysis. Thus the analysis serves as a basis for the formulation of specific relationships between the physical quantities involved in the experiment.

## REFERENCES

1. DuMond, J. W. M., and E. R. Cohen, "Least Squares Adjustment of the Atomic Constants 1952," *Review of Modern Physics*, **25**, 691–708 (July 1953).
2. Fisher, R. A., "Mathematical Probability in the Natural Sciences," *Technometrics*, **1**, 21–29 (Feb. 1959).
3. Linnig, F. J., J. Mandel, and J. M. Peterson, "A Plan for Studying the Accuracy and Precision of an Analytical Procedure," *Analytical Chemistry*, **26**, 1102–1110 (July 1954).
4. Randall, E. B., and J. Mandel, "A Statistical Comparison of the Wearing Characteristics of Two Types of Dollar Notes," *Materials Research and Standards*, **2**, 17–20 (Jan. 1962).
5. Toulmin, S., *The Philosophy of Science, An Introduction*, Harper, New York, 1960.

# chapter 3

# THE MATHEMATICAL FRAME-WORK OF STATISTICS, PART I

## 3.1 INTRODUCTION

The language and symbolism of statistics are necessarily of a mathematical nature, and its concepts are best expressed in the language of mathematics. On the other hand, because statistics is *applied* mathematics, its concepts must also be meaningful in terms of the field to which it is applied.

Our aim in the present chapter is to describe the mathematical framework within which the science of statistics operates, introducing the concepts with only the indispensable minimum of mathematical detail and without losing sight of the field of application that is the main concern of this book—the field of physical and chemical measurements.

## 3.2 PROBABILITY DISTRIBUTION FUNCTIONS

The most basic statistical concept is that of a *probability distribution* (also called *frequency distribution*) associated with a *random variable*. We will distinguish between *discrete* variables (such as counts) and *continuous* variables (such as the results of quantitative chemical analyses). In either category, a variable whose value is subject to chance fluctuations is called a *variate*, or a *random variable*.

For *discrete* variates, the numerical value $x$ resulting from the experiment can assume any one of a given sequence of possible outcomes: $x_1$, $x_2$, $x_3$, etc. The probability distribution is then simply a scheme associating with each possible outcome $x_i$, a probability value $p_i$. For example, if the experiment consists in determining the total count obtained in casting two dice, the sequence of the $x_i$ and the associated probabilities $p_i$ are given by Table 3.1.

**TABLE 3.1**

| $x_i$ | 2 | 3 | 4 | 5 | 6 | 7 | 8 | 9 | 10 | 11 | 12 |
|---|---|---|---|---|---|---|---|---|---|---|---|
| $p_i$ | $\frac{1}{36}$ | $\frac{2}{36}$ | $\frac{3}{36}$ | $\frac{4}{36}$ | $\frac{5}{36}$ | $\frac{6}{36}$ | $\frac{5}{36}$ | $\frac{4}{36}$ | $\frac{3}{36}$ | $\frac{2}{36}$ | $\frac{1}{36}$ |

Table 3.1 is a complete description of the probability distribution in question. Symbolically, we can represent it by the equation

$$\text{Prob } [x = x_i] = p_i \tag{3.1}$$

which reads: "The probability that the outcome of the experiment, $x$, be $x_i$, is equal to $p_i$." Note that the sum of all $p_i$ is unity.

At this point we introduce another important concept: the *cumulative probability function*. In the example of the two dice, Table 3.1 provides a direct answer to the question: "What is the probability that the result $x$ will be $x_i$?" The answer is $p_i$. We may be interested in a slightly different question. "What is the probability that the result $x$ will not exceed $x_i$?" The answer to this question is of course: the sum $p_1 + p_2 + \cdots + p_i$. For example, the probability that $x$ will not exceed the value 5 is given by the sum $1/36 + 2/36 + 3/36 + 4/36$, or $10/36$. Thus we can construct a second table, of *cumulative probabilities*, $F_i$, by simply summing the $p$ values up to and including the $x_i$ considered. The cumulative probabilities corresponding to our dice problem are given in Table 3.2.

**TABLE 3.2**

| $x_i$ | 2 | 3 | 4 | 5 | 6 | 7 | 8 | 9 | 10 | 11 | 12 |
|---|---|---|---|---|---|---|---|---|---|---|---|
| $F_i$ | $\frac{1}{36}$ | $\frac{3}{36}$ | $\frac{6}{36}$ | $\frac{10}{36}$ | $\frac{15}{36}$ | $\frac{21}{36}$ | $\frac{26}{36}$ | $\frac{30}{36}$ | $\frac{33}{36}$ | $\frac{35}{36}$ | $\frac{36}{36}$ |

Symbolically, the cumulative probability function is represented by the equation

$$\text{Prob } [x \leqslant x_i] = F_i = p_1 + p_2 + \cdots + p_i \tag{3.2}$$

Note that the value of $F_i$ corresponding to the largest possible $x_i$ (in this case this $x_i = 12$) is unity. Furthermore, it is evident that a $F_i$ value can never exceed any of the $F_i$ values following it: $F_i$ is a non-decreasing function of $x_i$.

Tables 3.1 and 3.2 can be represented graphically, as shown in Figs. 3.1a and 3.1b. Note that while the cumulative distribution increases over the entire range, it does so at an increasing rate from $x = 1$ to $x = 7$ and at a decreasing rate from $x = 7$ to $x = 12$.

In the case of *continuous* variates—which will be our main concern in this book—it is not possible to write down a sequence of possible outcomes $x_1$, $x_2$, etc. This is because any value between $x_1$ and $x_2$, or between $x_i$ and $x_{i+1}$, is also a possible outcome. In those cases use is made of a well-known device of differential calculus. We consider a small *interval* of values, extending from a given value $x_i$ to the nearby value $x_i + dx$, and define the frequency distribution in terms of the probability that the outcome of the experiment, $x$, will lie in this interval. Since the length of the interval, $dx$, is small, the probability that $x$ will lie in the interval may be considered proportional to $dx$, and we write accordingly

$$\text{Prob } [x_i \leqslant x \leqslant x_i + dx] = f_i dx \qquad (3.3)$$

where $f_i$ represents the probability corresponding to one unit of interval length. The suffix $i$ serves to indicate that this probability depends on the location of $x$. We can express this relationship more simply by using the functional notation $f(x)$, writing

$$\text{Prob } [x \leqslant x' \leqslant x + dx] = f(x)dx \qquad (3.4)$$

where $x'$ now stands for the *actual* outcome of the experiment, and $x$ is any one of the infinitely many values that *could* be the outcome of the experiment. Stated in words, Eq. 3.4 reads: "The probability that the value actually yielded by the experiment, $x'$, will be contained in the small interval extending from $x$ to $x + dx$ is equal to $f(x)dx$." In general, this probability is not constant over the entire range of possible values. Its behavior in this range is that of the function $f(x)$.

The function $f(x)$ is known as the *probability density function*. The function typified by Eq. 3.4 can be represented by a graph similar to that of Fig. 3.1a. However, because of the continuous nature of the variable, the probability distribution function is no longer represented by a finite number of bars. Instead, a continuous curve represents the probability density function for all values of $x$ (as in Fig. 3.2a). This curve may be visualized as one connecting the top points of an infinite number of vertical bars, such as those of Fig. 3.1a. From Eq. 3.4 it follows that the probability corresponding to a particular interval starting at $x$ and of length $dx$ is measured by the area of the rectangle of height $f(x)$ and base $dx$. This rectangle is, for sufficiently small $dx$, the same as the portion of the area under the curve between $x$ and $x + dx$. Thus, for continuous variates, the probabilities are represented by areas. This is an important notion of which we will make considerable use.

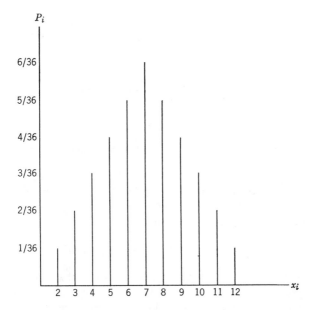

**Fig. 3.1a**   Probability distribution of a discrete variable: Prob $[x = x_i] = p_i$.

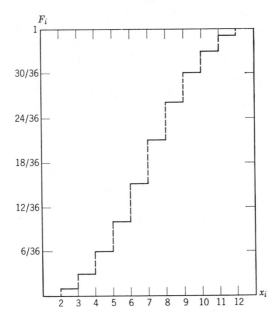

**Fig. 3.1b**   Cumulative distribution of a discrete variable: Prob $[x \leqslant x_i] = F_i$.

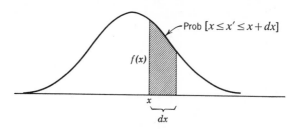

**Fig. 3.2a**  Probability distribution of a continuous variable.

As in the case of the discrete variable, we introduce a *cumulative probability function*, defined as the probability that the result $x'$ will not exceed a given value $x$.  Because of the continuous nature of the variable, the summation is replaced by an integral:

$$\text{Prob }[x' \leqslant x] = F(x) = \int_{L}^{x} f(y)dy \qquad (3.5)$$

In this equation $L$ represents the lowest value that $x$ can assume; it may be, but is not always, $-\infty$.  The upper limit of the integral is, of course, the value $x$ appearing in the first member of the equation.  In order to avoid confusion, the symbol for the variable in the integrand has been changed from $x$ to $y$.  The following two facts are easily apparent from the parallelism of Eq. 3.5 and Eq. 3.2.  In the first place, if $U$ represents the largest value that $x$ can assume (which could be, but is not necessarily, $+\infty$), $F(U) = 1$.  Secondly, the function $F(x)$ is a non-decreasing function of $x$.

In terms of Fig. 3.2a, the cumulative probability corresponding to a particular value $x$ is measured by the area under the curve to the left of $x$.

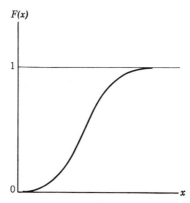

**Fig. 3.2b**  Cumulative distribution of a continuous variable.

If this area itself is plotted against $x$, the cumulative probability function, shown in Fig. 3.2$b$, is obtained.

Confining our attention to continuous random variables, we note that the complete characterization of their probability distribution can be made in two steps.

The first is the description of the *functional form* of the probability density function, together with the range of variation of the random variable. For example, it might be stated that a particular random variable $x$ varies from zero to plus infinity and that its density function $f(x)$ is of the form

$$f(x) = \frac{1}{k} e^{-x/k} \tag{3.6}$$

The information supplied by such a statement remains incomplete until a numerical value is supplied for the *parameter k* occurring in the equation. A *parameter* is any variable, other than the random variable $x$ itself, that occurs in the probability function $f(x)$. We will see that a distribution function may contain more than one parameter.

The second step, then, consists in providing the numerical values of all parameters occurring in the distribution function.

## 3.3  THE EXPECTED VALUE OF A RANDOM VARIABLE

The distribution of the count obtained in casting two dice, for which the probabilities are listed in Table 3.1, required 11 pairs of numbers for its complete description. In most practical problems, such as the distribution of the hourly number of telephone calls on a given trunk line, or the distribution of the velocity of molecules in a gas, the complete listing of all probabilities is either prohibitively laborious or completely impossible. It is true that in some cases a theoretical formula, such as Eq. 3.6, is available for the description of a frequency distribution. But while such a formula contains, theoretically, all the information about the distribution, it does not supply us with measures of immediate practical usefulness. It becomes, therefore, necessary to express frequency distributions in terms of a few well-chosen quantities that typify them for the purpose of practical applications. It turns out that the common notion of an *average value* is a particularly appropriate basis for the calculation of such quantities, regardless of the particular nature of the distribution function.

We begin by redefining the notion of an average, by giving it a mathematically more precise form, as well as a new name: *expected value*.

The *expected value* (or *mean*) of a random variable is obtained by multiplying each possible value of the variable by its probability and summing all the products thus obtained.    Thus, for Table 3.1, the expected value is  $(2 \times 1/36) + (3 \times 2/36) + (4 \times 3/36) + \cdots + (10 \times 3/36) + (11 \times 2/36) + (12 \times 1/36)$  which yields the value 7.

We denote the expected value of a random variable $x$ by the symbol $E(x)$.    Algebraically we have:

$$E(x) = \sum_i x_i p_i \qquad (3.7)$$

where the $\sum$ indicates a summation over all values of the associated index (in this case $i$).

In the case of a continuous random variable, the probability $p_i$ is replaced by $f(x)dx$ and the summation is replaced by an integral:

$$E(x) = \int x f(x) dx \qquad (3.8)$$

where the integral extends over the entire range of values of $x$.

For example, for the random variable with the distribution function 3.6, the expected value is

$$E(x) = \int_0^\infty x \frac{1}{k} e^{-x/k} dx$$

This is an elementary integral and is easily shown (2) to be equal to $k$.

For simplicity of notation we will often use the symbol $\mu$ rather than $E(x)$ to represent the mean of a frequency distribution.

In the case of Table 3.1 the expected value, 7, happens to be also the central value.    The reason for this is that the distribution in question is *symmetrical*: the set of $p_i$ values to the right of the value corresponding to $x = 7$ is the mirror image of the set occurring to the left of this value. This is by no means always the case.    Consider, for example, a shipment (or lot) of individual items of merchandise, for example, automobile tires, in which 10 per cent of the items are defective.    A random sample of 10 tires from this lot can conceivably contain 0, 1, 2, ..., 10 defective tires, but these values are not equally probable.    Table 3.3 lists the probabilities associated with each possible composition of the sample.

A distribution of this type is said to be *skew*, because the $p_i$ values are not symmetrical with respect to a central value of $x$.    The expected value for the distribution of Table 3.3 is equal to

$$(0 \times 0.349) + (1 \times 0.387) + \cdots + (10 \times 0.0000000001) = 1$$

Thus, in this case only one possible outcome (0 defective items) lies below the expected value, while nine possible outcomes lie above it.

**TABLE 3.3**

| Number of defectives in sample, $x_i$ | Probability, $p_i$ |
|---|---|
| 0 | 0.349 |
| 1 | 0.387 |
| 2 | 0.194 |
| 3 | 0.0574 |
| 4 | 0.0112 |
| 5 | 0.00149 |
| 6 | 0.000138 |
| 7 | 0.0000087 |
| 8 | 0.00000036 |
| 9 | 0.000000009 |
| 10 | 0.0000000001 |

The following equations state important properties of expected values. Their mathematical proof (1) is not difficult.

$$E(kx) = kE(x) \qquad (3.9)$$

$$E(x + y) = E(x) + E(y) \qquad (3.10)$$

$$E(x - y) = E(x) - E(y) \qquad (3.11)$$

## 3.4  MEASURES OF DISPERSION

The usefulness of average values (expected values) is well recognized. We will see that averages possess statistical properties that go well beyond their everyday utility.   Nevertheless, it is obvious that the expected value does not, by itself, reflect all the information supplied by a frequency distribution table (such as Table 3.1 or 3.3) or by a distribution function (such as Eq. 3.6).   One important aspect of a distribution about which the expected value tells us nothing is the amount of "crowding" of the values in the vicinity of the expected value.   Thus in the case of Table 3.1 the $p_i$ values decrease rather slowly on the left and on the right of $x = 7$, whereas in Table 3.3 the $p_i$ values fall off quite rapidly as $x$ deviates from 1.   This characteristic is generally referred to as *dispersion* or *spread*.

How is the dispersion of a frequency distribution to be measured ?   It is a remarkable fact that the best mathematical measure of the dispersion of a frequency distribution is again an expected value; not, of course, the expected value of the variable $x$, but rather *the expected value of the square of the deviation of x from its average value*.   Consider again Table 3.1. We know that $E(x) = 7$.   Corresponding to each value of $x$, there is its *deviation* from $E(x)$.   Table 3.4 lists: in the first row, $x_i$; in the second row, $p_i$; in the third row, the deviation $x_i - E(x)$, i.e., $x_i - 7$; and finally in the fourth row, the square of $x_i - E(x)$, i.e., $(x_i - 7)^2$.

**TABLE 3.4**

| $x_i$ | 2 | 3 | 4 | 5 | 6 | 7 | 8 | 9 | 10 | 11 | 12 |
|---|---|---|---|---|---|---|---|---|---|---|---|
| $p_i$ | $\frac{1}{36}$ | $\frac{2}{36}$ | $\frac{3}{36}$ | $\frac{4}{36}$ | $\frac{5}{36}$ | $\frac{6}{36}$ | $\frac{5}{36}$ | $\frac{4}{36}$ | $\frac{3}{36}$ | $\frac{2}{36}$ | $\frac{1}{36}$ |
| $x_i - E(x)$ | $-5$ | $-4$ | $-3$ | $-2$ | $-1$ | 0 | 1 | 2 | 3 | 4 | 5 |
| $[x_i - E(x)]^2$ | 25 | 16 | 9 | 4 | 1 | 0 | 1 | 4 | 9 | 16 | 25 |

We now form the expression for the expected value of this last quantity:

$$\left(25 \times \frac{1}{36}\right) + \left(16 \times \frac{2}{36}\right) + \cdots + \left(25 \times \frac{1}{36}\right) = \frac{210}{36} = 5.833$$

More generally, we define the *variance* of a random variable, which is a measure of its dispersion, by the formula

$$\text{Variance of } x = V(x) = \sum_i [x_i - E(x)]^2 p_i \tag{3.12}$$

The notation $V(x)$, to denote variance, will be used throughout this book. It is readily understood that Eq. 3.12 can also be written

$$V(x) = E[x - E(x)]^2 \tag{3.13}$$

since the variance is the expected value of the quantity $[x - E(x)]^2$.

Equation 3.13 applies both to discrete and to continuous variables. For the latter, the variance can also be expressed by replacing $p_i$ by $f(x)dx$ in Eq. 3.12, and the summation by an integral:

$$V(x) = \int [x - E(x)]^2 f(x)dx \tag{3.14}$$

where the integral extends over the entire range of values of $x$.

Applying Eq. 3.12 to Table 3.3, we obtain

$$V(x) = [(0 - 1)^2 \times 0.349] + [(1 - 1)^2 \times 0.387] + \cdots$$
$$+ [(10 - 1)^2 \times 0.0000000009] = 0.9$$

Why is variance defined in terms of the *squares* of the deviations?   Would it not be simpler to define it as the expected value of the deviation?   The answer is that the latter definition would yield the value *zero* for the variance of *any* frequency distribution.   This can easily be proved and may be verified for Tables 3.1 and 3.3.

The variance of a frequency distribution may also be described as the *second moment about the mean*. More generally, the $r$th *moment about the mean* is defined as:

$$r\text{th moment about the mean} = \mu^{(r)} = E\{[x - E(x)]^r\} \qquad (3.15)$$

It can be shown, for example, that the third moment about the mean is related to the skewness of the distribution. In this book we will make no use of moments beyond the second order (the variance). It should, however, be mentioned that the sequence of quantities consisting of the mean, the variance, and all successive moments about the mean provides, in a mathematical sense, a complete characterization of the frequency distribution from which they were derived.

Dimensionally, the variance of a random variable $x$ is expressed in terms of the square of the unit of measurement for $x$. Thus, if $x$ is a length expressed in centimeters, the variance has the unit centimeters-square. For practical reasons, it is useful to have a measure of dispersion expressed in the same units as the random variable. The *standard deviation* of a frequency distribution is defined as the square root of the variance. The common notation for standard deviation is $\sigma$. Thus

$$\sigma_x = \sqrt{V(x)} \qquad (3.16)$$

Applying this formula we obtain:

$$\text{for Table 3.1,} \quad \sigma = \sqrt{5.833} = 2.41;$$

$$\text{for Table 3.3,} \quad \sigma = \sqrt{0.9} = 0.949.$$

From Eq. 3.16 we infer that

$$\sigma_x{}^2 = V(x) \qquad (3.17)$$

Because of Eq. 3.17 the variance of a random variable $x$ is commonly represented by the symbol $\sigma_x{}^2$ or simply by $\sigma^2$.

In practical applications one is often interested in the relation between the standard deviation and the mean of a frequency distribution. The ratio of the standard deviation to the mean is known as the *coefficient of variation*. Multiplied by 100 it becomes the *per cent coefficient of variation*, often denoted by the symbol $\%CV$.

$$\%CV_x = 100 \, \frac{\sigma_x}{E(x)} \qquad (3.18)$$

For the data of Table 3.1, the per cent coefficient of variation is

$$\%CV = 100 \, \frac{2.41}{7} = 34.4$$

For the data of Table 3.3, we have

$$\%CV = 100\,\frac{0.949}{1} = 94.9$$

We give these values here merely as numerical illustrations.   Indeed, while a coefficient of variation can, of course, always be calculated, once the mean and the standard deviation are known, the concept expressed by this quantity is not always meaningful.   This matter will be discussed in detail in a subsequent chapter.

### 3.5   POPULATION AND SAMPLE

So far we have referred to random variables as essentially a mathematical abstraction.   Such a viewpoint is useful for an exposition of the mathematical concepts and relationships by which random phenomena are measured.   In applied statistics each value that a random variable can assume is, as it were, attached to a well-defined entity.   For example, if the random variable is the height of male adults in a given country, each height-value corresponds to an identifiable entity, in this case a male adult. The *totality* of all male adults considered in this application constitutes a *statistical population.*

In many cases the population does not consist of physical entities. Thus, in the example of Table 3.1, each member of the population was a *configuration* of two dice following a random throw.   Even less tangible are the members of the population of *all* replicate determinations of the sulfur content of a given piece of rubber.   In both these examples, the populations are infinite in size.   But whereas in the case of the dice we can readily visualize the unending sequence of throws, the example of the sulfur determination requires a further conceptual abstraction; for the piece of rubber is of finite dimensions and each determination requires a measurable portion of the piece.   Thus, the number of all possible determinations is actually finite.   Nevertheless we consider it as infinite by a purely conceptual process.   In this instance we would conceive of an infinitely large supply of rubber exactly analogous to the piece actually given, and of an infinite sequence of determinations made under essentially the same conditions of experimentation.   For the purpose of statistical treatment this conceptual abstraction is indispensable.   An abstraction of this kind is called a *hypothetical infinite population.*

It is, of course, impossible to obtain the value of the random variable for each member of an infinite population.   Even in the case of finite populations, such as all the tiles in a shipment of floor tiles, it very often is prohibitive to obtain a measurement value—for example, the thickness of the tile—for each individual item.　This applies with even greater force when the test that has to be carried out to obtain the measured value results in the physical destruction of the test specimen (*destructive tests*).　In all these situations one studies the frequency distribution of interest—height of people, sulfur content of rubber, thickness of tiles—on the basis of a finite selection of items from the population.　Such a selection is called a *statistical sample*, or more simply a *sample*.

Clearly one cannot expect to obtain a completely accurate representation of the population from the study of a sample.　The only exception to this statement is the case where the sample consists of *all* items of the population; this happens occasionally in industrial situations and is referred to as *100 per cent inspection*.　But wherever the sample is only a portion of the population, the information provided by it is only an *image*, sometimes a very imperfect image, of the population.　It is quite clear that one of the severest limitations of a sample, as a representation of the population, is its size, where by the *size of a sample* we mean the number of population items contained in it.　However, the size of a sample is not the only factor affecting its intrinsic merit as an image of the population.　Even a large sample may lead to erroneous conclusions.　What precautions must be taken in order to make sure that the sample is a satisfactory image of the population?

## 3.6  RANDOM SAMPLES

It is not our purpose to discuss the numerous aspects of the science of *sampling*, the selection of samples from populations.　Some facets of the subject will be dealt with in Chapter 10.　At this point, however, we will describe one basic method of sampling known as *random sampling*.

Rather than attempting a mathematical definition, we make use of the intuitive concept of randomness, exemplified by the shuffling of cards, the mixing of numbered chips in a bowl and similar manipulations.　Such procedures are known as *random processes*.　Their function is to exclude any form of bias, such as a conscious or even unconscious process of discriminatory selection on the part of the individual, or the effects of gradual shifts in the measuring apparatus.

From our viewpoint the importance of random sampling is that it lends itself to a mathematically precise formulation of the relationship between sample and population.　To understand this relationship we must first explain the concept of *sampling fluctuations*.

Consider again the example of a large lot of merchandise consisting of 90 per cent of satisfactory items and 10 per cent "defectives."　A random sample* of size 10 is taken from the lot.   If the composition of the sample were to be a perfect image of that of the lot, it should contain exactly *one* defective.   We know, however, that this will not always be the case.   In fact Table 3.3 tells us that such a "perfect" sample will be drawn in 38.7 per cent of all cases, i.e., not even in the majority of cases.   Other samples drawn from the same lot will contain 0, 2, 3, . . ., 10 defectives and the number of times that each such sample will be drawn, in the long run, is shown by the probabilities of Table 3.3.   If we consider a succession of samples, taken one after the other from the same population (the lot) and determine for each one the proportion of defectives, we will obtain a sequence of numbers.   The variability exhibited by these numbers demonstrates how the number of defectives *fluctuates* from sample to sample.   In short, samples are imperfect representations of the population in that they exhibit *random fluctuations* about the condition that would represent the exact image of the population.

While each individual sample of size 10 is but an imperfect image of the population, and therefore fails to give us full and accurate information about the population, the situation is entirely different when we consider the totality of all samples of size 10 from the population.   Indeed the fractions defective† of these samples constitute themselves a population, and it is precisely this population that is described by the probability distribution of Table 3.3.   This table is a typical example of the mathematical relationship between population and sample.   Knowing that the population contains 10 per cent defectives, it follows mathematically that the number of defectives in random samples of size 10 will be distributed according to the probabilities of Table 3.3.   We will not concern ourselves with the derivation of a table such as Table 3.3.

We emphasize, however, this fundamental point: *that Table 3.3 is incorrect and virtually valueless unless the selection of the samples from the lot satisfies* (at least to a good approximation) *the condition of randomness.* The science of statistics establishes relationships between samples and populations.   These relationships allow us to judge the reliability of a sample as a representation of the population.   But in practically all cases, such relationships are based on the assumption of randomness in the selection of the sample.   *It follows that the drawing of inferences from sample to population by statistical means is unjustified and may in fact lead to serious errors if the conditions of randomness are grossly violated.*

An analysis and interpretation of data by statistical methods should not

---

* The expression *random sample* is merely a short way of saying "a sample selected by a *random process*."   The idea of randomness implies an *act of selection*; the sample is the result of the act.

† See Chapter 10.

be attempted unless it has been established that the conditions of randomness necessary for the drawing of valid inferences have been properly fulfilled. The best way to accomplish this is to allow the statistician to participate in the planning stages of a sampling process or of a proposed experiment, so that in all instances in which samples are taken from populations—and this includes the making of measurements—the proper randomization processes are applied.

## 3.7  SAMPLE ESTIMATES

Suppose that in the situation represented by Table 3.3 a particular sample of 10 items was drawn by a random process and that 3 defectives were found in the sample. If the sample in question constitutes all the information available to us, we will have to conclude that, to the best of our knowledge, the proportion of defectives in the lot (the population) is 30 per cent. This value (or the proportion 0.3) is then our *sample estimate* of the "true" value or more precisely of the *population parameter*: fraction defective of the population.

In this example, using our knowledge that the true proportion of defectives is 10 per cent, we are painfully aware of the serious discrepancy between the population parameter and its sample estimate. In general, the true value of the population parameter is unknown to us. Accordingly an important aspect of statistical analysis is to provide means for evaluating the reliability of the sample estimate. This can be done in several ways, to be discussed in due course. Let us note at this point that the reliability of a sample estimate is necessarily related to the magnitude of its fluctuations from sample to sample.

Of particular interest in statistics are sample estimates of the mean, the variance (or of the standard deviation), and of the higher moments of a frequency distribution. Equations 3.7, 3.8, 3.12, and 3.13 provide us with definitions of the mean and the variance *considered as population parameters*. We still need means of *estimating* these quantities from samples taken from the populations.

Let $x_1, x_2, x_3, \ldots, x_N$ represent a random sample of size $N$ from any population. Then the following formulas express the *estimate* of the mean $\mu$ and variance $\sigma^2$, respectively:

$$\text{Estimate of } \mu = \hat{\mu} = \frac{x_1 + x_2 + \cdots + x_N}{N} = \frac{\sum_i x_i}{N} \tag{3.19}$$

$$\text{Estimate of } \sigma^2 = \hat{\sigma}^2 = \frac{\sum_i (x_i - \hat{\mu})^2}{N - 1} \tag{3.20}$$

The placing of a *caret* (^) above a population parameter to denote a sample estimate of the parameter is common in statistics, provided that the sample estimate belongs to a class known as "maximum-likelihood" estimates. In this book, we will ignore this restriction, though in most cases the estimates used will be of the maximum-likelihood class.

Common usage is responsible for the following additional symbols: (a) The arithmetic average of $N$ quantities $x_1, x_2, \ldots, x_N$ is generally represented by $\bar{x}$; thus Eq. 3.19 simply becomes:

$$\hat{\mu} = \bar{x} \tag{3.21}$$

(b) The sample estimate of the variance is generally represented by the symbol $s^2$; thus

$$\hat{\sigma}^2 = s^2 = \frac{\sum_i (x_i - \bar{x})^2}{N - 1} \tag{3.22}$$

The quantity $N - 1$ in the denominator of Eq. 3.22 is known as the number of *degrees of freedom* used for the estimation of the variance. Why not use $N$?

Consider the $N$ quantities: $x_1 - \bar{x}, x_2 - \bar{x}, \ldots, x_N - \bar{x}$. These quantities are the *deviations*, or *residuals*. Their sum is equal to

$$(x_1 + x_2 + \cdots + x_N) - N\bar{x}$$

But we have by definition of the average

$$\frac{x_1 + x_2 + \cdots + x_N}{N} = \bar{x}$$

Hence: $(x_1 + x_2 + \cdots + x_N) - N\bar{x} = N\bar{x} - N\bar{x} = 0$. Thus if any $N - 1$ of the $N$ residuals are given, the remaining one can be exactly calculated. Therefore the $N$th residual contains no information not already contained in the $N - 1$ residuals. We may also state this by saying that the residuals, though $N$ in number, are equivalent to only $N - 1$ *independent* quantities. (The concept of independence will be discussed in more detail later in this chapter.) It is therefore plausible that the denominator $N - 1$, rather than $N$, should be used.

The preceding explanation does not do justice to all pertinent aspects of the concept of degrees of freedom. A thorough discussion of this concept requires a more mathematical treatment than can be given in this book. We will, however, make considerable use of degrees of freedom and thereby gain further insight into this important concept.

The sample estimate of a standard deviation, generally denoted by $s$ or $\hat{\sigma}$, is the square root of the variance estimate:

$$\hat{\sigma} = s = \sqrt{s^2} = \left[ \frac{\sum_i (x_i - \bar{x})^2}{N - 1} \right]^{1/2} \tag{3.23}$$

The terms mean, variance, and standard deviation are in common usage for both the population parameters and their sample estimates. We will make clear in what sense the term is used wherever confusion could arise. To avoid serious misunderstandings the distinction between a population parameter and its sample estimate should always be kept in mind.

## 3.8  THE HISTOGRAM

In the preceding section we have seen how the mean and the variance (or the standard deviation) of a population can be estimated from a random sample. Now, it is apparent from a comparison of Tables 3.1 and 3.3, that frequency distributions vary widely in their over-all characteristics. If for each of these two sets of data, $p_i$ is plotted against $x_i$, a marked difference is noted in the *shape* of the two frequency distributions. Is it possible to learn something about the *shape* of a distribution curve from a random sample taken from it?

**TABLE 3.5**  A Set of 75 Measurements of the Sheffield Smoothness of Paper

| Laboratory | Smoothness values | | | | |
|:---:|:---:|:---:|:---:|:---:|:---:|
| 1 | 173 | 185 | 141 | 133 | 160 |
| 2 | 135 | 165 | 145 | 160 | 120 |
| 3 | 165 | 175 | 155 | 135 | 160 |
| 4 | 185 | 137 | 162 | 125 | 157 |
| 5 | 108 | 112 | 113 | 137 | 146 |
| 6 | 120 | 135 | 140 | 135 | 130 |
| 7 | 125 | 138 | 125 | 155 | 173 |
| 8 | 135 | 155 | 140 | 155 | 140 |
| 9 | 98 | 93 | 142 | 106 | 114 |
| 10 | 150 | 140 | 130 | 155 | 150 |
| 11 | 136 | 169 | 165 | 146 | 132 |
| 12 | 104 | 132 | 113 | 77 | 134 |
| 13 | 127 | 145 | 142 | 145 | 148 |
| 14 | 120 | 125 | 185 | 130 | 120 |
| 15 | 125 | 178 | 143 | 118 | 148 |

The data in Table 3.5 constitute a sample of 75 measurements of the smoothness of a certain type of paper. The measurements were made in accordance with the Sheffield method (3), in sets of 5 replicates by each of 15 laboratories. In the present discussion we ignore possible systematic differences between laboratories, and consider the entire set of 75 values as a random sample from a single population.* In Table 3.6 the data are

* This assumption, which is not strictly correct, does not invalidate the conclusions of this chapter.

**TABLE 3.6**   Frequency Tally of Sheffield Smoothness Data (Table 3.5)

| Class symbol | Interval | Number of measurements |
|:---:|:---:|:---:|
| A | 76–85 | 1 |
| B | 86–95 | 1 |
| C | 96–105 | 2 |
| D | 106–115 | 6 |
| E | 116–125 | 10 |
| F | 126–135 | 14 |
| G | 136–145 | 14 |
| H | 146–155 | 11 |
| I | 156–165 | 8 |
| J | 166–175 | 4 |
| K | 176–185 | 4 |
| | Total | 75 |

classified as indicated: eleven classes are defined in terms of "intervals" of smoothness values, and the number of measurements occurring in each class is recorded.   For example, class *H* contains smoothness values lying between 146 and 155, and an examination of Table 3.5 shows that exactly 11 measurements lie in this interval.   Figure 3.3 is a plot of the class frequencies against the intervals as listed in Table 3.6.   Such a plot is

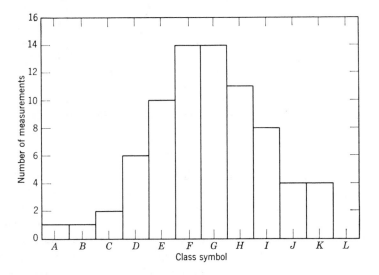

**Fig. 3.3**   Histogram for Sheffield smoothness data (data in Table 3.6).

called a *histogram* and represents, in fact, a sample-image of the frequency distribution of the population. It is plausible to assume that a histogram based on a much larger number of measurements and smaller intervals would show smoother transitions between the frequencies of adjacent class-intervals and that the limiting form of the histogram, for very large samples and very small intervals, is a smooth curve. The histogram based on 75 measurements gives us a reasonably accurate idea of the general shape of this smooth curve, but leaves us also with a number of unresolved questions. For example, there is a slight lack of symmetry in the histogram. Is this due to sampling fluctuations or does it reflect a similar situation in the population? We will not attempt, here, to answer this question or other questions of this type. We do observe, however, that 75 measurements of the same quantity constitutes a far larger number of replicates than is usually available and that even such a large sample gives a still imperfect picture of the population frequency distribution. This need not discourage us, for in general most of the pertinent questions can be satisfactorily answered without a complete knowledge of the frequency distribution. On the other hand, we may infer from Fig. 3.3 that the sampling fluctuations in small samples may attain serious proportions and this example should induce us to treat the conclusions drawn from small samples with a good deal of caution. We will have opportunities to express some of these matters in a more quantitative way.

**TABLE 3.7** Difference between Two Largest Values of Sets of Five Sheffield Smoothness Measurements

| Laboratory | Material | | | | |
|---|---|---|---|---|---|
| | $a$ | $b$ | $c$ | $d$ | $e$ |
| 1 | 12 | 5 | 2 | 9 | 5 |
| 2 | 5 | 20 | 20 | 5 | 5 |
| 3 | 10 | 15 | 20 | 5 | 20 |
| 4 | 23 | 12 | 1 | 5 | 15 |
| 5 | 9 | 6 | 2 | 4 | 11 |
| 6 | 5 | 5 | 15 | 20 | 5 |
| 7 | 18 | 12 | 10 | 1 | 23 |
| 8 | 0 | 3 | 10 | 32 | 20 |
| 9 | 28 | 3 | 2 | 4 | 2 |
| 10 | 5 | 0 | 5 | 40 | 0 |
| 11 | 4 | 11 | 5 | 9 | 3 |
| 12 | 2 | 12 | 1 | 16 | 8 |
| 13 | 3 | 3 | 3 | 8 | 10 |
| 14 | 55 | 0 | 25 | 15 | 10 |
| 15 | 30 | 6 | 14 | 14 | 1 |

To acquire more familiarity with frequency distributions and the histograms resulting from their sampling we now turn to a different example.

In Table 3.7, column *a* is derived from Table 3.5: each value in column *a* of Table 3.7 is the difference between the largest and next-to-largest value of the corresponding row of Table 3.5. For example, the value 9 listed for laboratory 5 in Table 3.7 is the difference between 146 and 137, the

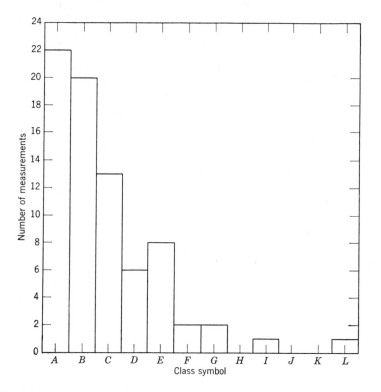

**Fig. 3.4**  Histogram for difference between two largest values of sets of five (data in Table 3.8).

two largest values listed for laboratory 5 in Table 3.5. Columns *b, c, d,* and *e* of Table 3.7 were obtained by a similar method, using the Sheffield smoothness values (in replicates of 5) obtained by the same laboratories for four other paper samples. In Table 3.8, the data of Table 3.7 are classified and counted and their histogram is shown in Fig. 3.4.

The difference between the histograms in Fig. 3.3 and 3.4 is striking. Whereas in Fig. 3.3 the most frequently occurring values are centrally located, in Fig. 3.4 the values of highest probability occur at the lower end of the range. Histograms should be constructed whenever enough data are available; they may then be considered as at least a rough approximation to the frequency distribution function of the population from which they are derived.

Applying Eqs. 3.21 and 3.23 to Tables 3.5 and 3.7 we obtain

for Table 3.5:*

$$\hat{\mu} = 140 \qquad \hat{\sigma} = 22 \qquad DF = 74$$

for Table 3.7:

$$\hat{\mu} = 10.4 \qquad \hat{\sigma} = 10.0 \qquad DF = 74$$

We can estimate these quantities much more rapidly—but with some loss of precision—using the "grouped data" of Tables 3.6 and 3.8. To

**TABLE 3.8** Frequency Tally for Difference of Two Largest Values of Sets of Five (Data of Table 3.7)

| Class symbol | Interval | Number of measurements |
|:---:|:---:|:---:|
| A | 0–4 | 22 |
| B | 5–9 | 20 |
| C | 10–14 | 13 |
| D | 15–19 | 6 |
| E | 20–24 | 8 |
| F | 25–29 | 2 |
| G | 30–34 | 2 |
| H | 35–39 | 0 |
| I | 40–44 | 1 |
| J | 45–49 | 0 |
| K | 50–54 | 0 |
| L | 55–59 | 1 |
| | | 75 |

this effect we consider all measurements in an interval as occurring at the center of the interval. For example, in Table 3.6 the midpoints of the intervals are 80.5, 90.5, ..., 180.5. Using this simplification, the estimate of the mean becomes:

$$\hat{\mu} = \frac{80.5 + 90.5 + (100.5 + 100.5) + \cdots + (180.5 + 180.5 + 180.5 + 180.5)}{75}$$

$$= \frac{(80.5 \times 1) + (90.5 \times 1) + (100.5 \times 2) + \cdots + (180.5 \times 4)}{1 + 1 + 2 + \cdots + 4} = 138.4$$

* The letters DF stand for "degrees of freedom."

We note that this calculation is expressed by the formula

$$\hat{\mu} = \frac{\sum_i m_i N_i}{\sum_i N_i} \tag{3.24}$$

where $m_i$ represents the midpoint of the $i$th interval and $N_i$ is the number of measurements in that interval. If we represent by $f_i$ the *relative frequency* corresponding to interval $i$, that is:

$$f_i = \frac{N_i}{\sum_i N_i} \tag{3.25}$$

we can write Eq. 3.24 in the form:

$$\hat{\mu} = \sum_i m_i f_i \tag{3.26}$$

This equation is analogous to Eq. 3.7, the definition of a population mean, and the observed frequency $f_i$ is the counterpart of the probability $p_i$.

Turning now to the estimation of the variance on the basis of the grouped data, we obtain by a similar argument:

$$\hat{\sigma}^2 = \frac{\sum_i (m_i - \hat{\mu})^2 N_i}{\sum_i N_i} = \sum_i (m_i - \hat{\mu})^2 f_i \tag{3.27}$$

Here again we observe the parallelism of Eqs. 3.27 and 3.12.

Using Eqs. 3.26 and 3.27 we obtain the following estimates:

for Table 3.6:

$$\hat{\mu} = 138.4 \qquad \hat{\sigma} = 21.3$$

for Table 3.8:

$$\hat{\mu} = 11.1 \qquad \hat{\sigma} = 10.1$$

## 3.9   THE NORMAL DISTRIBUTION

We have noted the distinct difference between the histograms shown in Figs. 3.3 and 3.4. The former suggests a symmetrical, or almost symmetrical, distribution function while the other is definitely skew. Even among symmetrical distribution functions there exist many different types, but one of these types was destined to acquire a central position in statistical theory and practice. Nor was this a mere coincidence, for that particular type, known as the *normal* or *Gaussian* curve (also as the *error function*) fully deserves its place of eminence. The reasons for this will be explained in due course. The normal distribution really comprises a family of curves differing from each other either in their mean or variance, or in both. If two normal distributions have the same mean and the same variance, they are completely identical. *Thus a distribution is fully characterized if it is known to be normal and its mean and its variance are given.*

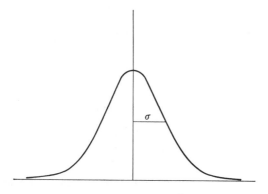

**Fig. 3.5** The normal curve.

The density function of a normal distribution of mean $\mu$ and variance $\sigma^2$ is given by the expression:

$$f(x) = \frac{1}{\sigma\sqrt{2\pi}} \exp\left\{-\frac{1}{2}\left(\frac{x-\mu}{\sigma}\right)^2\right\} \qquad (3.28)$$

It can be shown that the normal curve is symmetrical about its mean $\mu$ and that its standard deviation $\sigma$ is the distance between the location of the mean and that of the "point of inflection" of the curve, i.e., the point at which the curvature of the line changes (see Fig. 3.5).

We have seen previously that the probabilities for a continuous variate are measured by elements of area under the curve representing the probability density function. Consider an interval starting at $x$ and ending at $x + dx$. The corresponding probability is measured by the area of the rectangle of height $f(x)$ and of base $dx$.

$$\text{Probability} = f(x)dx$$

In view of Eq. 3.28, this probability can be written

$$\text{Probability} = \frac{1}{\sigma\sqrt{2\pi}} \exp\left\{-\frac{1}{2}\left(\frac{x-\mu}{\sigma}\right)^2\right\} dx \tag{3.29}$$

Suppose now that we change the scale of $x$, by measuring each $x$ value by its distance from the mean $\mu$ and adopting the standard deviation $\sigma$ as the unit of measurement. In this new scale, any value $x$ is replaced by the value

$$y = (x - \mu)/\sigma \tag{3.30}$$

From this equation we derive

$$x = \mu + y\sigma \tag{3.31}$$

If we now express the probability area in terms of the $y$ scale, we obtain from Eq. 3.29

$$\text{Probability} = \frac{1}{\sigma\sqrt{2\pi}} e^{-y^2/2} \, d(\mu + y\sigma)$$

Since $\mu$ and $\sigma$ are constants, we can write $d(\mu + y\sigma) = \sigma dy$. Consequently we obtain

$$\text{Probability} = \frac{1}{\sqrt{2\pi}} e^{-y^2/2} \, dy \tag{3.32}$$

The quantity $y$ is called the *standard normal deviate*. Of great importance is the fact that the probability distribution of the standard normal deviate contains no unknown parameters, as is seen by observing that the parameters $\mu$ and $\sigma$ have been eliminated in Eq. 3.32. Looking at it from a slightly different point of view, we can consider Eq. 3.32 as a particular case of Eq. 3.29, for $\mu = 0$ and $\sigma = 1$. Thus the standard normal deviate is a normal distribution with zero mean and unit standard deviation. We now see that *all* normal distributions are related to the standard normal deviate and can be reduced to it by a simple transformation in scale. This transformation is given by Eq. 3.30 or its equivalent, Eq. 3.31. If one possesses a table giving the probability values of the standard normal deviate, these probability values may be used for any normal distribution. To an interval $x_1$ to $x_2$ in the original normal distribution corresponds the "reduced" interval

$$\frac{x_1 - \mu}{\sigma} \qquad \text{to} \qquad \frac{x_2 - \mu}{\sigma}$$

of the standard normal deviate and the probabilities associated with these intervals are equal.

Suppose, for example, that it is required to find the probability that the normal variate $x$, of mean 140 and standard deviation 25, be located between 160 and 170. Construct the "reduced" interval:

$$\frac{160 - 140}{25} \quad \text{to} \quad \frac{170 - 140}{25}$$

or 0.80 to 1.20. The desired probability is therefore simply the probability that the standard normal deviate be situated in the interval 0.80 to 1.20.

The probability areas of a continuous distribution function can be tabulated in essentially two ways.

1. For a selected set of values, $x_1, x_2, \ldots$, the cumulative probabilities are tabulated, as illustrated in the following short table of the standard normal deviate:

| $x$ | 0 | 0.40 | 0.80 | 1.20 | 1.60 | 2.00 | 2.40 |
|---|---|---|---|---|---|---|---|
| Cumul. prob. | 0.500 | 0.655 | 0.788 | 0.885 | 0.945 | 0.977 | 0.992 |

From this table one infers, for example, that the probability for the interval 0.80 to 1.20 (the values of the example above) is $0.885 - 0.788 = 0.097$. In other words, 9.7 per cent of all values of a standard normal deviate fall between 0.80 and 1.20.

Tables of this type are contained in most handbooks of physics, chemistry, and engineering, as well as in practically every textbook on statistics. A fairly brief version is given in the Appendix as Table I.

2. A second way of tabulating a continuous distribution function is to select a sequence of probabilities of particular interest and to list the corresponding values of the variate. This method is illustrated by the following short table of the standard normal deviate.

| Cumul. prob. | 0.01 | 0.05 | 0.20 | 0.50 | 0.80 | 0.95 | 0.99 |
|---|---|---|---|---|---|---|---|
| $x$ | $-2.326$ | $-1.645$ | $-0.842$ | 0 | 0.842 | 1.645 | 2.326 |

This table may be used to answer questions of the following type: "Find that value of the standard normal deviate that will be exceeded only once in a hundred times." Since the value is to be exceeded one per cent of the time, the probability of *not* exceeding it is 99 per cent or 0.99. Thus, we are to find the value for which the cumulative probability is 0.99. From the table we see that this value is 2.326.

The values of $x$ in this second type of tabulation are known as *quantiles*. More exactly, a *p-quantile* is that value of the variate for which the cumulative probability is $p$. If $p$ is expressed in *per cent* probability, the quantile is called a *percentage point* or *percentile*. A quantile of particular importance is the *median* of a distribution. This is simply the 50 per cent quantile, or the 50 percentile. In other words, the median is that value which is as likely to be exceeded as not to be exceeded. Put differently, it is a value such that 50 per cent of all values fall below it and 50 per cent above. In any normal distribution the median coincides with the mean. The median of the distribution of the standard normal deviate is zero.

For many types of measurements, the experimental error tends to be distributed, at least approximately, in accordance with a normal distribution. At one time it was believed that this was a law of nature, and any series of measurements that failed to show the "normal" behavior was considered to be essentially faulty. Today we know this belief to be incorrect. Nevertheless the normal distribution is of the greatest importance in statistics, mainly because of the *central limit theorem* which will be explained in Section 4.3.

## 3.10 RELATION BETWEEN TWO RANDOM VARIABLES

Consider the following hypothetical experiment. From a homogeneous material of given density, a number of perfectly spherical specimens of different diameters are manufactured. For example the material might be steel and the specimens ball-bearings of different sizes. Each specimen is weighed, and its diameter is measured by means of a micrometer.

It is clear that apart from errors of measurement there is a mathematically precise relationship between the diameter and the weight:

$$\text{Weight} = (\text{specific gravity})(\pi/6)(\text{diameter})^3 \qquad (3.33)$$

Let $\varepsilon$ denote the experimental error in the measured diameter $d$, and $\delta$ the experimental error in the measured weight $w$. We then have: $w = \text{weight} + \delta$ and $d = \text{diameter} + \varepsilon$. It then follows from Eq. 3.33 that:

$$w - \delta = (\text{specific gravity})(\pi/6)(d - \varepsilon)^3 \qquad (3.34)$$

Thus there exists a functional relationship between the quantities $d$ and $w$. Nevertheless, we state that $d$ and $w$ are *statistically independent* measurements. By that we mean that the error $\varepsilon$ in the measurement of the diameter of any particular ball had no effect upon the error $\delta$, committed in weighing this ball, and vice-versa. In other words, if it were possible to determine the exact value of $\varepsilon$ for any particular ball, this knowledge would throw no light whatsoever on the value of $\delta$ for that ball. Conversely, knowledge of $\delta$ throws no light whatsoever on the corresponding value of $\varepsilon$. Thus the concept of statistical independence relates to the fluctuating parts $\varepsilon$ and $\delta$ of the measurements, not to their true values.

The true diameter of any given ball is equal to the expected value of all diameter measurements on that ball, provided that no systematic error exists in these length measurements. (In fact, this is the very definition of a *systematic error*\*.) Consequently, Eq. 3.33 is a relation between the *expected values* of populations of length and weight measurements. Functional dependence relates to the expected values of statistical populations, whereas statistical dependence relates to the *deviation* of observations from their expected values. As the example shows, statistical independence is by no means incompatible with the existence of a functional relationship. Conversely, one may have statistical dependence whether or not there exists a functional relationship between the expected values of the variables. A frequent cause of statistical dependence is the existence of a common part in the errors of related measurements. If, in an alloy of copper, lead, and antimony, the latter two constituents are directly determined, while the copper content is obtained by difference; i.e.,

$$\%\mathrm{Cu} = 100 - [\%\mathrm{Pb} + \%\mathrm{Sb}]$$

then any error in the determination of either lead or antimony will be reflected in the error of the copper content. The total error in per cent Cu is the sum of the errors in per cent Pb and per cent Sb. Consequently, there will be statistical dependence between the value obtained for per cent Cu and that of either per cent Pb or per cent Sb.

We will now discuss the matter in a more general way. Consider two random variables, $x$ and $y$, and their associated distribution functions $f_1(x)$ and $f_2(y)$. Let $x'$ and $y'$ represent the actual outcomes of an experiment in which both $x$ and $y$ are measured. Now consider the deviations $x - E(x)$ and $y - E(y)$. If the probability that the deviation $x - E(x)$ assume a certain given value is in no way affected by the value assumed by the deviation $y - E(y)$, and vice-versa, the random variables $x$ and $y$ are said to be *statistically independent*. For independent variables, the *joint* frequency distribution (i.e., the probability function for the joint occurrence of $x \leqslant x' \leqslant x + dx$ and $y \leqslant y' \leqslant y + dy$) is equal to the product of the distribution functions of $x$ and $y$, i.e., $f_1(x) \cdot f_2(y)$.

If, on the other hand, the probability for a certain deviation $x - E(x)$ is affected by the value taken by $y - E(y)$, or vice-versa, the variables $x$ and $y$ are statistically dependent. Their joint distribution is then no longer the product of $f_1(x)$ and $f_2(y)$.

The reader will note that we have defined statistical dependence in terms of the *deviations* $x - E(x)$ and $y - E(y)$, rather than in terms of $x$ and $y$. The distinction was made in order to avoid confusion with mathematical (or functional) dependence.

\* See Chapter 6.

A commonly used measure of statistical association is the *covariance* of two random variables. Formally the covariance of the random variables $x$ and $y$, denoted Cov $(x, y)$, is defined as the expected value of the product of $[x - E(x)]$ by $[y - E(y)]$.

$$\text{Cov}\,(x, y) = E[x - E(x)] \cdot [y - E(y)] \qquad (3.35)$$

To understand this definition, consider the data in the first two columns of Table 3.9, representing two random drawings of 10 one-digit numbers from a table of random numbers.* Column 3 represents the deviations

**TABLE 3.9**   The Concept of Covariance

| Column 1 | Column 2 | Column 3 | Column 4 | Column 5 |
|---|---|---|---|---|
| $x$ | $y$ | $x - \bar{x}$ | $y - \bar{y}$ | $(x - \bar{x})(y - \bar{y})$ |
| 2 | 6 | −2.4 | 1.9 | −4.56 |
| 0 | 2 | −4.4 | −2.1 | +9.24 |
| 9 | 4 | 4.6 | −0.1 | −0.46 |
| 9 | 7 | 4.6 | 2.9 | +13.34 |
| 4 | 7 | −0.4 | 2.9 | −1.16 |
| 7 | 3 | 2.6 | −1.1 | −2.88 |
| 5 | 3 | 0.6 | −1.1 | −0.66 |
| 0 | 2 | −4.4 | −2.1 | +9.24 |
| 7 | 0 | 2.6 | −4.1 | −10.66 |
| 1 | 7 | −3.4 | 2.9 | −9.86 |
| 44 | 41 | 0 | 0 | 1.6 |

Average =       Average =                      Estimate of covariance =

$\bar{x} = \dfrac{44}{10} = 4.4$   $\bar{y} = \dfrac{41}{10} = 4.1$         $\dfrac{1.6}{9} = 0.1778$

of the numbers of column 1 from their average. Column 4 is similarly obtained from column 2. In column 5, the product is formed of corresponding elements in columns 3 and 4. Finally, the "average" is formed of the numbers in column 5. For reasons analogous to those pertaining to estimates of standard deviations, the divisor is $N - 1$, not $N$. Thus in the present case, the divisor is 9. This average is the *sample estimate of the covariance* of $x$ and $y$. If the same calculations were carried out on

---

* Tables of random numbers are tables in which the digits 0 to 9 are assembled in rows and columns in a completely random fashion. A short table is given in the Appendix (Table VII).

larger and larger samples from the two populations represented by the symbols $x$ and $y$, the resulting covariance estimates would approach a limiting value, representing the *population covariance* of $x$ and $y$.    A little reflection shows that if an association had existed between $x$ and $y$, the covariance would have tended to be larger.    More specifically, if $y$ had tended to increase or decrease *along with* $x$, the covariance would have tended to be large and positive.    If, on the other hand, $y$ had tended to *increase* as $x$ *decreased*, or vice-versa, their covariance would have tended to be a large negative number.    It is also easily seen that the magnitude of the covariance depends not only on the degree of association of $x$ and $y$, but also on the magnitude of the deviations $x - \bar{x}$ and $y - \bar{y}$. For example, multiplication of all the numbers in the second column by 10 would have caused a ten-fold increase in those of column 4 and consequently also in the covariance.    Therefore the covariance must be "standardized" before it can be used as a generally applicable measure of association.    This is done by dividing the covariance by the product of the standard deviations of $x$ and $y$.    The quantity thus obtained is called the *correlation coefficient* of $x$ and $y$, denoted $\rho(x, y)$.    Thus:

$$\rho(x, y) = \frac{\text{Cov}(x, y)}{\sigma_x \cdot \sigma_y} \tag{3.36}$$

As is the case for the covariance, the correlations coefficient is a parameter pertaining to *populations*, but its value can be *estimated* from a sample.    For example, from the data of Table 3.9, we estimate the correlation coefficient between $x$ and $y$ to be:

$$\hat{\rho}(x, y) = \frac{\sum (x - \bar{x})(y - \bar{y})/9}{\sqrt{\frac{\sum (x - \bar{x})^2}{9}} \sqrt{\frac{\sum (y - \bar{y})^2}{9}}} = \frac{1.6/9}{3.53 \times 2.51} = \frac{0.1778}{8.86} = 0.020 \tag{3.37}$$

In the present case, the correlation coefficient is not far from zero.    It can be shown that a correlation coefficient always lies between $-1$ and $+1$. The former value, $-1$, denotes perfect *negative association* ($y$ decreases as $x$ increases).    The latter value, $+1$, denotes perfect *positive association* ($y$ increases as $x$ increases).    We will not dwell on this concept since the types of association currently encountered in the physical sciences can generally be treated by more satisfactory methods.    We merely mention the following theorem:

If two random variables $x$ and $y$ are statistically independent, their correlation coefficient is zero.    However, the converse is not always true; i.e., if $\rho(x, y) = 0$, this does not necessarily imply that $x$ and $y$ are statistically independent.

Finally, let us observe that the covariance of a random variable $x$ with itself is simply equal to the variance of $x$:

$$\text{Cov}(x, x) = V(x) \tag{3.38}$$

## 3.11  SUMMARY

The basic mathematical concept underlying statistical theory is a probability distribution associated with a random variable. Essentially this is a function describing the relative frequencies of occurrence of all possible values of a quantity that is subject to chance fluctuations. Probability distribution functions can follow many different mathematical patterns. All, however, can be described in terms of certain parameters called moments. Of primary importance are the first two moments of a distribution function, called the mean and the variance, both of which can be defined in terms of the concept of an expected value. The mean is basically a measure of location whereas the variance is a measure of spread (dispersion). The square root of the variance is called the standard deviation.

In practice there is seldom, if ever, enough information available to determine the probability distribution function of a random variable. A statistical sample is the collection of data actually obtained through observation or experiment. In contrast, the totality of *all* possible values of the random variable is denoted as the population. When the sample satisfies the condition of randomness, that is all avoidance of conscious or unconscious bias in the selection of values from the population, inferences can be made from the sample about the probability distribution function of the population. A useful tool for doing this is the histogram, which may be considered as an imperfect image of the population distribution function. The mean and variance can be estimated from a statistical sample by means of simple formulas. Such estimates are of course subject to uncertainty, and the smaller the sample, the greater is the uncertainty of the estimates.

Among the great variety of probability distribution functions, one plays a dominant role in statistical theory; this is the Gaussian, or normal distribution. A normal distribution is completely determined when its mean and variance are known.

One often deals with two or more random variables, the fluctuations of which are not independent of each other. A measure of association of considerable theoretical importance is the covariance of two random variables. As in the case for the mean and the variance, the covariance can be estimated, by means of a simple mathematical formula, on the basis of a joint sample from the two populations.

## REFERENCES

1. Brownlee, K. A., *Statistical Theory and Methodology in Science and Engineering*, Wiley, New York, 1960.
2. Courant, R., *Differential and Integral Calculus*, Nordeman, New York, 1945.
3. Lashof, T. W., and J. Mandel, "Measurement of the Smoothness of Paper," *Tappi*, **43**, 385–389 (1960).
4. *Symbols, Definitions, and Tables, Industrial Statistics and Quality Control*, Rochester Institute of Technology, Rochester, New York, 1958.

*chapter 4*

# THE MATHEMATICAL FRAME-WORK OF STATISTICS, PART II

## 4.1 THE COMBINATION OF RANDOM VARIABLES

Suppose that in the course of a physical experiment a time-duration is measured by means of a stopwatch. The correctness of the value obtained is limited in two ways: by the uncertainty in starting the watch at the correct instant and by the uncertainty in stopping it.* If $x$ represents the time of start and $y$ the time of stopping, the duration is measured by the quantity

$$d = y - x \tag{4.1}$$

From a statistical viewpoint, both $x$ and $y$ are random variables, and the quantity of interest $d$ is obtained by combining them a definite way, in this instance by simple subtraction. It is clear that a multitude of measured quantities are the result of combining two or more random variables. Even the simple operation of taking an average of several measurements involves a combination of random variables: from the relation

$$\bar{x} = \frac{x_1 + x_2 + \cdots + x_N}{N} = \frac{1}{N} x_1 + \frac{1}{N} x_2 + \cdots + \frac{1}{N} x_N \tag{4.2}$$

it is seen that an average is obtained by multiplying each of $N$ random variables by $1/N$ and adding the results.

* In the latter we include also any error committed in *reading* the stopwatch.

Both Eqs. 4.1 and 4.2 are examples of *linear combinations* of random variables. An example of non-linear combination is the determination of a density as the quotient of a measure of mass by a measure of volume.

Any combination of random variables, whether linear or non-linear, results in a new random variable, with its own probability distribution, expected value, variance, and moments of higher order. In this book, we will be concerned only with the expected values and variances of combinations of random variables.

Consider first a single variate $x$, for example a length expressed in centimeters. If a change of scale is made from centimeters to millimeters, the length in the new scale is expressed by the quantity $y = 10x$. More generally, we are interested in the derived quantity $y = kx$ where $x$ is a random variable, and $k$ a constant.

We first recall an important property of expected values (see Eq. 3.9):

$$E(kx) = kE(x)$$

From this relation, and the definition of a variance as the expected value of a squared deviation, it follows that:

$$V(kx) = k^2 V(x) \tag{4.3}$$

From Eq. 4.3 it follows that

$$\sigma_{kx} = k\sigma_x \tag{4.4}$$

where $k$ is a positive constant.

Consider now two random variables $x$ and $y$, and let $a$ and $b$ represent two constants. Consider the linear combination

$$L = ax + by$$

Then it can be shown (using Eqs. 3.9 and 3.10) that:

$$E(L) = aE(x) + bE(y)$$

More generally:

$$E[ax + by + cz + \cdots] = aE(x) + bE(y) + cE(z) + \cdots \tag{4.5}$$

From this general equation we can derive, among other things, an important theorem concerning the mean of a sample average. Consider a sample of size $N$ taken from a population of mean $\mu$. Represent the sample by $x_1, x_2, \ldots, x_N$, and its average by $\bar{x}$. Thus:

$$\bar{x} = \frac{x_1 + x_2 + \cdots + x_N}{N}$$

This relation can be written:

$$\bar{x} = \frac{1}{N} x_1 + \frac{1}{N} x_2 + \cdots + \frac{1}{N} x_N$$

Applying Eq. 4.5 to this relation, we obtain:

$$E(\bar{x}) = \frac{1}{N} E(x_1) + \frac{1}{N} E(x_2) + \cdots + \frac{1}{N} E(x_N)$$

But $E(x_1) = E(x_2) = \cdots = E(x_N) = \mu$, since the $x$ values all belong to the same population of mean $\mu$. Therefore:

$$E(\bar{x}) = N \left[ \frac{1}{N} \mu \right] = \mu$$

Thus:

$$E(\bar{x}) = E(x) \tag{4.6}$$

We state therefore that the population of the $\bar{x}$ has the same mean as that of the $x$ values. We now turn to the variance of linear combinations of random variables. Here it is necessary to distinguish two cases.

1. When two random variables $x$ and $y$ are *statistically independent*, their variances are additive:

$$V(x + y) = V(x) + V(y) \tag{4.7}$$

More generally, when $x$ and $y$ are statistically independent, we may write:

$$V(ax + by) = a^2 V(x) + b^2 V(y) \tag{4.8}$$

This rule applies to any number of *statistically independent* variables:

$$V(ax + by + cz + \cdots) = a^2 V(x) + b^2 V(y) + c^2 V(z) + \cdots \tag{4.9}$$

2. Let $x$ and $y$ be two random variables that are not necessarily statistically independent. Then it is no longer permissible to simply add their variances. The degree of interdependence is taken into account through introduction of their covariance. It can be shown that for non-independent variates:

$$V(x + y) = V(x) + V(y) + 2 \, \text{Cov} \, (x, y) \tag{4.10}$$

We know that for statistically independent variates the covariance is zero. Thus Eq. 4.10 includes Eq. 4.7 as a special case. Both these equations and many more can be derived from the following very general rule. Let $x$ and $y$ be two variates, each of which is a linear combination of other random variables, $u, v, w, \ldots$ ; $u', v', w', \ldots$ ;

$$x = au + bv + cw + \cdots$$
$$y = a'u' + b'v' + c'w' + \cdots \tag{4.11}$$

Then the covariance of $x$ and $y$ is obtained by a procedure similar to the familiar rule for the multiplication of two sums:

$$
\begin{aligned}
\text{Cov}(x, y) = {} & aa'\,\text{Cov}(u, u') + ab'\,\text{Cov}(u, v') + ac'\,\text{Cov}(u, w') + \cdots \\
& + ba'\,\text{Cov}(v, u') + bb'\,\text{Cov}(v, v') + bc'\,\text{Cov}(v, w') + \cdots \\
& + ca'\,\text{Cov}(w, u') + cb'\,\text{Cov}(w, v') + cc'\,\text{Cov}(w, w') + \cdots \\
& + \cdots
\end{aligned}
\tag{4.12}
$$

Now, some of the variates $u, v, w, \ldots$ could occur in both $x$ and $y$, in which case Eq. 4.12 would contain such terms as, for example,

$$aa'\,\text{Cov}(u, u)$$

We have already seen (Eq. 3.38) that $\text{Cov}(u, u) = V(u)$. Let us consider a few examples of the application of Eq. 4.12.

1. $x = u$, $y = u + v$.
   Then:
   $$
   \begin{aligned}
   \text{Cov}(x, y) &= \text{Cov}(u, u) + \text{Cov}(u, v) \\
   &= V(u) + \text{Cov}(u, v)
   \end{aligned}
   $$

If in this example $u$ and $v$ are statistically independent, $\text{Cov}(x, y) = V(u)$.

2. $x = u + v$, $y = u - v$.
   Then:
   $$
   \begin{aligned}
   \text{Cov}(x, y) &= \text{Cov}(u, u) - \text{Cov}(u, v) + \text{Cov}(v, u) - \text{Cov}(v, v) \\
   &= V(u) - V(v)
   \end{aligned}
   $$

3. $x = u + v$, $y = u + v$.
   Then:
   $$
   \begin{aligned}
   \text{Cov}(x, y) &= \text{Cov}(u, u) + \text{Cov}(u, v) + \text{Cov}(v, u) + \text{Cov}(v, v) \\
   &= V(u) + V(v) + 2\,\text{Cov}(u, v)
   \end{aligned}
   $$

But since in this case $x = y$, we have:

$$V(x) = V(u) + V(v) + 2\,\text{Cov}(u, v)$$

which is identical with Eq. 4.10.

4. $x = u - v$, $y = u - v$.
   Then, again $x = y$, and we obtain:
   $$\text{Cov}(x, y) = V(x) = V(u) + V(v) - 2\,\text{Cov}(u, v)$$

If, in this example, $u$ and $v$ are statistically independent, we obtain:

$$V(u - v) = V(u) + V(v)$$

Note that despite the negative sign in $u - v$, the variances of $u$ and $v$ are *added* (not subtracted). To understand why this is so, consider again the time-duration measured by means of a stopwatch. According to Eq. 4.1, $d = y - x$. If the errors in starting and stopping the watch are independent of each other, we have Cov $(x, y) = 0$, and according to example 4:

$$V(d) = V(y) + V(x)$$

Obviously, since both $x$ and $y$ are uncertain, the error in the duration $y - x$ is *larger* than that of either $x$ or $y$. We must therefore have an *addition* (not a subtraction) of variances, despite the fact that the duration is a difference of two time-instants. This reasoning would no longer be valid if the errors in starting and stopping the watch were *correlated* (i.e., if their covariance were different from zero). For example if the operator had a tendency to respond with a constant delay, both in starting and stopping, the covariance of $x$ and $y$ would be positive. In the formula

$$V(d) = V(y) + V(x) - 2 \operatorname{Cov}(x, y)$$

the last term would then tend to *reduce* the variance of $d$. This expresses the obvious fact that any part of the error which is the same for $x$ and $y$, both in magnitude and in sign, adds no uncertainty to the measurement of the difference $y - x$.

Equation 4.9, which applies to statistically independent variates, is particularly important. It represents a simple form of the *law of propagation of errors*, which we will study in more detail in Section 4.7.

## 4.2 THE VARIANCE OF A SAMPLE MEAN

We all have the intuitive feeling that an average of replicate observations is more reliable than an individual measurement. A quantitative measure for the improvement in precision due to replication is easily derived from the principles given in the preceding section.

Let $x_1, x_2, \ldots, x_N$ be observations taken at random from a particular statistical population (not necessarily normal), of variance $V(x)$. The random selection assures complete statistical independence of the $N$ observations. We now compute the variance of the average, $\bar{x}$, of the $N$ replicates. We have

$$\bar{x} = \frac{x_1 + x_2 + \cdots + x_N}{N} = \frac{1}{N}x_1 + \frac{1}{N}x_2 + \frac{1}{N}x_3 + \cdots + \frac{1}{N}x_N$$

Applying Eq. 4.9 to $\bar{x}$ (which is legitimate because of the statistical independence of the replicates) we obtain:

$$V(\bar{x}) = \left(\frac{1}{N}\right)^2 V(x_1) + \left(\frac{1}{N}\right)^2 V(x_2) + \cdots + \left(\frac{1}{N}\right)^2 V(x_N)$$

But $V(x_1) = V(x_2) = \cdots = V(x_N)$, since all replicates belong to the same statistical population, and this common variance is equal to $V(x)$, the variance of this population. Therefore:

$$V(\bar{x}) = \left(\frac{1}{N}\right)^2 V(x) + \left(\frac{1}{N}\right)^2 V(x) + \cdots + \left(\frac{1}{N}\right)^2 V(x)$$
$$= N\left[\left(\frac{1}{N}\right)^2 V(x)\right] = \frac{V(x)}{N}$$

Thus, the variance of a sample-mean is given by the equation:

$$V(\bar{x}) = \frac{V(x)}{N} \qquad (4.13)$$

From Eq. 4.13, it follows that

$$\sigma_{\bar{x}} = \frac{\sigma_x}{\sqrt{N}} \qquad (4.14)$$

The standard deviation $\sigma_{\bar{x}}$ is often referred to as the *standard error* of the mean.

Equations 4.13 and 4.14 are of fundamental importance. They are true for any population and require only that the $N$ values averaged be statistically independent observations from the same population. A careful examination of the proof shows that the formulas keep their validity even when the $N$ observations are taken from $N$ different populations, provided that the observations are statistically independent and that the $N$ populations, though different, all have the same variance, $V(x)$.

## 4.3   THE CENTRAL LIMIT THEOREM

In the preceding section we have observed that the formula expressing the variance of a mean of independent observations is of general validity and depends in no way on the nature of the distribution of the observations. We will now see that the usefulness of this theorem is further increased as the result of a most remarkable property of average values.

Suppose that we made a statistical study of the birthdays of a large assembly of people—for example of all persons attending a major political rally. If we retained only the day of the month, disregarding both the month and the year of birth, it is very likely that the number of people born on any particular day of the month, except the 31st, would be about the same. If we agree to disregard the entire group of people born on the 31st of any month, and represent by $x_i$ the day of birth for any person in the remaining sample, then the probability $p_i$ corresponding to $x_i$ is 1/30, for all $x_i = 1, 2, \ldots, 30$. The distribution of $x_i$ is said to be *uniform*. Its probability function is represented by 30 equidistant points situated on a horizontal line of ordinate 1/30. Suppose that we divide the entire group of people into sets of five persons *in a random manner*, and form the *averages* of birthdays (defined as before) for each group, rejecting any group containing one or more persons born on the 31st of any month. What is the frequency distribution of these averages? It turns out that this distribution is no longer uniform; the probability first increases as the variate (the average of random samples of 5) increases, reaches a maximum, and then decreases. This frequency distribution will show a noticeable resemblance with the Gaussian frequency distribution. If we had taken averages of 10, instead of 5, the resemblance would even be greater. More generally, as N, the number of observations in each average, increases, the frequency distribution of the averages actually approaches more and more the Gaussian distribution. These facts are known in statistical theory as the *central limit theorem*, which we may express as follows:*

*Given a population of values with a finite (non-infinite) variance, if we take independent samples from this population, all of size N, then the population formed by the averages of these samples will tend to have a Gaussian (normal) distribution, regardless of what the distribution is of the original population; the larger N, the greater will be this tendency towards "normality."* In simpler words: *The frequency distribution of sample averages approaches normality, and the larger the samples, the closer is the approach.*

Particularly noteworthy is the great generality of this theorem. The restriction that the population from which the samples are taken must have a finite (non-infinite) variance is of no practical importance, since all but a few very special populations possess finite variances.

Mathematically, the central limit theorem is an *asymptotic* law, i.e., the complete identity with the Gaussian distribution actually never takes place (unless the original population is itself Gaussian), but it is approached more and more as N increases. In practice, the approach is remarkably rapid. Even when we start with a population radically different from the Gaussian, for example the population illustrated by the data in Table 3.8 and Fig. 3.4, we obtain a reasonably close approximation to the Gaussian by taking averages of as few as four random observations.

The central limit theorem, when used in conjunction with the laws governing the mean and the variance of averages (Eqs. 4.6 and 4.13), is a powerful tool in practical statistical work.

Let $x$ represent a random variable with population mean $\mu$ and population variance $V(x)$. We need make no assumption about the nature of

---

* Our definition is not the most general possible; for a more complete account, see reference 1, at the end of this chapter.

the probability distribution of $x$. Consider the distribution of averages of $N$ randomly selected observations $x$. Let $\bar{x}$ represent any such average. Then we know that the mean of $\bar{x}$ is $\mu$, that its variance is $V(x)/N$ and that, in accordance with the central limit theorem, its distribution is approximately normal. We can make use of these facts in two ways. If $\mu$ and $V(x)$ are known, they provide a usable approximation for the distribution of $\bar{x}$, namely, the normal distribution of mean $\mu$ and variance $V(x)/N$. Conversely, if we can observe a sufficiently large sample from the population of $\bar{x}$ values we can deduce from it estimates of $\mu$ and $V(x)$.

## 4.4 ESTIMATION FROM SEVERAL SAMPLES

The following situation arises frequently in practice: a sample is taken from each of a number of statistical populations. The populations may differ from each other, but they have a common parameter. The problem consists in finding the best estimate for this parameter making appropriate use of the information provided by *all* the samples. We shall discuss two important examples.

1. Suppose that a particular quantity, for example a fundamental physical constant, has been measured independently by three different methods $A$, $B$, and $C$, all unbiased (i.e., free of systematic errors) but each having its own precision. Assume that the standard deviation that measures the variability among repeated measurements by method $A$ is $\sigma_A$, the standard deviations for methods $B$ and $C$ being $\sigma_B$ and $\sigma_C$, respectively. Finally, let $N_A$, $N_B$, and $N_C$ represent the number of replicate measurements made by the three methods respectively.

The average of the $N_A$ measurements made by method $A$ has a standard error equal to $\sigma_A/\sqrt{N_A}$. Similarly, the averages corresponding to methods $B$ and $C$ have standard errors equal to $\sigma_B/\sqrt{N_B}$ and $\sigma_C/\sqrt{N_C}$ respectively. The populations representing the values obtained by the three methods have, according to our assumption, all the same mean $\mu$, namely the "true value" of the quantity that is being measured. Let $\bar{x}_A$, $\bar{x}_B$, and $\bar{x}_C$ represent the averages of the replicate measurements obtained by the three methods. Each of these three quantities is an unbiased estimate of $\mu$. What is the best method for combining them in an over-all estimate of this quantity?

We will return to this problem in Chapter 7, where we will show that the solution to this problem is given by the expression

$$\hat{\mu} = \frac{w_A\bar{x}_A + w_B\bar{x}_B + w_C\bar{x}_C}{w_A + w_B + w_C} \qquad (4.15)$$

in which the "weights" $w_A$, $w_B$, and $w_C$ are the reciprocals of the variances of $\bar{x}_A$, $\bar{x}_B$, and $\bar{x}_C$; i.e.,

$$w_A = \frac{N_A}{\sigma_A^2}, \qquad w_B = \frac{N_B}{\sigma_B^2}, \qquad w_C = \frac{N_C}{\sigma_C^2} \qquad (4.16)$$

The quantity expressed by Eq. 4.15 is called a *weighted average* of $\bar{x}_A$, $\bar{x}_B$, $\bar{x}_C$, the weights of these three quantities being respectively, $w_A$, $w_B$, and $w_C$.

2. Our second example of the combination of samples from different populations is provided by the problem of the *pooling of variances*. The previous example dealt with different populations having a common mean. We now consider a number of populations of *different* means, but with a common variance. The problem here is to find the best estimate of this common variance.

To be more specific, let us assume that a routine analytical method has been in use for some time in a laboratory and that we wish to estimate its precision from the information that has been accumulated in its routine use. An inspection of the laboratory notebook reveals that the method has been applied to four chemical samples, as shown in Table 4.1.

**TABLE 4.1**

| Sample | Number of determinations | Results |
|--------|--------------------------|---------|
| $A$ | 4 | $x_1, x_2, x_3, x_4$ |
| $B$ | 2 | $y_1, y_2$ |
| $C$ | 2 | $z_1, z_2$ |
| $D$ | 3 | $u_1, u_2, u_3$ |

Let us assume that we have valid reasons for believing that the precision of the method is constant and is in no way affected by the fact that the magnitudes of the values $x$, $y$, $z$, and $u$ are distinctly different, nor by possible small differences in nature between the four samples. Then the statistical populations corresponding to the four samples all have a common variance $\sigma^2$, but they do not have the same mean. Our problem is to estimate the common parameter $\sigma^2$, using all four samples in the most effective way.

If only sample $A$ were available, the estimate of variance would be given by the quantity

$$\frac{(x_1 - \bar{x})^2 + (x_2 - \bar{x})^2 + (x_3 - \bar{x})^2 + (x_4 - \bar{x})^2}{4 - 1}$$

The numerator of the expression is appropriately called a *sum of squares* (more exactly, a sum of squares of deviations from the mean), and we denote it by the symbol $S_A$, where the subscript refers to the sample. The denominator $4 - 1$, is known as the *number of degrees of freedom*, or simply the *degrees of freedom* for sample $A$ (see also Section 3.7). Let us denote this number of degrees of freedom by $n_A$. Then, the estimate of the variance obtained from the values of sample $A$ is $S_A/n_A$. Similarly, the estimates of variance for the other samples are $S_B/n_B$, $S_C/n_C$, and $S_D/n_D$. But these four quantities are all independent estimates of the same quantity, $\sigma^2$, according to our assumption of the constancy of the precision of the method. It can be shown that the best estimate of $\sigma^2$, using the information supplied by all four samples, is

$$\hat{\sigma}^2 = \frac{S_A + S_B + S_C + S_D}{n_A + n_B + n_C + n_D} \tag{4.17}$$

This quantity is called the *pooled* estimate of variance. It is obtained by dividing the sum of the sums of squares by the sum of the degrees of freedom. The latter quantity is denoted as the (total) number of degrees of freedom of the pooled estimate.

Note that if we represent by $N_A$ the number of replicate measurements in sample $A$, we have $n_A = N_A - 1$, and similarly for the other samples. Hence:

$$n_A + n_B + n_C + n_D = (N_A - 1) + (N_B - 1) + (N_C - 1) + (N_D - 1)$$
$$= [N_A + N_B + N_C + N_D] - 4$$

Denoting the *total* number of observations by $N$, and the total degrees of freedom by $n$, we have in this case $n = N - 4$. More generally, if the number of samples is $k$, the degrees of freedom, $n$, of the *pooled* estimate of variance is given by the formula

$$n = N - k \tag{4.18}$$

where $N$ is the total number of observations, including all samples. In the example of Table 4.1, $N = 4 + 2 + 2 + 3 = 11$ and $k = 4$. Hence $n = 11 - 4 = 7$.

Equation 4.17 can be written in a slightly different, but highly instructive way. Let $\hat{\sigma}_A^2$ represent the estimate of $\sigma^2$ that we would have obtained from sample $A$ alone. Use a similar notation for sample $B$, $C$, and $D$. Then we have:

$$\hat{\sigma}_A^2 = \frac{S_A}{n_A} \tag{4.19}$$

and hence

$$S_A = n_A \hat{\sigma}_A^2$$

with similar formulas for the other samples. Using these relations, we can now write Eq. 4.17 as follows:

$$\hat{\sigma}^2 = \frac{n_A \hat{\sigma}_A^2 + n_B \hat{\sigma}_B^2 + n_C \hat{\sigma}_C^2 + n_D \hat{\sigma}_D^2}{n_A + n_B + n_C + n_D} \tag{4.20}$$

This formula shows that the pooled estimate of variance is a *weighted average* of the individual variance estimates, with weights equal (or proportional) to the individual degrees of freedom.

It should be emphasized at this point that the pooling of variance estimates in accordance with Eq. 4.17 or Eq. 4.20 is justified only if we can assume confidently that all the populations involved have a common variance $\sigma^2$. Should this not be the case, then the pooling procedure degenerates into a mere exercise in arithmetic, and leads to a quantity which at best is of nebulous meaning and at worst has no meaning at all. This note of warning is motivated by the much used pooling routine in the statistical technique known as analysis of variance. Because this technique is often used by persons unfamiliar with its theoretical background, or used carelessly by persons who do not take the trouble to investigate its applicability in any particular problem, pooling has often led to questionable or outright erroneous results. Statistical techniques of inference are more often than not of a delicate and vulnerable nature. The pooling of variances provides a striking example of the care that must be exercised in their use.

## 4.5 THE GENERAL PROBLEM OF DERIVED MEASUREMENTS

In Section 4.1 mention was made of the need of combining random variables in the computation of quantities of experimental interest. The situation is in fact quite general. Few experimentally determined quantities are measured directly. Most frequently they are derived from one or more direct measurements, and are indeed well-defined functions of these measurements. Thus, in gravimetric and volumetric chemical analyses, where the quantity of interest is the percentage of a constituent in a compound or in a mixture, it is a function of the size of the chemical sample and of the weight of a precipitate or the volume of a standard reagent. In spectrophotometric analysis, the measurement is an optical property, functionally related to the concentration of the unknown. In the determination of physical properties a similar situation generally applies.

A problem of fundamental importance in derived measurements of this type is the evaluation of the error of the derived quantity in terms of the errors of the direct measurements. The formulas in Section 4.1, particularly Eqs. 4.5 and 4.9 provide a partial answer to this problem, for all cases in which the derived quantity is a *linear* function of statistically independent measurements. Let us first examine the physical meaning of Eq. 4.5. Consider a quantity $u$ expressible as a *linear* function of measurements $x, y, z, \ldots$

$$u = ax + by + cz + \cdots \tag{4.21}$$

For example, $u$ might be the iron content of a copper-base alloy, determined by a back-titration method; iron in the ferrous state is first oxidized by an excess of standard potassium dichromate solution $D$, and the excess of $D$ is titrated by a standard solution of ferrous-ammonium sulfate $F$. Then:

$$\% Fe = aD - bF$$

where $a$ and $b$ depend on the "normality" of the reagents $D$ and $F$, and on the weight of the sample subjected to the analysis.

The quantities $x, y, z, \ldots$ are experimentally determined and therefore subject to random errors as well as possible systematic errors. If, for example, $x$ is subject to a systematic error, then its expected value $E(x)$ is not equal to its true value. Representing the true value by $X$, the difference

$$B(x) = E(x) - X \tag{4.22}$$

may be termed the *bias* (or *systematic error*) of $x$. Similarly we define $B(y)$, $B(z)$, .... We now have, for the true values:

$$U = aX + bY + cZ + \cdots \tag{4.23}$$

Furthermore, according to Eq. 4.5 we have:

$$E(u) = aE(x) + bE(y) + cE(z) + \cdots \tag{4.24}$$

Subtracting Eq. 4.23 from 4.24, we obtain

$$E(u) - U = a[E(x) - X] + b[E(y) - Y] + c[E(z) - Z] + \cdots \tag{4.25}$$

This equation may be written in terms of biases:

$$B(u) = aB(x) + bB(y) + cB(z) + \cdots \tag{4.26}$$

Thus, the systematic error (bias) of a *linear function* of random variables is equal to the same linear function of the systematic errors (biases) of the random variables. *In particular, if $x, y, z, \ldots$ are all free of systematic error, then so is the derived quantity $u$.* This result is by no means self-evident. In fact it is generally not true for non-linear functions, as we shall presently show.

## 4.6  BIASES IN NON-LINEAR FUNCTIONS OF MEASUREMENTS

A simple example of a *non-linear* function is provided by the determination of the area of a circle from a measurement of its radius. Let $A$ represent the area and $R$ the radius. Then:

$$A = \pi R^2 \tag{4.27}$$

Denote by $r$ a measured value of $R$.   The error of $r$ is by definition

$$e = r - R \qquad (4.28)$$

If we derive a value for the area, for example, $a$, from the measured value $r$, we have

$$a = \pi r^2 \qquad (4.29)$$

What is the error of $a$?   Representing the error of $a$ by $d$, we have:

$$d = a - A = \pi r^2 - \pi R^2 = \pi(r^2 - R^2) \qquad (4.30)$$

Now, from Eq. 4.28 we derive

$$r = R + e$$

Substituting this value in Eq. 4.30 we obtain

$$d = \pi[(R + e)^2 - R^2] = \pi(2eR + e^2) \qquad (4.31)$$

Let us now consider an unending sequence of measurements $r_1, r_2, r_3, \ldots$ with their corresponding errors $e_1, e_2, e_3, \ldots$.

If the distribution of these errors is symmetrical (as it would be if they were, for example, normally distributed) then in the infinite sequence there would correspond to each error $e$ an error $(-e)$.   The corresponding values of the errors $d$ (for the area) would, according to Eq. 4.31 be:

$$\text{for } e: \qquad d = \pi(2eR + e^2)$$
$$\text{for } (-e): \quad d' = \pi(-2eR + e^2)$$

The average of these two errors is

$$\tfrac{1}{2}(d + d') = \pi e^2 \qquad (4.32)$$

This is a positive quantity (unless $e = 0$).   The average of all errors in the area determination is of course simply the average of all the quantities $\tfrac{1}{2}(d + d')$ each corresponding to a pair of radius errors $e$ and $-e$.   This average is equal to

$$E[\tfrac{1}{2}(d + d')] = E(\pi e^2) \qquad (4.33)$$

Now this quantity cannot possibly be zero, unless each single error $e$ is zero, because an average of positive quantities cannot be equal to zero. We have therefore established that the average error of the area $a$ is positive.   In other words we have

$$E(a) > A$$

or in terms of the bias of $a$:

$$B(a) = E(a) - A > 0. \qquad (4.34)$$

This is true despite the fact that the radius measurement is free of bias. Indeed, since we assumed that to each $e$, there corresponds a $(-e)$, the average of all $e$ values is zero ($E(e) = 0$). It is easy to compute the exact value of $B(a)$. We have (using Eq. 3.10):

$$B(a) = E(a) - A = E(A + d) - A = E(d)$$

But, in view of Eq. 4.30 we have:

$$E(d) = \pi[E(r^2) - E(R^2)] = \pi[E(r^2) - R^2]$$

Hence:

$$B(a) = \pi[E(r^2) - R^2] \tag{4.35}$$

Now, by definition of the variance (see Eq. 3.13), we have:

$$V(r) = E[r - E(r)]^2 = E(r^2) - 2[E(r)]^2 + [E(r)]^2$$
$$= E(r^2) - [E(r)]^2$$

Hence:

$$E(r^2) = V(r) + [E(r)]^2$$

Introducing this result in Eq. 4.35 we obtain:

$$B(a) = \pi\{V(r) + [E(r)]^2 - R^2\}$$

Since $r = e + R$, the variance of $r$ is identical to that of $e$ and we may write:

$$B(a) = \pi\{V(e) + [E(r)]^2 - R^2\} \tag{4.36}$$

We have thus discovered a remarkable fact: for if the radius is measured without bias, we have $E(r) = R$ and then

$$B(a) = \pi V(e) \tag{4.37}$$

Thus, areas derived from measurements of radii that are free of systematic errors are themselves not free of systematic errors. The area values are systematically high, by an amount proportional to the variance of the radius error. In other words, the purely *random error* in the measurement of the radius has *induced* a *systematic* error in the derived area value. It is therefore a significant result that in the case of *linear* functions, no such situation can arise.

Note that the proof of Eq. 4.37 makes no use of the earlier assumption that the distribution of $e$ is symmetrical. The only condition required is that $E(r) = R$ or its equivalent, $E(e) = 0$. Evidently a distribution can be skew and nevertheless have a mean value equal to zero.

In general the biases of non-linear functions induced by the random errors of the direct measurements are small and may be neglected in most cases. For example, in the case of the area of the circle, if the radius is one meter and is measured with a standard deviation of 0.01 meters, the variance of the radius error $e$ will be $V(e) = 0.0001m^2$. Thus, $B(a) = \pi \cdot 10^{-4}$. But the value of the area is $\pi \cdot (1)^2$ or $\pi m^2$. Hence, the systematic error, expressed as a percentage of the value of the area is $100 \cdot \pi \cdot 10^{-4}/\pi$ or 0.01 per cent. Thus a random error of one per cent in the radius induces a positive bias of only 0.01 per cent in the area. Obviously, the random error of the radius measurement also induces, in addition to this systematic error, a *random* error in the derived area value. We will now examine, in a more general way, how the random errors in the direct measurements "propagate" themselves into the derived quantities. In other words, we wish to evaluate the random errors of functions of variables subject to random fluctuations.

### 4.7   THE LAW OF PROPAGATION OF ERRORS

Again we examine first the linear case. If $u = ax + by + cz + \cdots$ and $x, y, z, \ldots$ are statistically independent, we have by virtue of Eq. 4.9:

$$V(u) = a^2 V(x) + b^2 V(y) + c^2 V(z) + \cdots$$

As an application of this formula, consider again the determination of iron in a copper-base alloy (see Section 4.5). If $D$ represents the number of milliliters of 0.01 normal potassium dichromate and $F$ the number of milliliters of 0.01 normal ferrous ammonium sulfate, and if the analysis is made on 5 grams of the alloy, the per cent iron is given (2) by:

$$\%\text{Fe} = 0.01117(D - F) \qquad (4.38)$$

Suppose that $D = 21.0$ ml and $F = 0.8$ ml. Then

$$\%\text{Fe} = 0.01117(21.0 - 0.8) = 0.226$$

Now if the standard deviation of the titrations, with either $D$ or $F$, is 0.1 ml, then:

$$V(D) = V(F) = 0.01$$

Equation 4.9, applied to Eq. 4.38 gives:

$$V(\%\text{Fe}) = (0.01117)^2 V(D) + (0.01117)^2 V(F) = 2(0.01117)^2(0.01)$$

Thus the standard deviation of $\%\text{Fe}$ is

$$\sigma_{\%\text{Fe}} = \sqrt{2}(0.01117)(0.1) = 0.00156$$

The coefficient of variation of per cent Fe is equal to

$$\%\text{CV}_{\%\text{Fe}} = 100 \, \frac{\sigma_{\%\text{Fe}}}{\%\text{Fe}} = 100 \, \frac{0.00156}{0.226} = 0.7\%$$

We now turn to the non-linear case.   The exact calculation of standard deviations of non-linear functions of variables that are subject to error is generally a problem of great mathematical complexity.   In fact, a substantial portion of mathematical statistics is concerned with the general problem of deriving the complete frequency distribution of such functions, from which the standard deviation can then be derived.   In practice it is generally not necessary to solve these difficult problems.   There exists a device called *linearization* that allows us to replace any non-linear function by a linear one, for the purpose of obtaining approximate estimates of standard deviations.   The approximation is quite adequate for most applications.   Linearization is based on a Taylor expansion of the non-linear function with retention of only the linear portion of the expansion.

Consider first a function of a single random variable:

$$Z = f(X) \tag{4.39}$$

For example, $X$ might represent the radius of a circle and $Z$ its area as in Section 4.6.   However, in our present discussion we are interested in the *random error* of $Z$ rather than its bias, which was our concern in Section 4.6.

To an error $\varepsilon$ in $X$, corresponds an error $\delta$ in $Z$, given by

$$\delta = f(X + \varepsilon) - f(X)$$

If $\varepsilon$ can be considered small with respect to $X$, it may be treated as a differential increment.   Then

$$\frac{f(X + \varepsilon) - f(X)}{\varepsilon} \approx \frac{df(X)}{dX}$$

and consequently

$$\delta = \frac{df(X)}{dX} \cdot \varepsilon = Z_X{}' \cdot \varepsilon \tag{4.40}$$

where $Z_X{}'$ represent the derivative of $Z$ with respect to $X$, taken at or very near the measured value of $X$.   Equation 4.40 is a simple linear relation between the errors $\delta$ and $\varepsilon$.   Applying Eq. 4.9 we then obtain:

$$V(\delta) = (Z_X{}')^2 V(\varepsilon) \tag{4.41}$$

As an application of Eq. 4.41 consider again the determination of the area of a circle.   Here, the relation $Z = f(X)$ becomes

$$A = \pi R^2$$
$$Z_X{}' = A_R{}' = 2\pi R$$

Hence:

$$V(\delta) = (2\pi R)^2 V(\varepsilon)$$

where $\varepsilon$ is the random error in the measurement $r$ of the radius, and $\delta$ the *random* error, induced by $\varepsilon$, in the derived value $a$ of the area.    From the preceding equation we derive:

$$\sigma_\delta = 2\pi R \sigma_\varepsilon$$

Considering, as in Section 4.6, that $\sigma_\varepsilon = 0.01$ meters, for a radius $R = 1$ meter, we obtain

$$\sigma_\delta = 2\pi(1)(0.01)$$

Expressing this error as a coefficient of variation we have

$$\% CV_a = \frac{2\pi(0.01)}{\pi(1)^2} \cdot 100 = 2\%$$

We now understand why it would be pedantic to be concerned with the bias of $a$, as calculated in Section 4.6.    Indeed, an error of one per cent in measurement of the radius induces a random error of two per cent in the area, and a bias of only 0.01 per cent.    The latter is therefore entirely negligible.

Equation 4.41 expresses the *law of propagation of errors* for the case of a single independent variable.    Its proof is based on the assumption that the error $\varepsilon$ is reasonably small* with respect to the measured value of $X$. Should this not be the case, then the identification of $\varepsilon$ with a differential is not justified and the formula loses its validity.    In that case, the situation is somewhat aggravated by the fact that for large relative errors of the measured quantity, bias induced in the derived function may also become appreciable.    However, for most practical applications, the law in its stated form gives satisfactory results.

The law of propagation of errors for the general case, i.e., for functions of several random variables, is an immediate extension of Eq. 4.41.    Let $x, y, z, \ldots$ represent random variables whose true values are $X, Y, Z, \ldots$; let $u$ represent a derived quantity whose true value is given by

$$U = f(X, Y, Z, \ldots) \tag{4.42}$$

Let $\varepsilon_1, \varepsilon_2, \varepsilon_3, \ldots$ represent the statistically independent errors of $x, y, z, \ldots$, respectively.    Then the error induced in $u$, which we will denote by $\delta$, as a result of the errors $\varepsilon_1, \varepsilon_2, \varepsilon_3, \ldots$, has a variance equal to

$$V(\delta) = \left(\frac{\partial f}{\partial X}\right)^2 V(\varepsilon_1) + \left(\frac{\partial f}{\partial Y}\right)^2 V(\varepsilon_2) + \left(\frac{\partial f}{\partial Z}\right)^2 V(\varepsilon_3) + \cdots \tag{4.43}$$

The partial derivatives are taken at values equal or close to the measured

---

* In practice, if $\sigma_\varepsilon$ is of the order of 10% of $X$ or smaller, the law can be reliably used.

values $x, y, z, \ldots$. As an application of Eq. 4.43, consider the determination of a specific gravity as the ratio of a weight $P$ to a volume $W$. Representing the specific gravity by $\rho$, we have

$$\rho = \frac{P}{W} \tag{4.44}$$

If $\varepsilon_1$ is the error of $P$ and $\varepsilon_2$ that of $W$, the error, $\delta$, of $\rho$ has a variance equal to

$$V(\delta) = \left(\frac{\partial \rho}{\partial P}\right)^2 V(\varepsilon_1) + \left(\frac{\partial \rho}{\partial W}\right)^2 V(\varepsilon_2)$$

$$= \left(\frac{1}{W}\right)^2 V(\varepsilon_1) + \left(-\frac{P}{W^2}\right)^2 V(\varepsilon_2) \tag{4.45}$$

This expression can be written in a much more symmetrical form by factoring out the quantity $P^2/W^2$ in the right member of the equality:

$$V(\delta) = \frac{P^2}{W^2}\left[\frac{V(\varepsilon_1)}{P^2} + \frac{V(\varepsilon_2)}{W^2}\right]$$

or

$$\frac{V(\delta)}{\rho^2} = \frac{V(\varepsilon_1)}{P^2} + \frac{V(\varepsilon_2)}{W^2} \tag{4.46}$$

Equation 4.46 can be simplified further through the introduction of relative errors, i.e., coefficients of variation. Indeed, the coefficient of variation of $\rho$ is $(1/\rho)\sigma_\rho$, the square of which is precisely the first member of Eq. 4.46. Similarly $V(\varepsilon_1)/P^2$ and $V(\varepsilon_2)/W^2$ are the squares of the coefficients of variation of $P$ and $W$, respectively. Hence

$$\left(\frac{\sigma_\rho}{\rho}\right)^2 = \left(\frac{\sigma_{\varepsilon_1}}{P}\right)^2 + \left(\frac{\sigma_{\varepsilon_2}}{W}\right)^2$$

or

$$(CV_\rho)^2 = (CV_P)^2 + (CV_W)^2 \tag{4.47}$$

It is easily shown that for any function of a purely *multiplicative* form, the law of propagation of errors takes the simple form Eq. 4.47. More specifically, if

$$u = \frac{x \cdot y \cdot z \cdots}{p \cdot q \cdot r \cdots} \tag{4.48}$$

then

$$(CV_u)^2 = (CV_x)^2 + (CV_y)^2 + (CV_z)^2 + \cdots$$
$$+ (CV_p)^2 + (CV_q)^2 + (CV_r)^2 + \cdots \tag{4.49}$$

For functions containing sums or differences of random variables, this simple rule is no longer valid, and it is then necessary to derive the error variance from Eq. 4.43. The remark made in connection with the validity of Eq. 4.41 as a usable approximation applies also to Eqs. 4.43 and 4.49.

## 4.8   A DIFFERENT APPROACH

In the last three sections we have considered functions of measurements as *derived* measurements, and we have attempted to infer the errors of derived measurements from those of the direct measurements on which they are based.   One can approach the problem from an entirely different viewpoint.   Thus, in the determination of the area of a circle, $A = \pi R^2$, one may choose to recognize only one measured value, namely $R$.   In this approach the area $A$ is not considered as a measurement at all, but rather as a fixed, though unknown constant.   Such a constant is called a *parameter*.   The problem now consists in finding that value of the parameter that is most consistent with: (*a*) whatever prior knowledge we possess about the problem and (*b*) the value or values obtained in the measurement of the radius, $R$.   The significance of this approach becomes clearer when several parameters are involved.   In Section 2.4 we described two problems illustrating this situation: the determination of fundamental physical constants, and the study of a method of chemical analysis. Either example involved a number of unknown constants, the values of which were to be inferred from those of related measured quantities. In a later chapter we will develop more fully what might be called the measurement-parameter approach, and present its solution by the method of least-squares.

## 4.9   SUMMARY

One often deals with quantities that are the sums, differences, products, or any other mathematical functions of two or more random variables. A basic problem in mathematical statistics is to derive the probability distributions of such combinations of random variables.   From a practical viewpoint, the important task is to obtain values for the mean and the variance of these combined quantities.   This is accomplished by means of certain mathematical formulas, some of which are only approximately correct.   In particular, the formulas of this type that pertain to the variance of a function of several random variables are known as the law of propagation of errors.   This law is exact for linear combinations of random variables, but only approximate for non-linear combinations. An important special case consists in the formula giving the variance of a mean of observations taken from the same population (the variance of a sample mean).

The pre-eminence of the normal distribution in statistical theory is due in large part to a basic statistical law known as the central limit theorem. This important theorem states essentially that the arithmetic average of the values of a statistical random sample taken from any population has a probability distribution function that rapidly approaches the Gaussian form, as the size of the sample increases.   As a result of this theorem, it is generally safe to regard averages as being normally distributed, even if the averages contain as few as four or five observations.   Taken in conjunction with the law of propagation of errors, the central limit theorem provides a powerful tool of statistical inference.

When random variables are combined in a non-linear fashion, the resulting quantities are generally subject to biases; i.e., the combined effect of the fluctuations of the individual variables will cause the derived quantity to be *systematically* larger or smaller than it would have been in the total absence of fluctuations.   In most cases, however, these systematic effects are negligible in comparison with the random fluctuations of the derived quantities.   Since the random fluctuations of derived quantities are governed by the law of propagation of errors, the latter is generally all that is required for the description of the uncertainty of such quantities.

It is often fruitful to look upon the mathematical relationship between calculated and observed quantities in a different way.   The true, though unknown values of the calculated quantities are defined as parameters. The problem then consists in deriving estimates of the parameters that will be most consistent with the observed values of the directly measured quantities.   This approach underlies the method of least squares, to be discussed in Chapter 7.

## REFERENCES

1. Parzen, E., *Modern Probability Theory and Its Applications*, Wiley, New York, 1960.
2. Snell, F. D., and F. M. Biffen, *Commercial Methods of Analysis*, McGraw-Hill, New York, 1944, p. 331.

# chapter 5

# HOMOGENEOUS SETS OF MEASUREMENTS

## 5.1 INTRODUCTION

We have seen that the endlessly repeated application of a measuring process on a physical system, for the determination of a particular property of this system, generates a hypothetical infinite population. In practice one is, of course, always limited to finite sets. To indicate that all the measurements of the finite set belong to the same population, we will call it a *homogeneous* set.

The idea that the mathematical laws of chance could be used to advantage in the interpretation of scientific measurements is of relatively recent origin. The inconsistencies encountered in measurement data caused early experimenters a great deal of concern. Many ingenious ways were proposed to resolve them, but it was not until the end of the 18th century that a clear formulation was given to the relation between errors of measurement and the mathematical laws of chance. There then followed a period during which attempts were made to prove mathematically that all experimental errors must necessarily obey the Gaussian distribution function. This trend has left its mark on our present terminology, in which the Gaussian law is still referred to as *the normal law* of errors or even simply as *the error function*. We now know that errors of measurement are not necessarily distributed in accordance with the Gaussian function. On the other hand, this function rightfully plays a dominant role in the theory of errors. The reasons for this are twofold. In the first place, many situations exist in which the normal curve is a good approximation to the law that actually governs the errors of measurement. The second reason for the privileged position of the Gaussian curve in measurement statistics is contained in the remarkable law known as the central limit theorem, discussed in Section 4.3. Essentially this law assures us that the averages of random samples of size $N$ from *any* statistical population (provided that it has a finite variance) tend to be distributed in accordance with the Gaussian curve and that the approximation improves rapidly as $N$ increases. We will see, furthermore (Chapter 13), that it is possible, and in fact desirable for more than one reason, to express certain types of data in a transformed scale (for example, as logarithms) for the purpose of analysis and interpretation. The fluctuations of the measurements in the transformed scale are then quite often in much better agreement with the Gaussian distribution than were the measurements in the original scale.

A population of measurements is really a mathematical abstraction; i.e., its elements are conceptual rather than actual. This is true because the totality of *all* replicate measurements that *could* be made will never *actually* be made. The distinction between actual and hypothetical populations may be related to that between *mass* phenomena and *repetitive* events. The former include, among others, human and animal populations, the populations underlying the statistical tables used by insurance companies, and even the populations of molecules considered in thermodynamic studies. Repetitive phenomena, on the other hand, occur in the evaluation of measurements, in studies of radioactive decay, in investigations of the load distribution on telephone trunk lines, and in many other situations. Whereas the populations of mass phenomena are *actual* (physically real), those occurring in the study of repetitive events are mostly *potential*; each element in such a population is a *possibility*, though not necessarily a *reality*. We need not concern ourselves, in this book, with the philosophical questions regarding the legitimacy of using such concepts. Suffice it to say that abstractions of this type are very common both in mathematics and in the physical sciences.

The concept of a homogeneous set of measurements is important, both in scientific applications and in the quality control of manufactured products. In both cases, two problems arise: (1) to judge whether a given set of data is really homogeneous, and in the affirmative case, (2) to make inferences about the population from which the set constitutes a sample.

To illustrate these questions we mention two examples. Consider first a series of determinations of the velocity of light. Today it is possible to make such measurements in the laboratory and to make a fairly large number of determinations in a single day. Now, in order that we may have confidence in such a set of values, we must satisfy ourselves that the values, *in the order in which they were obtained*, do not show systematic changes, such as a gradual increase or decrease, or a cycling effect, or a sudden shift. In other words, we must have reasonable assurance that the fluctuations shown by the data exhibit the same kinds of patterns that a set of values *picked at random* from a single population would exhibit. It is easy to see why this requirement is of such primary importance. Suppose, for example, that 30 determinations were made between 9:00 a.m. and noon, and that an examination of the values reveals a gradually increasing trend. Can we then state that the average of these 30 values is a better value than, say, the first or the last? Obviously not, since a continuation of the experiment in the afternoon might show a continuation of the trend, and the average of all determinations of the day would then be closer to a single measurement obtained around noon than to the average of the first set of 30.

A similar problem arises in the field of quality control. Uniformity of product requires that the conditions maintained during the manufacturing process be kept as constant as possible. Some fluctuations are of course unavoidable but what must be guarded against is the presence of systematic trends, cycles, or shifts. This is done by examining a series of samples taken periodically during the manufacturing process and by studying the test values for these samples *in the chronological order in which they were taken.*

The second problem, drawing inferences from a homogeneous set of data, is of course related to the objectives of the investigation. It is often possible to express these objectives in terms of the parameters of statistical populations, such as averages, standard deviations, and others. We will return to this matter in Section 5.3, but first we discuss in more detail the problem of judging the homogeneity of statistical samples.

## 5.2 CONTROL CHARTS

There exist many statistical methods for judging the homogeneity of a set of values. Many of these methods are specific, in that they examine randomness against the alternative of a *specific type of non-randomness.* Thus, in some cases one might suspect a gradual increase or decrease in the values, whereas in others it is more likely that the data may exhibit a sudden shift to a different level. The latter situation may arise in manufacturing processes involving batches. It may also arise in series of measurements made on different days.

In this book we will not discuss tests of randomness as such.* We will, however, explain a general principle underlying these matters and discuss two aspects of this principle, the first in this chapter and the second in Chapter 9.

The principle is simple: divide the set into a number of subsets and examine whether the variability *between* subsets is greater than *within* subsets. In order that this principle make any sense, it is necessary that the subsets be chosen in a rational way. The choice depends very much on the nature of the data.

There are situations in which the subsets clearly exist as an obvious consequence of the way in which the data were obtained. We will encounter a problem of this type in Chapter 9, in connection with the determination of the faraday. There the subsets correspond to different samples of silver and to different treatments of these samples (annealing and etching). A method of analysis for such data is given in Chapter 9.

In other cases no obvious subsets exist. For example, the values shown in Table 5.1, constituting 39 determinations of the velocity of light, were

**TABLE 5.1**  Measurement of the Velocity of Light

| Determination number | Value of $c$ | Determination number | Value of $c$ |
|---|---|---|---|
| 1 | 299799.4 | 21 | 299799.1 |
| 2 | 799.6 | 22 | 798.9 |
| 3 | 800.0 | 23 | 799.2 |
| 4 | 800.0 | 24 | 798.5 |
| 5 | 800.0 | 25 | 799.3 |
| 6 | 799.5 | 26 | 798.9 |
| 7 | 799.6 | 27 | 799.2 |
| 8 | 799.7 | 28 | 798.8 |
| 9 | 800.0 | 29 | 799.8 |
| 10 | 799.6 | 30 | 799.5 |
| 11 | 799.2 | 31 | 799.6 |
| 12 | 800.9 | 32 | 799.8 |
| 13 | 799.2 | 33 | 799.7 |
| 14 | 799.4 | 34 | 799.7 |
| 15 | 799.0 | 35 | 799.2 |
| 16 | 798.6 | 36 | 799.5 |
| 17 | 798.7 | 37 | 799.7 |
| 18 | 799.0 | 38 | 799.8 |
| 19 | 798.6 | 39 | 800.1 |
| 20 | 798.7 | | |

* Some tests of randomness are discussed in reference 3.

obtained consecutively, and no natural grouping of the data is suggested by the experiment.

In 1924, Walter Shewhart, a physicist employed by the Bell Telephone Laboratories, suggested an ingenious way of studying the homogeneity of sets of data. Shewhart's interest was primarily in the quality control of manufactured products but his method, known as the *Shewhart control chart*, is applicable to a wide variety of problems (8). In Shewhart's method, the data are partitioned into *rational subgroups*, if such groups exist, or, in case no *a priori* grouping exists, into groups of a small number (3 to 6) of consecutive values. Two quantities are calculated for each subgroup: its average and its range.* The average value of the ranges of all subgroups provides a measure for the within-subgroup variability.

**TABLE 5.2**  Control Chart Analysis of Velocity of Light Data

| Subset number[a] | | Average | Range |
|:---:|:---:|:---:|:---:|
| 1 | | 299799.666 | 0.6 |
| 2 | | 9.833 | 0.5 |
| 3 | | 9.766 | 0.4 |
| 4 | | 9.900 | 1.7 |
| 5 | | 9.200 | 0.4 |
| 6 | | 8.766 | 0.4 |
| 7 | | 8.800 | 0.5 |
| 8 | | 8.866 | 0.7 |
| 9 | | 9.133 | 0.4 |
| 10 | | 9.366 | 1.0 |
| 11 | | 9.700 | 0.2 |
| 12 | | 9.466 | 0.5 |
| 13 | | 9.866 | 0.4 |
| | Average | 299799.41 | 0.592 |
| Two-sigma control limits[b] | Upper | 9.81 | 1.214 |
| | Lower | 9.01 | 0 |

> [a] Each subset consists of three consecutive measurements (see Table 5.1).
> [b] For calculations, see references 1 and 9.

This measure is then used as a criterion to judge whether the averages of the subgroups vary by an amount larger than can be accounted for by within-group variability. To illustrate this technique, consider now the data of Table 5.1. In Table 5.2 are listed the averages and ranges of

* The range is defined as the difference between the largest and the smallest of a set of observations (see Section 6.6).

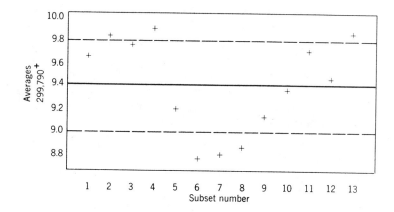

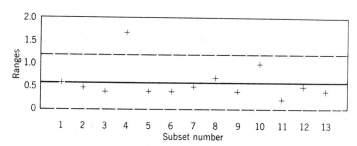

**Fig. 5.1** Control charts for averages and ranges.

subgroups of 3 consecutive measurements. These values are plotted on two charts, known as the *control chart for averages* and the *control chart for ranges*, as shown in Fig. 5.1. The dashed lines on these charts are known as *control-lines*, and their purpose is to examine whether the values, *in the order in which they were obtained*, behave like a set of random drawings from a single population. To show how this is done, let us first denote the population of the individual measurements by $P(c)$. Now, consider the chart for averages in Fig. 5.1. These values may also be considered as belonging to a statistical population. Let us denote this population by $P(\bar{c})$. According to Eq. 4.14, the standard deviation of $P(\bar{c})$ is equal to that of $P(c)$ divided by $\sqrt{3}$. But how do we estimate the standard deviation of $P(c)$? Here we anticipate some results to be obtained in the following chapter: from the average value of the range, we obtain an estimate of the standard deviation of $P(c)$, by dividing the average range by a factor depending on the size of subgroups. For ranges of three values, the factor is 1.693 (see Table 6.3). Thus, the standard deviation of $P(c)$ is estimated by $0.592/1.693 = 0.350$. Consequently, the estimate for the standard deviation of $P(\bar{c})$ is $0.350/\sqrt{3} = 0.20$. We now assume, as is customary in control chart work, that this estimate is close enough to be taken as the true standard deviation of $P(\bar{c})$. We also assume that the average of the $\bar{c}$ values, in this case 299799.41, is close enough to the population average of $P(\bar{c})$ to be identified with it. Using the theory of the normal distribution for $P(\bar{c})$,* we may then assert that:

1. About 2/3 of all $\bar{c}$ values should fall within "one-sigma limits," i.e., within $299799.41 \pm 0.20$.

2. About 95 per cent of all $\bar{c}$ values should fall within "two-sigma limits," i.e., within $299799.41 \pm 0.40$.

3. About 99.7 per cent of all $\bar{c}$ values should fall within "three-sigma limits," i.e., within $299799.41 \pm 0.60$.

Generally only one of these facts is used, the common one being the statement involving three-sigma limits. We will, however, use two-sigma limits for our illustration. The dashed lines on the average chart are at a distance of two standard deviations from the central line, which represents the mean. We see that out of 13 values, only 7 fall within the two-sigma limits, whereas we expected 95 per cent of 13, or approximately 12 values to fall within these limits. The obvious conclusion is that the averages show more scatter than can be accounted for by the within-subgroup variability. The control chart technique has the great advantage of not only revealing such a situation, but also providing clues about its cause. A glance at the chart for averages shows that these values, instead of fluctuating in a random manner, tend to follow a systematic pattern suggesting a cycling effect: high, low, high. A lack of randomness of this type, i.e., a situation in which consecutive values are, on the average, less different (or more different) than non-consecutive values is called a *serial correlation*.

Let us now examine the chart for ranges. In a set of values generated by a true random process, the ranges of subgroups form themselves a population. Statistical theory allows one to estimate the standard deviation of this population, using only the average of the ranges, and to set two-sigma limits (or any other probability limits), as was done for the average chart. The formulas can be found in references 1 and 9. In Fig. 5.1, two-sigma limits are indicated for the range chart. The lower limit is zero. Only one of the 13 points falls outside the control lines. There is therefore no evidence that the ranges behave in a non-random fashion.

---

* The central limit theorem is also used here.

Summarizing the previous discussion, we have found that the set of measurements on the velocity of light does not appear to satisfy the requirements of randomness that are necessary for the drawing of valid inferences about an underlying population of measurements. The control chart technique has been found particularly valuable for this investigation since it combines the advantages of visual appraisal with those of a simple theory and simple calculations. The great value of the control chart technique, even outside the field of quality control, is still not sufficiently appreciated, despite the fact that G. Wernimont, in a paper published in 1946 (10), illustrated its successful use in many problems arising in chemical analysis.

## 5.3 ORDER STATISTICS

When, as in the example of the light velocity measurements, a strong suspicion exists regarding the randomness of the data, it is wise to investigate the causes of non-randomness before proceeding with any further analysis of the data. In the following we will assume that we are dealing with a set of measurements that has not been found wanting in regard to homogeneity. Statistically speaking this means that we are dealing with a random sample from a single statistical population. The estimation of the mean and of the standard deviation of this population can be made in accordance with Eqs. 3.19 and 3.23 on the sole assumption of the randomness of the sample. These estimates are valid, regardless of the type of the underlying population. But we are now concerned with a different problem: to learn something about the nature of this population. Here again a simple graphical procedure is most useful. This time the data are considered, not in the order in which they were obtained, but rather in order of increasing magnitude.* When a sample has been thus rearranged, the values, in their new order, are called *order statistics*.† We will now see what can be learned from an examination of a sample in the form of its order statistics. As is often the case in statistical theory, the nature of a particular method of analysis of data is better illustrated in terms of artificial data than with real experimental values. The reason for this is that by using artificial data we know exactly what we are dealing with and can therefore evaluate the appropriateness and the power of the analysis. We will use this method for the presentation of the basic theory

---

* It is equally valid to order the data in the reverse order, i.e., in order of decreasing magnitude. For the sake of consistency, we will use the increasing order exclusively.

† The word "statistic," when used in the singular, denotes a function of the observations. A statistic involves no unknown parameters. Thus, an order statistic, as here defined, is truly a statistic.

of order statistics and begin therefore by constructing a set of normally distributed "data."

Column 1 in Table 5.3 contains 20 values *randomly selected* from a table of standard normal deviates, i.e., values from a normal population of mean zero and unit standard deviation. Column 2 is derived from

**TABLE 5.3**   The Use of Order Statistics

| | 1 | 2 | 3 | 4 | 5 | 6 | 7 | 8 |
|---|---|---|---|---|---|---|---|---|
| $i$ | $t$ | $y$ | $y'$ | $\dfrac{i}{N+1}$ | $t'$ | $\dfrac{i-3/8}{N+1/4}$ | $t'$ | Exponential $t'$ |
| 1 | 1.2 | 5.12 | 4.80 | 0.0476 | −1.66 | 0.0309 | −1.87 | 0.0500 |
| 2 | 1.2 | 5.12 | 4.82 | 0.0952 | −1.31 | 0.0802 | −1.41 | 0.1026 |
| 3 | 0.2 | 5.02 | 4.91 | 0.1428 | −1.07 | 0.1296 | −1.13 | 0.1582 |
| 4 | 1.2 | 5.12 | 4.94 | 0.1905 | −0.87 | 0.1790 | −0.92 | 0.2170 |
| 5 | −0.2 | 4.98 | 4.95 | 0.2381 | −0.71 | 0.2284 | −0.75 | 0.2795 |
| 6 | 0.8 | 5.08 | 4.97 | 0.2857 | −0.57 | 0.2778 | −0.59 | 0.3461 |
| 7 | −0.5 | 4.95 | 4.98 | 0.3333 | −0.43 | 0.3272 | −0.45 | 0.4175 |
| 8 | −0.2 | 4.98 | 4.98 | 0.3809 | −0.30 | 0.3765 | −0.31 | 0.4944 |
| 9 | 0.3 | 5.03 | 5.00 | 0.4286 | −0.18 | 0.4259 | −0.19 | 0.5777 |
| 10 | 0 | 5.00 | 5.01 | 0.4762 | −0.06 | 0.4753 | −0.06 | 0.6686 |
| 11 | 0.2 | 5.02 | 5.02 | 0.5238 | 0.06 | 0.5247 | 0.06 | 0.7686 |
| 12 | −2.0 | 4.80 | 5.02 | 0.5714 | 0.18 | 0.5741 | 0.19 | 0.8797 |
| 13 | −0.3 | 4.97 | 5.03 | 0.6190 | 0.30 | 0.6235 | 0.31 | 1.0047 |
| 14 | 0.5 | 5.05 | 5.04 | 0.6667 | 0.43 | 0.6728 | 0.45 | 1.1476 |
| 15 | 0.4 | 5.04 | 5.05 | 0.7143 | 0.57 | 0.7222 | 0.59 | 1.3143 |
| 16 | −1.8 | 4.82 | 5.08 | 0.7619 | 0.71 | 0.7716 | 0.75 | 1.5143 |
| 17 | 0.1 | 5.01 | 5.12 | 0.8095 | 0.87 | 0.8210 | 0.92 | 1.7643 |
| 18 | −0.9 | 4.91 | 5.12 | 0.8571 | 1.07 | 0.8704 | 1.13 | 2.0976 |
| 19 | 2.2 | 5.22 | 5.12 | 0.9048 | 1.31 | 0.9198 | 1.41 | 2.5976 |
| 20 | −0.6 | 4.94 | 5.22 | 0.9524 | 1.66 | 0.9691 | 1.87 | 3.5976 |

$$\bar{y} = 5.009 \qquad \hat{\sigma}_y = 0.1001$$

1: a random sample of 20 standard normal deviates
2: $y = 5 + 0.1t$
3: $y' = y$, rearranged in order of increasing magnitude
4: $i/(N + 1)$, in this case $i/21$
5: $t' =$ the value of the standard normal deviate for which the area to the left is equal to $i/(N + 1)$
6: a modified value to be used instead of $i/(N + 1)$
7: $t' =$ the value of the standard normal deviate for which the area to the left is equal to $(i - 3/8)/(N + 1/4)$
8: expected value of the order statistic of the exponential variate of parameter one

column 1 by multiplying the values in the latter by 0.1 and then adding 5 to the product. Thus, if $t$ denotes a standard normal deviate, the values in the second column, denoted by $y$, are represented by the relation

$$y = 0.1t + 5 \qquad (5.1)$$

This equation shows that the $y$ values are linearly related to the $t$ values. It can be shown that as a consequence of this linear relationship, the $y$ values are also normally distributed. Using Eqs. 4.5 and 4.3 we see that the expected value of $y$ is given by

$$E(y) = 0.1E(t) + 5 = 5$$

and the variance of $y$ by

$$V(y) = 0.01V(t) = 0.01$$

Thus, the standard deviation of $y$ equals $\sigma_y = \sqrt{0.01} = 0.1$. Now, when $y$ is plotted against $t$, then according to Eq. 5.1 the slope of the line is 0.1 and the $y$ intercept is 5. *Thus, the slope of this line represents the standard deviation and the y intercept represents the mean of y*, and a graph of the line would therefore allow us to compute these parameters.

Let us now imagine that the $y$ values in column 2 have been obtained experimentally as a set of replicate measurements of some physical quantity. Under those conditions, the corresponding $t$ values (column 1) are of course unknown to us. We cannot, therefore, use the plot of $y$ versus $t$ for the estimation of the mean and the standard deviation of $y$. We can, of course, estimate these parameters by the usual formulas. Such estimates are given in the lower part of the table. We now propose to derive estimates of these parameters by a graphical method, admittedly less precise than the one based on the usual formulas, but having the advantage of giving us new and useful insight into the structure of homogeneous samples.

In column 3 of Table 5.3 the $y$ values of column 2 have been rewritten in order of increasing magnitude. They are now denoted as $y'$ values. These are of course none other than the order statistics. Thus, the first $y'$ is the lowest value of the set of $y$ values, the second $y'$ is the next-to-lowest, and so on. Figure 5.2 shows the positions of the 20 order statistics, $y'$, on the normal curve of mean 5 and standard deviation 0.1.

Let us draw vertical lines through the positions occupied by the 20 order statistics as shown in Fig. 5.2. In this way the total area under the curve (whose measure we know to be unity) is divided into a number of parts. If all 20 "measurements" (the $y'$ values), had been different from each other, there would have been 21 such parts. In the present case, the value 4.98 occurs twice, as the seventh and eighth order statistics. So does

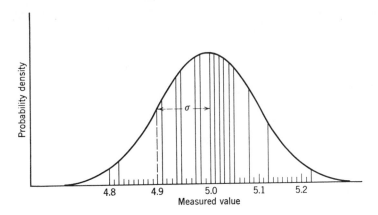

**Fig. 5.2**   Positions of the 20 order statistics (column (3) of Table 5.3) on the normal curve of mean 5 and standard deviation 0.1.

the value 5.02, as the eleventh and twelfth order statistics.   Similarly, the value 5.12 occurs three times, as the seventeenth, eighteenth and nineteenth order statistics.   For the sake of generality we will consider that *each* pair of consecutive order statistics, whether or not they are identical, define an area under the curve; if the pair in question consists of two identical values, the corresponding area has simply shrunk to zero measure. With this convention, a sample of $N$ values *always* divides the area under the probability curve into $N + 1$ parts.   In general, these parts will be of unequal area, and some may even have zero-area, as explained above. For the data of Table 5.3, there are 21 areas, 4 of which have zero measure.

We now state the following theorem: for any continuous distribution curve (not only the Gaussian), the *expected values* of the $N + 1$ probability *areas* determined by a random sample of $N$ values are all equal to each other; and consequently their common value is $1/(N + 1)$.*

Operationally, the expected value of the area located between, say, the third and the fourth order statistic has the following interpretation. Suppose that we draw a large number of samples, each of size $N$, from the same population.   For each sample we measure the area between the third and the fourth order statistics.   The sequence of areas thus obtained define a statistical population of areas.   The average value of the population is the expected value in question.   According to our theorem, this

* This theorem explains the peculiar pattern of the vertical lines exhibited by Fig. 5.2; the order statistics tend to crowd in the center and spread out towards the tails.   This occurs because the height of the curve is greatest in the center, thus requiring a shorter base for the approximately rectangular portion to achieve equal area.

average area is the same, and equal to $1/(N + 1)$, regardless of what pair of consecutive order statistics were chosen.

We can state our theorem in an alternative and more useful way. Consider the following set of areas: the area *to the left* of the first order statistic, the *total* area to the left of the second order statistic (thus *including* the preceding area) and so on up to, and including the $N$th order statistic. It then follows from the theorem that the expected values of the $N$ areas in this sequence are respectively equal to $1/(N + 1)$, $2/(N + 1)$, $3/(N + 1), \ldots, N/(N + 1)$.

The practical importance of this theorem is this: for any random sample of size $N$, the total area to the left of the $i$th order statistic is equal to $i/(N + 1)$, except for a random fluctuation whose average value is zero. We have already stated that this theorem is true for any continuous distribution and is therefore in no way limited to Gaussian distributions. If we now determine $N$ points on the abscissa of a given distribution function such that the total areas to their left are equal respectively to $1/(N + 1)$, $2/(N + 1)$, $3/(N + 1), \ldots, N/(N + 1)$, then these points are in some sense an *idealized* set corresponding to our *actual* set of $N$ order statistics. But *for any known distribution function*, this idealized set is readily obtained. To show this, we have listed in column 4 of Table 5.3, the area values $1/21 = 0.0476$, $2/21 = 0.0952, \ldots, 20/21 = 0.9524$. Using a table of the areas of the normal distribution we then associate with each of these area-values, a corresponding value of the standard normal variate. For example, the value in the normal distribution of mean zero and unit standard deviation for which the area to the left is 0.0476 is found to be $t' = -1.66$ (see column 5). This value corresponds, in some sense, to the $t$ value associated with the first order statistic of our sample. The latter is 4.80 and the associated $t$ value is, according to Eq. 5.1: $t = (4.80 - 5)/0.1 = -2.00$. Similarly, $t'$ for the second order statistic is $-1.31$, while the $t$ value associated with the observed value of the second order statistic is $t = (4.82 - 5)/0.1 = -1.80$. We have assumed that the $t$ values are unknown to us. On the other hand, *the $t'$ values are known*. If we could substitute $t'$ for $t$, our plot of $y'$ versus $t'$ would allow us, as explained above, to estimate the mean and the standard deviation of $y$, as respectively equal to the intercept and the slope of the straight line that best fits the $(t', y')$ points.

To recapitulate: the observed measurements $y$ are rearranged as order statistics $y'$; the values $i/(N + 1)$ are calculated; finally, using a table of the standard normal distribution, a value $t'$ is associated with each $i/(N + 1)$. A plot of $y'$ versus $t'$ should yield an approximate straight line, whose slope is an estimate of the standard deviation, and whose intercept is an estimate of the mean.

For approximate values this method is adequate. It is however subject to some inaccuracy, due to the following circumstances. The theorem on which the procedure is based is rigorously correct in terms of areas. We have, however, shifted the discussion from the *areas* associated with the sample values to these values themselves. It is true that to any given value of an area there corresponds a unique abscissa such that the area to its left has this particular value, but it is *not* true that what is stated for the expected values of areas also applies to the expected values of the associated abscissas. Expected values are essentially averages and it is readily seen that, for example, the average value of the squares of any set of numbers is *not* equal to the square of the average value of these numbers (unless all the numbers are alike). By passing from areas to the corresponding abscissas we have introduced a bias, similar to the difference between the average of squares and the square of the average. There are two ways to meet this difficulty. Blom (2) has shown that in the case of the Gaussian distribution the bias can be practically eliminated by simply substituting for the sequence of areas

$$\frac{1}{N+1} \quad \frac{2}{N+1} \quad \frac{3}{N+1} \quad \cdots \quad \frac{N}{N+1}$$

the modified sequence:

$$\frac{1 - 3/8}{N + 1/4} \quad \frac{2 - 3/8}{N + 1/4} \quad \frac{3 - 3/8}{N + 1/4} \quad \cdots \quad \frac{N - 3/8}{N + 1/4}$$

These values, for $N = 20$, are listed in column 6 of Table 5.3 and the corresponding values of the standard normal deviate, $t'$, are listed in column 7. It is now legitimate to plot the $y'$ values versus this new set of $t'$, and to estimate the mean and the standard deviation of $y$ from the intercept and the slope of a straight line fitted to the plotted points. Figure 5.3 exhibits such a plot. Even a visual fit would provide usable estimates for the slope (standard deviation) and the intercept (mean).

The second method of avoiding bias in the plotting of order statistics is to use directly the expected values of order statistics, either in tabulated form or as computed by means of analytical expressions. Fisher and Yates (4) have tabulated the expected values of order statistics for the reduced normal distribution for sample sizes up to $N = 50$. A shorter table is given in reference 7.

By consulting these references, the reader may verify that the expected values of order statistics for the case $N = 20$ are identical to within the second decimal, with the values listed in column 7 of Table 5.3, which were obtained by Blom's formula.

For some simple distributions, such as the rectangular or the exponential,

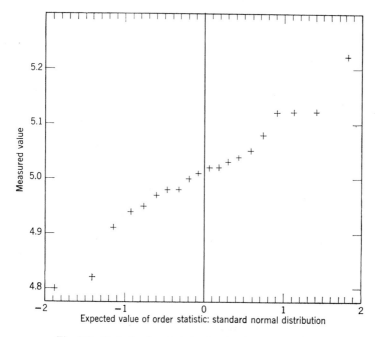

**Fig. 5.3** Plot of order statistics for the data of Table 5.3.

no tables are required for the expected values of order statistics; in these cases, the expected values of order statistics can be computed by means of simple algebraic formulas (7).

What use can be made of a plot of order statistics such as Fig. 5.3? Let us suppose that a sample is drawn from some population $P_1$ and that, ignoring the nature of this population, we postulate that it is of a certain type, say $P_2$. If $P_2$ is different from $P_1$, the $t'$ values derived from the $P_2$ distribution will not produce a straight line with the observed $y'$ values. To illustrate this point we have replotted the data of Table 5.3 (for which the population $P_1$ is Gaussian), referring them this time to the expected values of the order statistics for the exponential distribution. The latter were obtained by the formula given in reference 7 and are listed in column 8 of Table 5.3. It is apparent from the plot shown in Fig. 5.4, that the expected straight line did not materialize, indicating that the data do not conform to the exponential distribution. *Thus, a plot of order statistics is useful in drawing inferences from a sample about the nature of the parent population.*

A striking feature of Fig. 5.3 is its wavy appearance: the plotted points do not fluctuate in a purely random fashion about the theoretical straight

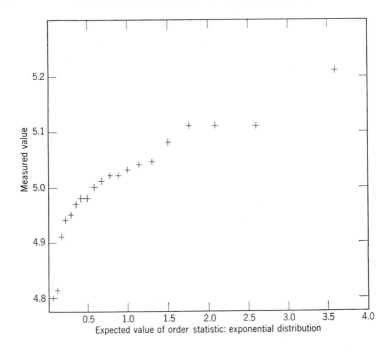

**Fig. 5.4**  Curvature resulting from an erroneous assumption about the underlying distribution function.

line, but tend to stay close to their neighbors.  We have here another typical cause of serial correlation: the mere act of ordering the data destroys their statistical independence.  The drawing of inferences from correlated data presents serious difficulties.  For example one might mistake the wavy character of the plot for real curvature and erroneously conclude that the underlying population is not normal.  In a later chapter we will be concerned with still another type of serial correlation that arises in some types of measurements, and we will see how misinterpretations can be avoided.  For our present purposes, a comparison of Figs. 5.3 and 5.4 shows that the serial correlation does not appear to seriously interfere with the ability to choose between widely different populations, such as the normal and exponential.  However, the true form of an underlying distribution function can be determined only on the basis of very large samples.

A second important application of a plot of order statistics versus their expected values is its use as an additional method for judging the *homogeneity* of a statistical sample.  Suppose that in a sample of 20 observations, 19 belong to a normal population of mean 5 and standard deviation 0.1, while the 20th (not necessarily the *last* observation) is taken from a normal population of mean 5.5 and standard deviation 0.1.   Then it is likely that the measurement originating from the second population will be the largest of the set of 20, and it may in fact turn out to be appreciably larger than the largest of the 19 observations from the first population.   A plot of the 20 order statistics versus their reduced expected values (such as in Fig. 5.3) will then be likely to reveal the foreign nature of the outlying observation, *not* because the latter differs appreciably from the average (the smallest or the largest observation in a sample must generally be expected to differ appreciably from the average), but rather because it departs significantly from the straight line that best fits all the other points.

More generally, if a sample consisted of portions from distinctly different populations, there is a likelihood of detectable breaks in the

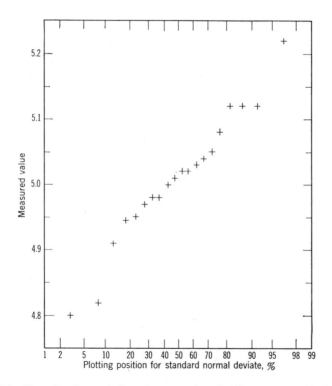

**Fig. 5.5**  Plot of order statistics using normal probability paper.  This figure is identical with Fig. 5.3.

ordered sequence of observations. A plot such as Fig. 5.3 is helpful in detecting such breaks and may therefore be used as another visual aid in judging the homogeneity of samples.

The plotting of order statistics is often done on special types of graph paper, called probability paper. In terms of Table 5.3, the conversion of column 4 into column 5 can be accomplished by constructing a scale whose *distances* are proportional to the reduced variable $t'$ (column 5) but which is *labeled* in terms of the area values (column 4). A graph paper having this scale in one dimension and an ordinary arithmetic scale in the other direction serves the purpose. The observed values of the sample are measured along the arithmetic scale, while the area values are laid out in accordance with the labeled values of the probability scale. Figure 5.5 is a plot of the "observations" of Table 5.3 on *normal probability paper*.

Note that all that is needed, in addition to the ordered observations (column 3), are the appropriate areas (column 6). The latter are denoted as a set of *plotting positions* for the order statistics. The graph is identical with that of Fig. 5.3. Its advantage is that it obviates the use of a table of the probability distribution. Of course, each type of distribution necessitates its own probability paper. Also, the use of probability paper is unnecessary when the expected values of the order statistics are directly available.

## 5.4   SAMPLES FROM SEVERAL POPULATIONS WITH DIFFERENT MEAN VALUES

While considerations based on order statistics are valid and useful for the analysis of samples of simple replicate measurements, they suffer from a major weakness: large samples are required before one can get a reasonably firm grip on the type of distribution underlying the sample. Inferences that depend on the postulation of a particular underlying distribution, say the Gaussian, are therefore of a tentative nature. It is, however, not uncommon that experimenters possess, in their notebooks, sets of data obtained on different occasions, but belonging to the same measuring process. For example, an analytical chemist may have a record of triplicate determinations of a given type of chemical analysis, made on a large number of compounds of a similar nature.

In Section 4.4 we have seen how the common variance of a number of populations with different means can be estimated from data of this type. We will now show that further useful information can be obtained by applying to such data the theory of order statistics. It is, however, necessary to assume in the following that the samples from the different populations all contain the same number of measurements. Table 5.4

**TABLE 5.4**    Logarithm of Bekk Smoothness

| Material | Ordered Observations | | | | | | | |
|---|---|---|---|---|---|---|---|---|
| | 1 | 2 | 3 | 4 | 5 | 6 | 7 | 8 |
| 1 | 1.724 | 1.878 | 1.892 | 1.914 | 1.919 | 1.919 | 1.929 | 1.952 |
| 2 | 0.580 | 0.681 | 0.699 | 0.699 | 0.748 | 0.763 | 0.778 | 0.778 |
| 3 | 0.944 | 1.000 | 1.033 | 1.093 | 1.100 | 1.114 | 1.114 | 1.176 |
| 4 | 1.072 | 1.114 | 1.114 | 1.114 | 1.121 | 1.170 | 1.182 | 1.193 |
| 5 | 1.584 | 1.587 | 1.606 | 1.611 | 1.638 | 1.683 | 1.687 | 1.699 |
| 6 | 2.083 | 2.097 | 2.137 | 2.143 | 2.172 | 2.236 | 2.259 | 2.268 |
| 7 | 1.946 | 1.982 | 1.984 | 1.989 | 2.088 | 2.126 | 2.129 | 2.177 |
| 8 | 2.106 | 2.112 | 2.125 | 2.127 | 2.143 | 2.152 | 2.164 | 2.170 |
| 9 | 1.072 | 1.079 | 1.079 | 1.114 | 1.121 | 1.164 | 1.176 | 1.199 |
| 10 | 0.342 | 0.716 | 0.732 | 0.732 | 0.748 | 0.778 | 0.778 | 0.845 |
| 11 | 2.205 | 2.207 | 2.224 | 2.255 | 2.263 | 2.276 | 2.280 | 2.285 |
| 12 | 0.681 | 1.146 | 1.220 | 1.225 | 1.225 | 1.230 | 1.230 | 1.250 |
| 13 | 1.428 | 1.508 | 1.534 | 1.546 | 1.564 | 1.649 | 1.651 | 1.688 |
| 14 | 2.185 | 2.191 | 2.193 | 2.222 | 2.236 | 2.250 | 2.259 | 2.354 |
| Average | 1.425 | 1.521 | 1.541 | 1.556 | 1.578 | 1.608 | 1.615 | 1.645 |

exhibits 14 sets of 8 values each, obtained by measuring the Bekk smoothness* of 14 different types of paper (materials).    For reasons that will be discussed in a subsequent chapter, the original measurements of Bekk seconds have been converted to a logarithmic scale.    Suffice it to say, here, that though the Bekk smoothness varied considerably among these 14 papers, the *spread* of *replicate* measurements on the same sample, when measured in the logarithmic scale, is about the same for all 14 materials.    For each material, the 8 measurements have been rearranged in increasing order of magnitude (left to right).    Now it is readily understood that if the populations corresponding to these 14 samples have distribution functions of the same geometrical shape, differing only in location (as indicated, for example, by the location of their mean values), then the order statistics should also show the same general pattern for all 14 samples.    Consequently, the typical relations between the order statistics for samples from this type of distribution function apply also to the *averages* of corresponding order statistics, i.e., to the column averages at the bottom of Table 5.4.    Furthermore, the averaging should result in greater stability, so that the resulting averages may be expected to behave more closely in accordance with the laws established for the expected values of order statistics than do the ordered values of individual sets.

* The Bekk method is a procedure for measuring the smoothness of paper; see reference 3 of Chapter 3.

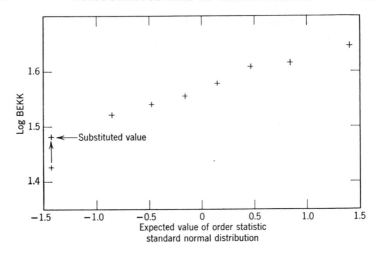

**Fig. 5.6**  Plot of order statistics for the column averages of Table 5.4.

It is therefore possible to use such accumulated sets of data for more reliable inferences regarding the underlying distribution than one could ever achieve with a single set of data, and this can be done despite the fact that the measured value varies widely from set to set.    Figure 5.6 shows a plot of the column averages of Table 5.4 against the expected values of the order statistics of the reduced normal deviate.    A striking feature of this graph is the anomaly of the first point, the only one that shows considerable departure from an otherwise excellent straight line fit.    An examination of Table 5.4 reveals that two values in the first column are abnormally low: the tenth and the twelfth.    Any doubt regarding the anomaly of these two values is removed when one computes the differences between the first two order statistics for each of the 14 rows of Table 5.4.    The values obtained are: 0.154, 0.101, 0.056, 0.042, 0.003, 0.014, 0.056, 0.006, 0.007, 0.374, 0.002, 0.465, 0.080, 0.006.*    It is likely that the two values in question result either from faulty measurements or from erroneous recordings of the measured values.    One may attempt to substitute for these two faulty data, values obtained by subtracting from the second value in rows 10 and 12, the average of the remaining 12 differences above. This average is equal to 0.044.    Subtracting 0.044 from 0.716 and from

* The reader who remains skeptical is advised to compute and tabulate the difference between *all* pairs of successive order statistics for each of the 14 rows.    It is of interest to add, at this point, that the unpleasant feature of serial correlation between order statistics is largely removed when *differences* of consecutive order statistics are considered, rather than the order statistics themselves.

1.146, we obtain 0.672 and 1.102. When these two values are substituted for 0.342 and 0.681 respectively, the new average of the first order statistics (column 1 of Table 5.4) becomes 1.479. When this value is used in Fig. 5.6, as a replacement for 1.425, all eight points can be satisfactorily fitted by a single straight line. The slope of this line is seen to be approximately equal to 0.059. We have already shown that this slope represents an estimate for the standard deviation of the measurements. In the present case, it is an estimate of the standard deviation of the logarithm of Bekk seconds. Let us denote a Bekk value by $X$ and its common logarithm by $Z$. Applying Eq. 4.41 to the function $Z = \log_{10} X$, for which the derivative $Z_x' = 1/2.30X$, we obtain

$$V_e(Z) = V(\log_{10} X) = \left(\frac{1}{2.30X}\right)^2 V_e(X)$$

where the symbol $V_e$ denotes the variance of the experimental error. Thus,

$$\sigma_e(Z) = \frac{1}{2.30X} \sigma_e(X)$$

and hence

$$\frac{\sigma_e(X)}{X} = 2.30\sigma_e(Z)$$

Substituting for $\sigma_e(Z)$ the value 0.059, found for the standard deviation of log Bekk, we obtain:

$$\frac{\sigma_e(X)}{X} = (2.30)(0.059) = 0.14$$

Thus, using the data of Table 5.4, our estimate of the coefficient of variation of the Bekk measuring process is 14 per cent.

We note parenthetically that the analysis by order statistics has allowed us to detect with considerable assurance the presence of two faulty measurements. While a number of statistical tests are available for this purpose, the present method has the unique advantage of being based on an underlying distribution that is *shown* to fit the data rather than one that is merely postulated.


## 5.5  EXTREME-VALUE THEORY

Many situations exist in which the property of interest for a particular system is related to the largest or to the smallest of a set of measurable values. For example, a "flood" is defined as the largest discharge of water of a river during a year. Clearly the consideration of floods is of critical importance in the construction of dams. Another example is

given by the life of light bulbs.    The quality control engineer is interested
in the shortest observed life for a series of bulbs under test.    It is therefore
of interest to study the statistical properties of the *extreme values* of a
sample, i.e., of the smallest and the largest values of the sample.    In the
language of order statistics, the extreme values are of course simply the
first and the last order statistics of a statistical sample.

To gain some insight into the behavior of extreme values, consider a
sample of 10 observations taken from a normal population.    From
Section 5.3 we know that the last order statistic will be situated approximately at that point of the measurement axis (see Fig. 5.2) for which the
area to the *right* is $1/(N + 1) = 1/11$.    Using the refinement given by
Blom's formula, the exact average location is that for which the area to
the *left* is $(N - 3/8)/(N + 1/4) = (10 - 3/8)/(10 + 1/4) = 77/82 = 0.939$.
From a table of the cumulative normal distribution we easily infer that
this point is located at 1.55 standard deviations above the population
mean.    Thus, the largest of a set of 10 observations from a normal
universe is located, on the average, at 1.55 standard deviations above the
mean.

The above calculation can of course be made for any sample size.
For example, for $N = 100$, we obtain the area $(100 - 3/8)/(100 + 1/4) =$
0.99376 which corresponds to a point situated at 2.50 standard deviations
above the mean.    Similarly, for $N = 1000$, we obtain for the average
location of the largest value: 3.24 standard deviations above the mean.
The striking feature of these values is the slowness with which the extreme
value increases with sample size.    This property of extreme values has
far-reaching consequences; it implies for example that the probability of
observing an extreme that is very far removed from the mean of the
original distribution is quite small.    Now when an event has a very small
probability, it takes a large number of trials to obtain an occurrence of the
event.    We can even calculate the *average* number of trials required to
obtain an occurrence.    For example, the probability that the topmost
card of a well-shuffled deck be a king of spades is 1/52.    Therefore it
takes, *on the average*, 52 shufflings to obtain the king of spades as the topmost card.    More generally, if the probability of an event is $p$, it takes on
the average $1/p$ trials to obtain an occurrence of the event.    Let us apply
this theorem to the theory of floods.    Suppose that for the Colorado
River at Black Canyon, a flood of 250,000 cubic feet per second has been
observed to have a probability of occurrence of 5 per cent.    Then, it will
take $1/0.05 = 20$ years, on the average, to obtain a flood of this magnitude.
This fact is expressed by stating that the *return period* of a flood of 250,000
cubic feet per second is 20 years.    For the construction of dams, isolated
facts of this type are of course insufficient.    What is needed here is a

knowledge of the probabilities of occurrence of floods of various magnitudes, in other words, of the frequency distribution of floods at the location of the prospective dam.

The study of the frequency distributions of extreme values such as floods has become a topic of great importance in statistics. Applications occur in climatology, in aeronautics, in demographic studies, in bacteriology, and in the study of the strength of materials. It has been found that a particular type of distribution function, known as the *double exponential*, represents to a good approximation the behavior of many types of extreme values. Theoretically, the double exponential can be shown to be the *limiting* distribution of extreme values from *large* samples taken from many types of populations, including the Gaussian and the exponential. The mathematical properties of the double exponential are well-known and tables have been prepared to facilitate its practical use (5). A particularly useful device consists in plotting the order statistics of a sample of extreme values on a special type of probability paper. As explained in Section 5.3, a scale can be constructed for any known distribution function such that when the order statistics of a sample from this distribution are plotted against the appropriate *plotting positions* on this scale, a straight line will be obtained. For further details on the interesting subject of extreme value theory the reader is referred to references 5 and 6.

## 5.6  SUMMARY

The simplest set of data that may be subjected to a statistical investigation is a random sample from a single population. Such a set is said to be homogeneous. In practice one can seldom be sure that a set of data is homogeneous, even when the data were obtained under what was believed to be a constant set of conditions.

In the control of the quality of mass-produced items, it is important to detect unsuspected changes in the manufacturing process that might cause a gradual change in the properties of the product. A semi-graphical statistical technique was developed by Shewhart in 1924 to accomplish this aim. This technique, known as the control chart, turns out to be equally valuable for the detection of lack of homogeneity in a set of repeated measurements of any type. It is illustrated here in terms of a set of repeated measurements of the velocity of light and reveals, in this case, the presence of a cyclic pattern of change in the values. When such non-randomness exists the classical formula for the variance of a mean cannot be safely applied to the average of the measurements of the set.

Another powerful tool for the detection of non-homogeneity in a set of repeated measurements is given by the examination of order statistics.

These are simply the observed values rearranged in increasing order.   In a random sample from any statistical population the order statistics tend to occupy, on the average, fixed positions.   These fixed positions, called plotting positions, can be calculated for any given distribution function, and they have been tabulated for the normal distribution as well as for a few other important frequency distributions.   A plot of the actually observed order statistics versus the corresponding plotting positions allows one to draw statistical inferences about the probability distribution function underlying the data.   Such a plot may also be used to detect outlying observations, i.e., measurements that are very probably subject to additional errors other than the usual random fluctuations characterizing the population to which the data belong.   Furthermore, the plot of order statistics provides a simple method for estimating the standard deviation of the measurements.   An interesting application of the theory of order statistics is found in the examination of several homogeneous samples of equal size when the populations corresponding to the various samples differ from each other only in terms of their mean value.   This technique is illustrated in terms of the values reported for 14 materials in a study of a method for measuring the smoothness of paper.   For each of the 14 materials, 8 replicate smoothness measurements were made.   The statistical analysis reveals the presence of some faulty measurements and allows one to estimate the replication error of the measuring process.

Situations exist in which the property of interest is the largest or the smallest of a set of values (extreme values).   Examples are provided by the theory of floods of rivers and by the study of the useful life of manufactured items, such as light bulbs.   The statistical theory underlying such phenomena has been studied, and a particular frequency distribution, known as the double exponential, has been found particularly useful for the description of extreme values.   Of particular practical interest is a technique combining the theory of order statistics with that of the distribution of extreme values.

## REFERENCES

1. *American Society for Testing Materials Manual on Quality Control of Materials*, Special Technical Publication 15-C, American Society for Testing Materials, Philadelphia, 1951.
2. Blom, G., *Statistical Estimates and Transformed Beta-Variables*, Wiley, New York, 1958.
3. Brownlee, K. A., *Statistical Theory and Methodology in Science and Engineering*, Wiley, New York, 1960.
4. Fisher, R. A., and F. Yates, *Statistical Tables for Biological, Agricultural, and Medical Research*, Oliver and Boyd, Edinburgh, 1953.

5. Gumbel, E. J., "Probability Tables for the Analysis of Extreme-Value Data," *National Bureau of Standards Applied Mathematics Series*, **22**, 1953. (For sale by the Superintendent of Documents, U.S. Government Printing Office, Washington 25, D.C.)

6. Gumbel, E. J., "Statistical Theory of Extreme Value and Some Practical Applications," *National Bureau of Standards Applied Mathematics Series*, **33**, 1954. (For sale by the Superintendent of Documents, U.S. Government Printing Office, Washington 25, D.C.)

7. Sarhan, A. E., and B. G. Greenberg, eds., *Contributions to Order Statistics*, Wiley, New York, 1962, p. 43.

8. Shewhart, W. A., *Economic Control of Quality of Manufactured Product*, Van Nostrand, New York, 1931.

9. *Symbols, Definitions, and Tables, Industrial Statistics and Quality Control*, Rochester Institute of Technology, Rochester, New York, 1958.

10. Wernimont, G., "Use of Control Charts in the Analytical Laboratory," *Industrial and Engineering Chemistry*, Analytical Edition, **18**, 587–592 (October 1946).

# chapter 6

# THE PRECISION AND
# ACCURACY OF MEASUREMENTS

## 6.1 INTRODUCTION

The question of how experimental error arises in measurements has been briefly discussed in Chapter 2 and we will return to it in greater detail in Chapters 13 and 14. In the present chapter, we are concerned with *measures* of experimental error. An analogy will help in understanding the need for such measures. The mass of an object, or its color, are physical properties of the object and susceptible of physical methods of measurement. Similarly the experimental errors arising in a measurement process are statistical properties of the process and susceptible of statistical methods of measurement. Mass can be expressed by a single number, measured for example in grams; but the measurement of color is somewhat more complicated and the results of such measurements are best expressed in terms of a spectrum of wavelengths. The problems raised by the measurement of experimental error are somewhat similar to those arising in the measurement of color; only seldom will a single number suffice for the characterization of experimental error. The matter is further complicated by the need to specify exactly of what one is measuring the experimental error. To state merely that one is measuring the experimental error of a particular measuring process is ambiguous, since in order to obtain experimental errors the process must be carried out on a particular system. For example, the process of measuring tensile strength can be described in terms of a particular instrument, but the experimental errors obtained in making a tensile strength determination with this instrument will depend on the particular material of which the tensile strength is being measured.   It matters a great deal whether this material is a piece of leather, rubber, or steel.   Even specifying that it is rubber is insufficient because of the great differences that exist in the tensile strength of rubber sheets vulcanized in different ways.   Similarly, the nature of the experimental errors arising in the determination of nickel in an alloy may depend not only on the nature of the alloy but also on the actual nickel content of the alloy.

To a large extent the confusion surrounding the measurement of experimental error is due to a failure to give proper attention to the matters we have just discussed.   The desire to obtain simple measures of experimental error has led to two concepts: *precision* and *accuracy*. When properly defined these concepts are extremely useful for the expression of the uncertainty of measurements.   We will see, however, that contrary to common belief, it is generally not possible to characterize measuring methods in terms of two numbers, one for precision and one for accuracy.   We now proceed to a definition of these two concepts.

## 6.2   THE CONCEPT OF PRECISION

It is easier to define *lack of precision* than precision itself.   For convenience we will refer to lack of precision as *imprecision*.   *Given a well-described experimental procedure* for measuring a particular characteristic of a chemical or physical system, *and given such a system*, we can define *imprecision* as the amount of scatter exhibited by the results obtained through repeated application of the process to that system.   This scatter is most adequately expressed in terms of a probability distribution function, or more exactly in terms of parameters measuring the spread of a probability distribution.

This definition implies that the repeated application of a given measuring technique to a given system will yield results that belong to a *single* statistical population.   This assumption is not supported by a detailed examination of actual measuring processes.   Data obtained in the study of measuring techniques lead almost invariably to populations *nested* within populations (also referred to as hierarchical situations).   The reasons for this will become apparent in Chapter 13 where we will study the systematic evaluation of measuring processes.   For the present we will ignore these complications, but in Section 6.4 we will introduce the statistical techniques most commonly used for the *expression* of the precision of measured values.

## 6.3   THE CONCEPT OF ACCURACY

To define accuracy, we refer, once more to a *well-described measuring process, as applied to a given system.*   Again we consider the statistical population of measurements generated by repeated application of the process to that system.   We must now introduce a new concept: the *reference value* of the measured property for the system under consideration.

Reference values can appear in at least three different forms: (*a*) A reference value can be defined as a real, though unknown "true" value, conceived in some philosophically accepted sense.   An example of this type is provided by the basic physical constants, such as the velocity of light *in vacuo*.   (*b*) The reference value can be an "assigned" value, arrived at by common agreement among a group of experts.   This type of reference value is commonly used in technological measurements, and its choice is dictated primarily by considerations of convenience.   For example, in conducting an interlaboratory study of a particular method of chemical analysis, the reference values of the samples included in the study may simply be the values provided for these samples by a recognized standard institution.   Such a choice does not necessarily imply that these values are considered absolutely correct, but it does imply a reasonable degree of confidence in the quality of testing done at this institution. (*c*) Finally, the reference value may be defined in terms of a *hypothetical* experiment, the result of which is *approximated* by a sequence of actual experiments.   For example, in establishing the appropriateness of a standard yard stick, the real reference value is the length measurement that *would have been* obtained for the stick *had it been subjected* to a comparison, in accordance with a well-described experimental procedure, with the wavelength of the orange-red line of krypton 86.   The cost involved in making such a comparison would be completely unwarranted by the modest requirements of exactness for the yard stick, and this experiment would therefore not be performed.   This does not, however, preclude an estimation of the exactness of a yard stick in reference to the primary length standard.   The estimation is accomplished in practice by performing a simpler and cheaper comparison, involving secondary length standards.   Since the determination of the exactness of the latter involves at least one new experiment, we generally obtain a sequence of experiments, leading from the primary standard to a secondary one, and so forth, down to the yard stick in question.

It should be noted that of the three types of reference values described above, only the second type lends itself to an operationally meaningful statement of the *error of a particular measurement*: the error of the measurement is simply the difference between the value obtained and the assigned value.   On the other hand, the use of reference values of the first type necessitates, in addition to the measurement, a *judgment* of its error.   Thus, any measurement of the velocity of light *in vacuo* is in error by an unknown quantity.   Attempts can be made to estimate an *upper bound* for this unknown error, on the basis of a detailed analysis of all the steps involved in the measurement, and of their possible shortcomings. Comparisons of the chain-type involving standards of decreasing degrees of perfection lead to an accumulation of errors of which the true values are unknown, but which can generally be guaranteed not to exceed certain limits.

Whichever way we have defined the reference value, let us denote it, for a particular property of a particular system, by the symbol $R$ and let $\mu$ denote the mean of the population of repeated measurements of the system.   We now define the *bias* or *systematic error* of the process of measurement for that system by $\mu - R$, i.e., as the difference between the population mean of repeated measurements $\mu$ and the reference value $R$.

Regarding the concept of accuracy there exist two schools of thought. Many authors define accuracy as the more or less complete absence of bias; the smaller the bias, the greater is the accuracy.   Complete accuracy, according to this view, simply means total absence of bias.   This of course implies that $\mu - R = 0$, or that $\mu = R$.   According to this school of thought, a measurement can be accurate even though it has an experimental error.   If $x$ represents the measurement, the total error is $x - R$, which can be written:

$$\text{error} = x - R = (x - \mu) + (\mu - R)$$

Accuracy implies that $\mu = R$, but this still leaves the quantity $x - \mu$ as experimental error.

The second school of thought defines accuracy in terms of the *total* error, not merely the bias $\mu - R$.   According to this view, a measurement can never be accurate unless its total error, $x - R$, is zero, and the *process* is accurate only if this is true for *all* repeated measurements.

If we observe that $x - \mu$ is really a measure for the imprecision of the measurement, we see that the second school of thought includes imprecision in the measurement of accuracy while the first school of thought separates imprecision from the measurement of accuracy.

Without going into the merits of the two conflicting views on the definition of accuracy we will, for the sake of consistency, adopt the first view only.   Thus, accuracy for us means the absence of bias.   *Inaccuracy* is then measured by the magnitude of the bias.

## 6.4  STATISTICAL MEASURES OF PRECISION

Among the methods most commonly used for the expression of precision we mention the standard deviation (or its square, the variance), the mean deviation, the range, the standard error, the probable error, the coefficient of variation, confidence intervals, probability intervals, and tolerance intervals.  Such a profusion of measures requires that some criteria be given to select that measure which is most adequate for any given situation. The choice depends mainly on the objective of the experiment.  We will discuss the various measures in terms of two sets of data.  Throughout our discussion we must keep in mind the distinction between a population parameter and its estimate derived from a sample.

**TABLE 6.1**   Isotopic Abundance Ratio of Natural Silver

| Measurement number | Ratio | Measurement number | Ratio |
|---|---|---|---|
| 1 | 1.0851 | 13 | 1.0768 |
| 2 | 834 | 14 | 842 |
| 3 | 782 | 15 | 811 |
| 4 | 818 | 16 | 829 |
| 5 | 810 | 17 | 803 |
| 6 | 837 | 18 | 811 |
| 7 | 857 | 19 | 789 |
| 8 | 768 | 20 | 831 |
| 9 | 842 | 21 | 829 |
| 10 | 786 | 22 | 825 |
| 11 | 812 | 23 | 796 |
| 12 | 784 | 24 | 841 |

Average $= 1.08148$
Variance $= 667 \times 10^{-8}$
Standard Deviation $= 0.00258$
Standard Error of the Mean $= \dfrac{0.00258}{\sqrt{24}} = 0.00053$
Per Cent Coefficient of Variation of the Mean $=$
$$100 \, \frac{0.00053}{1.081} = 0.049\%$$

The data in Table 6.1 are measurements of the isotopic abundance ratio of natural silver, i.e., the ratio of $Ag^{107}$ to $Ag^{109}$ in a sample of commercial silver nitrate (5).   Those in Table 6.2 are tensile strength measurements made on 30 tanned leather hides (3).   Both sets can be considered as random samples from well defined, though hypothetical populations.

## 6.5  STANDARD DEVIATION AND STANDARD ERROR

The variances and standard deviations given at the bottom of each set are the sample estimates calculated in accordance with Eqs. 3.22 and 3.23. They are proper estimates of the corresponding population parameters, regardless of the nature of the frequency distribution.  It is only when we attach certain interpretations to these estimates that the nature of the distribution becomes a pertinent issue.

**TABLE 6.2**  Measurement of the Tensile Strength of Leather

| Hide number | Tensile strength | Hide number | Tensile strength |
|---|---|---|---|
| 1 | 449 | 16 | 398 |
| 2 | 391 | 17 | 472 |
| 3 | 432 | 18 | 449 |
| 4 | 459 | 19 | 435 |
| 5 | 389 | 20 | 386 |
| 6 | 435 | 21 | 388 |
| 7 | 430 | 22 | 414 |
| 8 | 416 | 23 | 376 |
| 9 | 420 | 24 | 463 |
| 10 | 381 | 25 | 344 |
| 11 | 417 | 26 | 353 |
| 12 | 407 | 27 | 400 |
| 13 | 447 | 28 | 438 |
| 14 | 391 | 29 | 437 |
| 15 | 480 | 30 | 373 |

Average = 415.6
Variance = 1204
Standard Deviation = 34.7
Standard Error of the Mean = $\dfrac{34.7}{\sqrt{30}}$ = 6.3

In the case of the isotopic ratio, measurement of precision may take either one of two aspects: (*a*) our interest may be focused on the measuring technique by which the data are obtained, for example in order to compare it with an alternative technique; or (*b*) we may be concerned with the reliability of the best value of the ratio derived from this set of data.   In case *a*, the standard deviation that characterizes the statistical population underlying these measurements is a logical criterion for the expression of precision.   In case *b* if we take as the best value for the ratio the arithmetic average of the values obtained, we are particularly interested in the uncertainty of this average.    In that case, the appropriate measure is the standard error of the average, which, in accordance with Eq. 4.14 equals the standard deviation of the population of the measurements, divided by the square root of the actual number of measurements.    Finally, if we wish to express the standard error of the mean as a fraction of the measured value, we obtain the coefficient of variation of the mean.    For the data of Table 6.1, this measure, expressed as per cent, is equal to 0.049 per cent, or approximately 1 part in 2000.    It should be emphasized that this measure is derived entirely from the internal variability of the given set of data, and should not be taken as an index of the over-all uncertainty of the average.

We must also give careful consideration to the following point.    There is a subtle duality to the concept of precision as applied to a single value, such as the average of a series of measurements of the isotopic abundance ratio.    Operationally, all we mean by the standard error of the average in our example is a measure of scatter of a hypothetical infinite population, namely the population of averages of similar experiments, each of which consists of 24 individual measurements.    Yet we use this standard error as if it applied, not to a population, but to a single average.    This is justified in the sense that the standard error, by measuring the scatter of a population of averages all of which are obtained in the same way as our single average, gives us a tangible idea of the *uncertainty* of this average.    The passage from a measure of *scatter of a population of values* to that of the uncertainty of a single value involves perhaps only a philosophical difficulty.    In any case the matter cannot be ignored if one is to interpret correctly the use of a standard error as a measure of uncertainty.    Further clarification of this concept will be obtained when we discuss the subject of confidence intervals; we will also consider another aspect of this matter in Section 6.13.

Turning now to the data of Table 6.2, we again begin by asking ourselves what objectives we have in mind in measuring precision.    The estimated standard deviation derived from the data is in this case not necessarily an index of the precision of the measurement technique.    In fact the errors of measurement are probably considerably smaller than the fluctuations in tensile strength from hide to hide.    Each of the 30 measurements of Table 6.2 is actually the average of 21 breaking strength values corresponding to 21 locations on the hide.    It is therefore as good a representation of the tensile strength of the hide as one can get.    Primarily, then, the standard deviation derived from the 30 values is a measure of hide to hide variability in tensile strength.    It is unlikely that the use of a more refined instrument for the measurement of tensile strength would result in an improved standard deviation.    It is clear that the standard deviation derived from our measurements is but one of many pertinent measures for the variability of the tensile strength of leather.    In addition to variability among hides, we might be interested in the variations of tensile strength on the hide area.    Is this variation random, or are there strong and weak positions on the hide?    If the pattern of variation is systematic, the use of a standard deviation to describe it is not adequate, since the locations of the hide do not constitute a random sample from a larger collection of locations.    Even though the standard deviation calculated from the data in Table 6.2 does not measure the precision of a measuring technique, it is nevertheless a useful quantity.    For example, it might be used as a measure of the uniformity of the population of hides from which the 30 hides were taken.    Or, if tensile strength is strongly influenced by any of the treatments to which the hides were subjected (such as tanning), it would be an appropriate measure of the effectiveness of the experimental control of the techniques used in applying these treatments.

Analogous to our treatment of the data of Table 6.1, we can calculate a standard error of the average of the measurements in Table 6.2.    In the case of the isotopic abundance ratio, the standard error was a useful concept, related to the reliability of the best value derived from the data. Is the standard error of the average a meaningful quantity for the data of Table 6.2?    The answer depends, as always, on the objective of the experiment.    If the 30 hides in question were tanned by a given procedure, and a number of other hides from the same cattle stock were tanned by a different procedure, the comparison of the average tensile strengths corresponding to the two tanning procedures might well be of interest. But a quantitative appraisal of the difference between the two averages cannot be made unless one has a measure of the reliability of each average. Such a measure is provided by the standard error of the average.    Thus, the standard error is a meaningful concept for the hide data as well as for measurements of the isotopic abundance ratio.

It should be emphasized that in either case, the isotopic abundance ratio or the leather data, the standard error is not a measure of the *total* error of the average (i.e., of its deviation from the true or reference value). We have seen that the evaluation of the total error is sometimes impossible and mostly difficult and uncertain.    However, in many cases, the standard error of the average is all that is required.    Thus, a comparison of tannages based on a comparison of average values and their standard errors might well be completely satisfactory.    If, as a result of deficiencies in the measuring technique, systematic errors are present, they are likely to be the same for all tannages, and the *differences* between the average values corresponding to the different tannages will then be entirely free of these systematic effects.    Let $\bar{x}$ and $\bar{y}$ represent the averages of respectively, $N_1$ measurements for tannage $A$ and $N_2$ measurements for tannage $B$. If $\sigma_1$ and $\sigma_2$ are the (population) standard deviations for the scatter among single measurements (or hides) for $A$ and $B$, respectively, then the standard errors for $\bar{x}$ and $\bar{y}$ are $\sigma_1/\sqrt{N_1}$ and $\sigma_2/\sqrt{N_2}$.    Consequently, the standard error of the difference $\bar{x} - \bar{y}$ assuming statistical independence between these averages, is

$$\sigma_{\bar{x} - \bar{y}} = \left[ \frac{\sigma_1^2}{N_1} + \frac{\sigma_2^2}{N_2} \right]^{1/2} \tag{6.1}$$

This, then, is a measure of the *total* uncertainty of $\bar{x} - \bar{y}$, despite the fact that the *individual* standard errors, $\sigma_1/\sqrt{N_1}$ and $\sigma_2/\sqrt{N_2}$ fail to account for systematic errors in the measuring process.

It has been pointed out already that in the case of the isotopic abundance ratio, the standard error of the average, derived from the differences among replicate measurements, is far less satisfactory as an index of uncertainty.    For here we are concerned not with a mere comparison, but rather with a quantity of fundamental intrinsic interest.    Possible systematic errors in the measuring process are not reflected in the standard error of the average.    They must be explored on the basis of evidence *external* to the data, rather than from their internal structure.

## 6.6    THE RANGE AND THE MEAN DEVIATION

It has already been mentioned that if the population standard deviation $\sigma$ is unknown, a sample estimate $s$ calculated in accordance with Eq. 3.23, is used in its place.    For small samples, say samples consisting of less than 15, a simple formula can be used to provide a good approximation of $s$.    Defining the *range* as the difference between the largest and the smallest of a sample of $N$ observations, the approximate value of the standard

**TABLE 6.3**    Relation Between Range and Standard Deviation

| Sample size, $N$ | Factor, $d_2$[a] | Sample size, $N$ | Factor, $d_2$[a] |
|---|---|---|---|
| 2 | 1.128 | 9 | 2.970 |
| 3 | 1.693 | 10 | 3.078 |
| 4 | 2.059 | 11 | 3.173 |
| 5 | 2.326 | 12 | 3.258 |
| 6 | 2.534 | 13 | 3.336 |
| 7 | 2.704 | 14 | 3.407 |
| 8 | 2.847 | 15 | 3.472 |

[a] To obtain an estimate of $\sigma$, divide the range by $d_2$.

deviation is obtained by dividing the range by the factor $d_2$ given in Table 6.3.  The notation $d_2$ for this factor is standard in quality control literature (1, 6).

Denoting the range by $R$, the new estimate of the standard deviation, which we may denote by $\tilde{\sigma}$, is given by

$$\tilde{\sigma} = R/d_2 \tag{6.2}$$

It is interesting to observe that the value $d_2$ is very closely approximated by $\sqrt{N}$.  Thus, without recourse to any table one can estimate the standard deviation of small samples by dividing the range of the sample by the square root of the number of observations.  It follows that the standard error of the average is approximately equal to the range divided by the number of observations, a useful device for a rapid evaluation of the uncertainty of an average.

Equation 6.2 has the following useful extension.  Suppose that several samples of equal size, $N$, are available, and that it may be assumed that the populations from which the samples were taken, all have the same standard deviation, $\sigma$.  An estimate of this common $\sigma$ is then obtained by computing the *average range*, i.e., the average of the ranges of the various samples, and dividing it by the factor $d_2$ corresponding to the common sample size $N$.  Denoting the average range by $\bar{R}$, we thus obtain:

$$\tilde{\sigma} = \bar{R}/d_2 \tag{6.3}$$

If the samples are not all of the same size, an estimate of $\sigma$, the common standard deviation, can still be obtained by dividing the range of each sample by the factor $d_2$ corresponding to its size, and averaging the quotients thus obtained.  If $k$ denotes the number of samples, the formula is:

$$\tilde{\sigma} = \frac{1}{k}\left(\frac{R_1}{d_{2,N_1}} + \frac{R_2}{d_{2,N_2}} + \cdots + \frac{R_k}{d_{2,N_k}}\right) \tag{6.4}$$

where $d_{2,N_i}$ represents the $d_2$ factor for a sample of size $N_i$.

Equation 6.4 is the analogue of Eq. 4.20.  It is appreciably simpler than the latter but not as precise.  It should not be used if any of the $N_i$ values (the sizes of the samples) is greater than 15.

Many generations of physicists and chemists have used a measure of precision (or rather imprecision) known as the *average deviation*, or *mean deviation*.  The average deviation is the average value of the absolute values of the deviations from the mean:

$$\text{Average deviation} = \frac{\sum_i |x_i - \bar{x}|}{N} \tag{6.5}$$

Defined in this manner, the average deviation is a *statistic*,* not a population parameter, although it is possible to define it also as the latter. Despite its popularity, the use of this measure cannot be recommended. Youden (7) has pointed out that if the precision of a process is estimated by means of the average deviation of a number of replicate measurements, a subdivision of the measurements into smaller subsets will result in a systematic under-estimation of the true variability of the process. If the experimental errors of the measuring process follow the normal distribution, average deviations calculated from sets of 2 measurements will tend to be about 18 per cent smaller than average deviations calculated from sets of 4 measurements from the same population. These latter average deviations still underestimate the population standard deviation by about 31 per cent. The degree of underestimation depends not only on the sample size but also on the nature of the probability distribution of the errors.

It must be pointed out that the sample standard deviation is not entirely free from the blemish that has just been described for the average deviation. On the other hand, the square of the standard deviation, i.e., the variance, is always an unbiased estimate† of the population variance, provided that the proper number of degrees of freedom is used. In the case of the normal population, the sample standard deviation is appreciably less biased than the average deviation as a measure of the population standard deviation. Thus, for samples of size 2, the sample standard deviation has a bias‡ of −20 per cent as compared to −44 per cent for the average deviation; for samples of size 4, the bias of the sample standard deviation is −8 per cent, as compared to −31 per cent for the average deviation; and for samples of size 20, the bias of the sample standard deviation is only −1.3 per cent, as compared to −22 per cent for the average deviation.

Another measure of precision that has enjoyed great popularity among physical scientists is the *probable error*. This value is generally defined as the product of the standard deviation by the numerical quantity 0.6745. The definition is ambiguous, since no distinction is usually made between the population parameter and the sample estimate. A discussion of the probable error is more meaningful after introduction of the concept of probability intervals.

---

* See footnote on page 85.

† An estimate of a parameter is said to be unbiased if its expected value is equal to the true value of the parameter. Thus an unbiased estimate is essentially one that is free of systematic errors.

‡ Bias is here defined as the difference between the expected value of the estimate and the true value of the parameter, expressed as a percentage of the latter.

## 6.7  PROBABILITY INTERVALS

One of the basic problems in the theory of errors is to determine the frequency with which errors of any given size may be expected to occur.

Suppose that a particular measuring process leads to results with normally distributed errors.  For a particular sample, the measurements can be expressed as

$$x = \mu + \varepsilon \tag{6.6}$$

where $\varepsilon$ is a random variable of zero mean and standard deviation $\sigma$. Alternatively, we can say that $x$ itself is a normally distributed variable with mean $\mu$ and standard deviation $\sigma$.  If we limit ourselves to considerations of precision (i.e., if we exclude the effects of possible systematic errors in the measuring process), we will be particularly interested in the quantity

$$\varepsilon = x - \mu \tag{6.7}$$

Now suppose that a certain positive quantity, $D$, is given, and the question is asked how frequently the error $\varepsilon$, or $x - \mu$, will be larger than $D$.

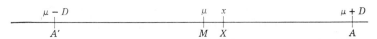

**Fig. 6.1**  If $x$ is normally distributed, and $D = 1.96\sigma$, the point $X$ will be between $A'$ and $A$ with a probability of 0.95.

Consider Fig. 6.1 in which the points $X$ and $M$ represent the quantities $x$ and $\mu$.  The deviation $x - \mu$, taken without regard to sign, will exceed $D$ whenever the point $X$ lies either to the left of $A'$ or to the right of $A$. (If $D = 1.96\sigma$, this will happen 5 times in 100.  In that case the probability that the point $X$ will lie between $A'$ and $A$ is 0.95.)  Algebraically, the event that $X$ lies between $A'$ and $A$ can be expressed by the inequality $|x - \mu| < D$, or alternatively, by the double inequality

$$\mu - D < x < \mu + D \tag{6.8}$$

The interval extending from $\mu - D$ to $\mu + D$ ($A'$ to $A$) is a *probability interval* for the random variable $x$.  For $D = 1.96\sigma$, the probability associated with the interval is 0.95.  More generally, if $D = k\sigma$, the probability interval for the normally distributed variable $x$ is

$$\mu - k\sigma < x < \mu + k\sigma \tag{6.9}$$

and the probability associated with this interval depends only on $k$.

Note that the limits of the probability interval are, in the present case, symmetrically located with respect to $\mu$. For skew distributions this symmetrical choice is not necessarily the most desirable one, and a probability interval may be considered of the form:

$$\mu - k_1\sigma < x < \mu + k_2\sigma \qquad (6.10)$$

For any given probability, it is possible to choose the pair of values $k_1$ and $k_2$ in infinitely many ways. For variables with skew distributions, the concept of a probability interval lacks the intuitive simplicity of the symmetrical case.

Referring again to the double inequality Eq. 6.9, we deduce from a table of the normal curve that for normally distributed $x$, a value of $k$ equal to 0.6745 corresponds to a probability interval of 0.50. Thus, for $D = 0.6745\sigma$, the probability that $x$ will lie between $\mu - D$ and $\mu + D$ is 0.50. It is for this reason that the quantity $0.6745\sigma$ has been defined as the *probable error*, in the sense that such a quantity is as probable to be exceeded as it is not to be exceeded. Three circumstances militate against the use of the probable error as a measure of precision, even for symmetric distributions. In the first place, the numerical factor 0.6745 is appropriate only for the normal distribution, if the 50 per cent probability is to be maintained. Secondly, if the standard deviation is estimated from a sample of small or moderate size, the factor 0.6745 is incorrect even for the normal distribution, as we will see in the following section. Finally, no convincing case can be made for the adoption of a 50 per cent probability interval as being necessarily the most appropriate choice in all situations.

## 6.8 CONFIDENCE INTERVALS

The concept of the probability interval has given rise to two developments that are of particular interest to the science of measurement. The first of these is the notion of confidence intervals and, more generally, of confidence regions.

Consider again the expression 6.8, and its illustration in Fig. 6.1. Writing $D$ as $k\sigma$, we obtain expression 6.9

$$\mu - k\sigma < x < \mu + k\sigma$$

Assuming a normal distribution for $x$ there corresponds to each value of $k$ an associated probability. The latter can be interpreted as the relative frequency with which $x$ will lie in the interval $\mu - k\sigma$ to $\mu + k\sigma$ ($A'$ to $A$ is Fig. 6.1). It is, however, not often that we are interested in problems in which all pertinent values—in this case $\mu$ and $\sigma$—are known.

The purpose of measurement in most cases is to *estimate* these values. Let us assume for the moment that we are dealing with a measuring process for which $\sigma$, the standard deviation of experimental error, is known with great accuracy. A measurement is made and leads to a value $x$. Realizing that $x$ is not necessarily equal to the true value $\mu$ (we ignore systematic errors in the present discussion), we wish to make a pertinent statement, *not* about the location of $x$ but rather about the location of $\mu$. Consider Fig. 6.2.

The top part of Fig. 6.2 is similar to Fig. 6.1, and can be taken to represent the event

$$\mu - k\sigma < x < \mu + k\sigma$$

This is equivalent to stating that $X$ lies between $A'$ and $A$.

The bottom part of Fig. 6.2, on the other hand, represents the relation:

$$x - k\sigma < \mu < x + k\sigma \tag{6.11}$$

which states that $M$ lies between $B'$ and $B$.

Now, as can be readily verified, relations 6.9 and 6.11 are algebraically completely equivalent. But whereas the former expresses the event that the *random* variable $x$ lies between the *fixed* quantities $\mu - k\sigma$ and $\mu + k\sigma$, the latter states that the *fixed* quantity $\mu$ lies between the two *random* variables $x - k\sigma$ and $x + k\sigma$. The latter statement is not immediately meaningful from the probability viewpoint, for the position of $\mu$, although it may be unknown to us, is not subject to random fluctuations. An interesting interpretation of relation 6.11 has been given by Neyman and Pearson (4) in their theory of *confidence intervals*. The quantity $\mu$ being fixed, the relation expresses the probability that the *random interval* extending from $x - k\sigma$ to $x + k\sigma$ will *bracket* $\mu$. In terms of Fig. 6.2,

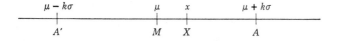

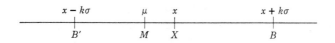

**Fig. 6.2** Top: illustrates the relation $\mu - k\sigma < x < \mu + k\sigma$. Bottom: illustrates the relation $x - k\sigma < \mu < x + k\sigma$.

this is the probability that the random interval $B'B$ will contain the fixed point $M$.   Of course, this probability is none other than that associated with relation 6.9, which is readily deduced from a table of the normal curve for any given value of $k$.   The true significance of Neyman and Pearson's theory is more readily understood by considering, Fig. 6.3. Each "trial" represents an independent measurement of the quantity $\mu$. Because of random errors, the measurements $x$ fluctuate about the fixed value $\mu$.   The vertical segments are intervals analogous to $B'B$ in Fig. 6.2: the lower limit of the segment represents the quantity $x - k\sigma$ and the upper limit, $x + k\sigma$.   The length of each segment is $2k\sigma$.   For each trial, the corresponding segment either does or does not intersect the horizontal line representing the true value $\mu$.   The probability of the former occurrence is precisely the probability associated with $k$ in relation 6.9, and it

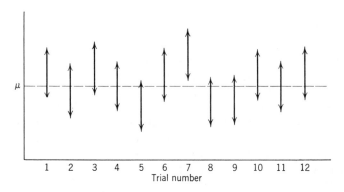

**Fig. 6.3**   Confidence intervals for known $\sigma$; the length of the segments is constant, but the position of their midpoints is a random variable.

measures the relative frequency of segments that contain $\mu$ in the population of all segments.   Figure 6.3 clearly shows that the varying element is the *interval* constructed around $x$, not the quantity $\mu$.   Each such segment represents a *confidence interval* for the unknown quantity $\mu$, based on the measured quantity $x$.   The probability associated with the interval (i.e., the probability that it contains $\mu$) is called the *confidence coefficient*. The introduction of the new expression, "confidence interval," is necessary because of the apparently paradoxical association of a probability statement with a fixed quantity; the paradox is resolved by observing that the probability refers to the random interval, not to the fixed quantity. On the other hand, the practical value of the theory lies, of course, in the bracketing of the unknown quantity by a known interval.   To avoid semantic confusion, the word *confidence* is substituted for *probability* in statements pertaining to the position of unknown, but fixed quantities.

So far we have considered single measurements. A "trial" may consist of $N$ replicate measurements $x$ from a population whose mean is $\mu$ (unknown) and whose standard deviation is $\sigma$ (known). In this case, we obtain a confidence interval for $\mu$, based on the average, $\bar{x}$, of the $N$ measurements by first writing the *probability* statement

$$\mu - k \frac{\sigma}{\sqrt{N}} < \bar{x} < \mu + k \frac{\sigma}{\sqrt{N}} \qquad (6.12)$$

which is then transformed into the *confidence* statement

$$\bar{x} - k \frac{\sigma}{\sqrt{N}} < \mu < \bar{x} + k \frac{\sigma}{\sqrt{N}} \qquad (6.13)$$

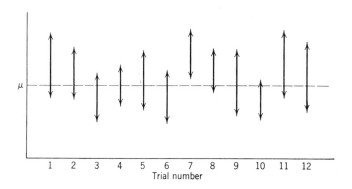

**Fig. 6.4** Confidence intervals when $\sigma$ is unknown; $s$ varies from trial to trial; the length of the segments as well as the position of their midpoints are random variables.

The length of each segment is now $2k\sigma/\sqrt{N}$. The association of the confidence coefficient with any given value of $k$, or vice-versa is identical with that described for single measurements.

Unlike most presentations of the subject, we have described the concept of confidence intervals for the case of known standard deviation $\sigma$. The extension to cases in which $\sigma$ is unknown, and must be estimated from the sample of replicate measurements, is not elementary from a mathematical viewpoint, but *conceptually* it requires very little beyond the simple case of known $\sigma$. In the general case, a *trial* is defined as a set of $N$ replicate measurements. Both the mean $\mu$ and the standard deviation $\sigma$ are unknown. The objective is to bracket the unknown $\mu$ by means of a confidence interval based on the sample.    By analogy with relation 6.13, we consider an interval defined by

$$\bar{x} - \frac{k's}{\sqrt{N}} < \mu < \bar{x} + \frac{k's}{\sqrt{N}} \tag{6.14}$$

The estimate $s$ has been substituted for the unknown value of $\sigma$.    This results in two modifications, the first of which becomes apparent by comparing Fig. 6.4 with Fig. 6.3.    Since $s$ is a function of the $N$ observations $x$, it will vary from trial to trial.    Consequently, the length of the confidence interval, equal to $2k's/\sqrt{N}$, is no longer constant from trial to trial.    This constitutes the first basic difference between the general case and the special one of known $\sigma$.    Had we retained the same value of $k$, the probability of obtaining a segment that intersects the horizontal line through $\mu$ would no longer be equal to the confidence coefficient formerly associated with $k$.    It can be shown mathematically that through a proper choice of the modified multiplier $k'$, the requirement of a pre-assigned probability (that the segment contains $\mu$) can be maintained.    Without going into any of the mathematical details, let us note, nevertheless, that relation (6.14) is equivalent to

$$|\bar{x} - \mu| < k' \frac{s}{\sqrt{N}} \tag{6.15}$$

or

$$\left| \frac{\bar{x} - \mu}{s/\sqrt{N}} \right| < k' \tag{6.16}$$

Thus, $k'$ is associated with a new distribution function: that of the ratio of an observed deviation from the mean, $\bar{x} - \mu$, to its estimated standard error, $s/\sqrt{N}$.    This distribution function has been tabulated and is known as *Student's t*.    We will therefore substitute the letter $t$ for $k'$, and write Eq. 6.14 as

$$\bar{x} - t \frac{s}{\sqrt{N}} < \mu < \bar{x} + t \frac{s}{\sqrt{N}} \tag{6.17}$$

and Eq. 6.16 as

$$\left| \frac{\bar{x} - \mu}{s/\sqrt{N}} \right| < t \tag{6.18}$$

Unlike the standard normal distribution Student's $t$ is not a single distribution.    It depends on the number of degrees of freedom associated with the estimate $s$.    Thus, tables of Student's $t$ always contain a parameter, denoted "degrees of freedom," and represented by the symbol DF or $n$.    If a confidence interval is based on a single sample of size $N$, the standard deviation is based on $N - 1$ degrees of freedom.    In such cases, the appropriate value for Student's *t* is taken from the table, using $N - 1$ degrees of freedom, and whatever level of probability is desired.

Tables of Student's *t* appear in a variety of forms.   Because their most frequent use is in the calculation of confidence intervals or in connection with significance testing (see Chapter 8), they generally appear in the form of percentiles.   Table II of the Appendix gives, for various values of the number of degrees of freedom, the *t* values corresponding to cumulative probabilities of 0.75, 0.90, 0.95, 0.975, 0.99, 0.995, and 0.999.   In using this table one must remember that the construction of a symmetrical confidence interval involves the elimination of *two* equal probability areas, one in each tail of the distribution curve.   Thus, a 95 per cent confidence interval extends from the 2.5 per centile to the 97.5 per centile.   The *t* values corresponding to these two points are equal in absolute value, but have opposite algebraic signs.

As an illustration of the use of confidence intervals based on Student's *t* distribution, consider the data of Table 6.1.   The standard deviation is estimated with $24 - 1 = 23$ degrees of freedom.   If a 95 per cent confidence interval is desired—i.e., a confidence interval associated with a probability of 0.95—we find the appropriate value of Student's *t* in Table II, for 23 degrees of freedom *and a probability of* 0.975 (*not* 0.95). This value is found to be $t = 2.07$.   Application of relation 6.17 then yields the following confidence interval for the mean:

$$1.0804 < \mu < 1.0826 \qquad (6.19)$$

It should be kept in mind that this interval pertains only and specifically to the mean of the population of which the data of Table 6.1 are a sample. The statement contained in relation 6.19 gives us *no* assurance, at any confidence level, that the *true* value of the isotopic abundance ratio is contained within the given interval.

Returning now to the concept of the probable error, let us find in Student's *t* table, the multipliers corresponding to 50 per cent confidence intervals.   It is seen that the celebrated value 0.6745 applies to an infinite number of degrees of freedom.   For $n = 20$, the correct value is 0.6870; for $n = 10$, it is 0.6998; and for $n = 5$, it is 0.7267.   Thus, a 50 per cent confidence interval requires a different multiplier for each sample size, and the value 0.6745, when used for small samples, will lead to a probability appreciably smaller than 50 per cent.

## 6.9  GENERALIZED APPLICATION OF STUDENT'S *t*

Relation 6.18 suggests the following more general formulation for problems to which Student's *t* distribution can be applied.

Let $z$ be an observed quantity subject to experimental error, and let $E(z)$ be the expected value of $z$. It is immaterial whether $z$ is a single direct measurement, an average of several such measurements, or any numerical quantity derived in some specified manner from one or more measurements. Let $s_z$ represent an estimate of the standard error of $z$, and let this estimate be based on $n$ degrees of freedom.

Now, if the error of $z$ obeys the Gaussian distribution, then the quantity

$$\frac{z - E(z)}{s_z} \tag{6.20}$$

obeys Student's $t$ distribution with $n$ degrees of freedom. Note that expression 6.20 denotes simply the deviation of $z$ from its expected value divided by its estimated standard error. Thus formulated, Student's $t$ distribution becomes a powerful tool for calculating confidence intervals. Suppose for example that in some particular application $n$ is equal to 16. Table II then shows us that the probability is 95 per cent that $t$ will be contained between $-2.12$ and $+2.12$. Thus the probability is 95 per cent that

$$-2.12 < \frac{z - E(z)}{s_z} < +2.12 \tag{6.21}$$

Now $z$ and $s_z$ are numerical values in any given application. Our aim is to set an uncertainty interval around $E(z)$. From Eq. 6.21 we derive

$$z - 2.12s_z < E(z) < z + 2.12s_z \tag{6.22}$$

This then constitutes a 95 per cent confidence interval for the unknown quantity $E(z)$.

Several remarks are in order. Note, in the first place, that the confidence interval concerns $E(z)$, the mean of the population from which the value $z$ was selected. This is not necessarily the "true value" for $z$. Only if the measuring process giving the $z$ values is unbiased, will $E(z)$ equal the true value. Thus, confidence intervals should never be considered as giving necessarily an interval of uncertainty for the *true* value. Expressed in a different way, confidence intervals pertain to precision, not to accuracy.

A second remark concerns the length of the confidence interval. From relation 6.22 we derive for the length of the interval $(z + 2.12s_z) - (z - 2.12s_z)$, or $2(2.12)s_z$. More generally, we write for the length $L$ of a confidence interval for $E(z)$:

$$L = 2ts_z \tag{6.23}$$

Here $t$ is the critical value of Student's $t$, for the chosen confidence coefficient. Instead of choosing a 95 per cent confidence coefficient, we could select a 99 per cent, or a 99.9 per cent confidence coefficient. Our confidence in having bracketed $E(z)$ would then be greater, but for this advantage we would have paid a price: the length $L$ of the confidence interval would also have increased since $t$ increases as the confidence coefficient increases. On the other hand, in order to reduce $L$, without losing the desired degree of confidence, it is necessary to reduce $s_z$. If $z$ is an average of $N$ replicate measurements $x_1, x_2, \ldots, x_N$, and if $\sigma_x$ is the standard deviation of $x$, then the standard error of $z$ is

$$\sigma_z = \frac{\sigma_x}{\sqrt{N}}$$

and the estimate $s_z = s_x/\sqrt{N}$ also reflects the division by $\sqrt{N}$. Thus, by increasing $N$, we shorten the confidence interval. This is, however, an uneconomical way to reduce $L$, for in order to reduce $L$ by, say, a factor of 4, one has to increase the number of measurements by a factor of $4^2$ or 16. The best way to achieve short confidence intervals is to start out with a small standard deviation of error, i.e., with a measuring process of high precision. The same conclusion would, of course, also result from mere common sense reasoning. The foregoing analysis has the advantage of giving it quantitative significance.

The method of confidence intervals is a logical choice in the case of the isotopic ratio data of Table 6.1, provided that it is remembered that it relates to precision only, not to accuracy. For the data of Table 6.2, on the other hand, the nature of the problem requires a different approach. Indeed, the objective of the measurements of the isotopic ratio is to estimate a single constant. It is then appropriate to surround the value $\bar{x}$ actually obtained by an uncertainty interval, such that, barring systematic errors in the measuring process, we may have some confidence that the true value $\mu$ lies within the interval. On the other hand, no single value is of particular interest in the case of the tensile measurements on leather hides. Since the variability in these data is a reflection of heterogeneity of the material rather than measurement error, the more pertinent problem here is to estimate what *fraction* of all values shall fall within a given interval, rather than where the *average* of all values is likely to be located. The statistical technique used for the solution of this problem is known as the setting of *tolerance limits*, or the determination of a *tolerance interval*.

## 6.10 TOLERANCE INTERVALS

We assume, as before, that the measurements exhibited in Table 6.2 belong to a single normal population. If the mean and the standard deviation of this population were both accurately known, we could readily calculate the relative number of measurements that will be encountered in the long run in any given interval. Consider, for example, the interval extending from 350 to 500. Let us assume, for the moment, that both $\mu$ and $\sigma$ are known and that $\mu = 416$ and $\sigma = 35$. Then the lower limit of the proposed interval when expressed as a deviation from the mean in units of standard deviations, is $(350 - 416)/35 = -1.89$ or 1.89 standard deviations below the mean. Similarly, the upper limit corresponds to a point $(500 - 416)/35 = +2.40$, i.e., 2.40 standard deviations above the mean. The probability associated with this interval is now readily obtained from the table of the normal curve, and equals $0.4706 + 0.4918 = 0.9624$. Thus, the fraction of measurements of the populations that will be in the proposed interval is 0.9624 or 96.24 per cent.

The above calculation is a straightforward application of the concept of the probability interval. Its limitation lies in that it requires that both $\mu$ and $\sigma$ be accurately known. For the data of Table 6.2, these parameters are both unknown, though estimates are available for both. It is easily understood that the substitution of these estimates for the unknown true values, in carrying out the above described calculation, will introduce an element of uncertainty in the value obtained for the fraction of values in the interval. In the theory of tolerance intervals, this uncertainty appears in two ways: (1) The fraction of measurements lying in the interval is not completely defined; only its lower limit is given. (2) The statement that the interval will contain this fraction of values is made with less than complete certainty; a probability value is associated with the statement. Thus, a tolerance interval is of the following form: "The probability is $P$ (a chosen number) that the interval extending from $A$ to $B$ (two specific numbers) contains a fraction of the population not less than $\gamma$ (a chosen number)." Evidently the values chosen for $P$ and $\gamma$ are positive numbers less than unity. When the population of the measurements is Gaussian, Table VI of the Appendix may be used for the computation of tolerance intervals for values of $P$ equal to 0.90, 0.95, or 0.99 and values of $\gamma$ equal to 0.90, 0.95, 0.99, or 0.999. In addition to $P$ and $\gamma$, the table involves one other parameter, the number of degrees of freedom available for the estimation of the standard deviation of the population.

To illustrate the use of tolerance intervals consider again the data of Table 6.2. Let us construct a tolerance interval that will include, with probability $P = 0.95$, a fraction of the population not less than 90 per cent. The estimated mean of the population is $\hat{\mu} = 416$ and the sample standard deviation is $\hat{\sigma} = 35$, and is based on 29 degrees of freedom. The tolerance interval is defined by its limits, given by

$$\hat{\mu} \pm l\hat{\sigma}$$

The value of $l$ is found in the table for $n = 29$, $P = 0.95$, and $\gamma = 0.90$; $l = 2.14$.   Thus, the tolerance interval is given by

$$416 \pm (2.14)(35) = 416 \pm 75$$

or

$$341 \quad \text{to} \quad 491$$

In concluding this discussion, let us state once more the fundamental distinction between confidence intervals and tolerance intervals. A confidence interval expresses our uncertainty about the true location of the mean of a population.   A tolerance interval, on the other hand, is not an expression of uncertainty, though it is itself subject to uncertainty; it is a measure of the range of values comprising a given fraction of the population.

## 6.11   THE MEASUREMENT OF ACCURACY

We have pointed out already that the concept of accuracy is inextricably tied to that of a reference value.   When the reference value is numerically known, the measurement of accuracy is relatively simple: it then involves merely the determination of the difference between this known value and the mean of the population of measurements.   This is an important and frequently occurring situation.

Consider for example the data given in Table 6.4, representing 10

**TABLE 6.4**   Photometric Titration of Beryllium[a]

| Beryllium taken, mg. | Beryllium found, mg. |
|:---:|:---:|
| 3.179 | 3.167 |
| 3.179 | 3.177 |
| 3.179 | 3.177 |
| 3.179 | 3.169 |
| 3.179 | 3.173 |
| 3.179 | 3.177 |
| 3.179 | 3.177 |
| 3.179 | 3.177 |
| 3.179 | 3.171 |
| 3.179 | 3.169 |

Average $= 3.1734$
Standard Deviation, $\hat{\sigma}_x = 0.00409$
Standard Error of Average, $\hat{\sigma}_{\bar{x}} = 0.00129$

[a] See reference 2.

determinations of beryllium by a photometric titration method (2).    In this case the true amount of beryllium in the sample was known to be exactly 3.179 mg.    The average of the observed values, 3.1734, differs from this value, but the difference is, of course, partly due to sampling fluctuations of this sample mean.    We can estimate the uncertainty due to these fluctuations.

The estimate of the standard deviation of the population of individual measurements is 0.00409.    Consequently the estimated standard error of the mean is $0.00409/\sqrt{10} = 0.00129$.    How does the observed difference, 0.0056, compare with this uncertainty value?

Let $B$ represent the bias of the method.    Then $B = \mu - 3.179$, and an estimate of this bias is given by

$$\hat{B} = \hat{\mu} - 3.179 = 3.1734 - 3.179 = -0.0056$$

The variance of $\hat{B}$ is

$$V(\hat{B}) = V(\hat{\mu})$$

and its standard error is:

$$\sigma_{\hat{B}} = \sigma_{\hat{\mu}}$$

We do not know $\sigma_{\hat{\mu}}$ but we have an estimate for this quantity, with 9 degrees of freedom:

$$\hat{\sigma}_{\hat{\mu}} = 0.00129; \quad DF = 9$$

Hence

$$\hat{\sigma}_{\hat{B}} = 0.00129$$

Using the results of Section 6.8, we can now construct a confidence interval for the bias $B$.    We have:

$$\hat{B} - t\hat{\sigma}_{\hat{B}} < B < \hat{B} + t\hat{\sigma}_{\hat{B}}$$

Using a 95 per cent confidence coefficient, we have for 9 degrees of freedom: $t = 2.262$.    We therefore obtain:

$$-0.0056 - (2.262)(0.00129) < B < -0.0056 + (2.262)(0.00129)$$

or

$$-0.0085 < B < -0.0027$$

This interval does not include the value "zero"; we thus conclude that the bias is "significantly different from zero."    The exact meaning of the expression "statistical significance" will be discussed in Chapter 8. For the present we interpret the above statement by saying that, since the confidence interval for $B$ does not include the value zero, the latter is not a possible choice for $B$.    In other words, the data provide evidence for the existence of a real bias.*     Using a 95 per cent confidence coefficient we state that this bias has some value between $-0.009$ and $-0.003$. It is therefore negative, which means that the analytical method, as applied to this chemical sample, tended to give slightly low results.

It is worth emphasizing that this entire line of reasoning, in order to be valid, requires that the variability shown by the experimental values be a correct representation of the population variance of the measuring process. This would not be the case, for example, if the ten values were all obtained on a single day and if an appreciable day-to-day variability existed for the measuring process.     For in that case, the bias found would only represent the bias of the method for the particular day on which the measurements were made, and the possibility would exist that, averaged over many days, the method is really unbiased.     We prefer to look upon a bias that varies from day-to-day but averages to zero over all days, not as a lack of accuracy, but rather as a *factor of precision* (or more exactly, as a *factor of imprecision*).

It should also be noted that the data used in our analysis pertained to a single solution of beryllium.     The results of the analysis do not necessarily apply to other concentrations or to solutions of beryllium containing interfering substances.     We will return to this point in Section 6.12.

So far, we have assumed that the reference value for the measured system (or material) is exactly known.     The determination of accuracy becomes a far more difficult problem when such knowledge is not at hand.     Suppose, for example, that we have determined the velocity of light by a given measuring method.     How do we judge the accuracy of the determination?

Precision offers, of course, no basic problems; by sufficient repetition of the measuring process we can readily estimate its precision.     But the evaluation of accuracy in the absence of an exactly known reference value is reduced to an educated guess.

There exist essentially two methods for making such a guess.     The first consists in a careful survey of all systematic errors that, to the best of one's knowledge, might affect the measuring process.     Exact values for these systematic errors are of course unknown, for if they were known, one could make appropriate corrections in the data.     But it is frequently possible to estimate "upper bounds" for the systematic errors.     By combining the effects of all known systematic errors, with their respective upper bounds, it is then possible to find an upper bound for the total bias, or in other words, for the inaccuracy of the method.     When this method has been used, the statement of accuracy may be of the following form:

---

* This does not mean that this bias is necessarily of *practical* importance.     In the present case, the bias is probably small enough to be ignored for most practical applications.

"It is believed that the combined effect of all systematic errors of this method does not exceed ...." The careful wording of this statement is not due to an overly timid attitude; it merely reflects the physical impossibility of determining a difference of two values (the bias) one of which is totally unknown. Reflection will show that no "confidence interval" can express the bias of a method in the absence of a known reference value: the method of confidence intervals involves only the uncertainty due to statistical fluctuations, not that due to lack of knowledge of a reference value.

The second method that is available for the evaluation of accuracy in the absence of exactly known reference values consists in the comparison of results obtained by different and independent measuring processes. This is the method most frequently employed by physicists in the determination of basic physical constants. Its use, of course, does not preclude use of the other method (estimation of upper bounds of systematic errors). The principle underlying this second method is the plausible belief that if several measuring methods, unrelated to each other, yield the same value (or values close to each other) for a given quantity, the confidence we may have in this common value is greatly increased. Here again we cannot make use of confidence intervals or other exact statistical techniques. Physical intuition and background knowledge in the field of the measuring technique are the determining factors. Statistical methods are still useful, but they are concerned more specifically, in these cases, with questions of propagation of errors, and of course with all matters pertaining to precision.

## 6.12  FACTORS OF PRECISION AND OF ACCURACY

In the preceding sections we have introduced statistical measures that are commonly used for the expression of precision and accuracy, and we have related them to some specific questions that they are designed to answer. We have also had several occasions to point out that the general problem of determining the precision and the accuracy of a measuring process is far too broad to be expressible in terms of a single standard deviation (for precision) and a single bias (for accuracy). More specifically, whatever measures one uses for expressing the precision and accuracy of a process, these measures will, in general, be *functions* rather than fixed values. Among the variables on which precision and accuracy often depend, we mention:

1. The "level," or magnitude, of the measured value: thus a method for the determination of iron in an alloy may have appreciably different measures of precision and accuracy for alloys with low iron content than

for alloys with high iron content.   If the standard deviation increases proportionally with iron content, the coefficient of variation will be constant.   Thus it is often possible, *through a particular choice of the measure* of precision (or accuracy), to obtain a constant value for characterizing the precision or the accuracy of the measuring process, but no general rules can be given to achieve this.

2. The nature of the measured system: frequently, differences in precision and in accuracy are observed when the same measuring process is applied to two different systems, even when the level of the measured value is the same for both systems.   The reasons for this are that the process responds not only to differences in level of the measured characteristic, but in many cases to other characteristics of the system.   Thus, two solutions, even of equal concentration in the measured component, may contain different amounts of other components ("interferences") which influence the precision and accuracy, or both, of the process.

3. The exact conditions under which the measurements are made: these conditions should, of course, be standardized as much as possible in order to achieve uniformity among results obtained at different times or in different laboratories.   To a large extent the development of the measuring process consists in determining the optimum conditions for the environmental variables affecting the process.

4. The nature of the comparisons that are made as a result of the measurements: this matter, as well as many others relating to precision and accuracy, is discussed in detail in Chapters 13 and 14.

From the preceding discussion it follows that the determination of the precision and the accuracy of a measuring process involves far more than the calculation of simple indices for these measures.   It involves a careful study of the dependence of precision and accuracy on the *factors* that influence them.   The final *expression* of the precision and accuracy of a process depends then on the findings of this study.   Occasionally a single value may be found to characterize the precision and another single value to characterize its accuracy, but in most cases, a statement of precision and accuracy must carefully specify the conditions under which the measurements were made and the comparisons to which they apply.

## 6.13   THE UNCERTAINTY OF CALIBRATION VALUES

When a standardizing institution, such as the National Bureau of Standards, issues a certificate for a *standard sample* (such as a sample of standard steel) or a *physical standard* (such as a standard kilogram weight), it attributes a numerical value to this standard, together with a statement about the uncertainty of this value.   Evidently such a statement is of the utmost importance to the user of the standard.    How is it obtained and how is it to be interpreted?

The concepts of the precision and accuracy, as we have defined them, pertain to *measuring processes* as applied to particular systems; i.e., they describe properties of *populations* of measurements, not of individual measurements.    In general the buyer of a standard has no need to distinguish between the reproducibility and the bias of the measuring process by which the standard was calibrated.    Therefore the concepts of precision and accuracy are generally not of *direct* interest to him.    They are, however, of great importance to the institution that issues the certificate. The only way in which this institution can offer some guarantee about the "certified value" is through a thorough *knowledge of the process by which the standard was calibrated.*    The standardizing laboratory must have evaluated both the precision and the accuracy (the latter generally in the form of an "upper bound" for the total bias) of the process.    Strictly speaking, the laboratory can never be entirely sure that the certified value satisfies the requirements stated in the certificate, but from a practical viewpoint it is quite justified in deriving from its knowledge of the calibration process, confidence in the calibration value, and in transmitting this confidence in the form of an "uncertainty statement."

The user of the standard needs primarily an estimate of the total uncertainty (imprecision plus upper bound for bias) of the certified value. Occasionally he is also interested in a measure for its precision, which is related to the precision of the measuring (or calibration) process.    His needs are dictated, in each case, by the specific use he intends to make of the standard.

## 6.14    SUMMARY

Experimenters have always felt the need for expressing the uncertainty of their measurements in the form of a few well-chosen quantities.    To distinguish between random and systematic errors, the concepts of precision and accuracy have been developed.    To avoid confusion, it is best to define these terms first in relation to a given process of measurement carried out repeatedly on a given physical or chemical system or material.    When such a measuring process generates a well-defined statistical population, the variance of this population is an (inverse) measure of the precision of the measuring process.    The degree of agreement of the mean of this population with a reference value for the measured system or material measures the accuracy of the process.    By reference value is meant a value defined either theoretically or as a result of an agreement between experts for the system or material in question.

Many measures have been proposed for the expression of precision. Most of them are directly related to the standard deviation. The range, defined as the difference between the largest and the smallest of a set of observations, is a readily calculated measure of precision. For small samples (less than 15 observations) it is almost as good a measure as the standard deviation itself. The use of the average (or mean) deviation is not recommended.

A useful concept is the probability interval and its generalization, the confidence interval. The latter is an interval derived from the estimated values of the mean and standard deviation of the measurements; the interval is calculated to insure that the probability that it contains the population mean of the measurements is equal to a predetermined value. The calculation of confidence intervals for normally (or approximately normally) distributed measurements is made possible by the introduction of a new probability distribution function known as "Student's $t$." This distribution depends on the size of the sample, or more generally, on a parameter defined as the number of degrees of freedom. For the case of a homogeneous sample, the degrees of freedom are $N - 1$, where $N$ is the sample size. Student's $t$ may be used for deriving confidence intervals for any quantity for which an estimate is available, provided that an estimate of its standard error, based on a definite number of degrees of freedom, is also available. Strictly speaking, the estimate should be normally distributed, but since the procedure is generally applied to an average (or linear function) of a number of measurements, the central limit theorem then generally insures approximate compliance with this requirement.

For any given distribution function it is theoretically always possible to calculate what percentage of all measurements will fall between stated limits. The procedure of tolerance intervals is concerned with the same problem when the distribution function is not known but estimates are available for its mean and standard deviation. An interval is derived from these estimates; it is such that the probability that it contains not less than a stated fraction of the population is equal to a given value. This concept is important in quality control since it allows one to estimate what fraction of a lot of merchandize will comply with specified requirements.

The measurement of accuracy is not possible without a reference value. When such a value is available, one can obtain, by application of Student's $t$, a measure for the bias of a measuring process for a given system. The bias is defined as the difference between the population mean of the measurements and the reference value.

When the concepts of precision and accuracy are used to describe a measuring process as a whole, i.e., as applied to an entire class of systems or materials rather than just a specific one, difficulties are generally encountered. They are due to the fact that the precision and accuracy depend not only on the measuring technique but also on the thing that is measured. Often the precision and accuracy are, at least in part, mathematical functions of the magnitude of the property that is measured. The systematic study of these matters is presented in Chapter 13.

In developing a method of measurement, great stress is laid on determining the best possible steps of procedure for the use of the method. This requires a study of the factors that may influence the precision and the accuracy of the method. We refer to these quantities as factors of precision and accuracy. The study of factors of accuracy and precision includes that of the dependence of these measures on the type of material that is measured, on the magnitude of the measured value, on environmental conditions (temperature, relative humidity, etc.), as well as matters of calibration of the measuring equipment.

The certification of the value assigned to a physical standard or to a chemical standard sample raises general questions relating to the expression of uncertainty of a certified value. Any such expression must be understood in terms of the confidence that the certifying agency has in the measuring process by which the value was obtained. Statements relating to the accuracy of a certified value must generally be made in the form of beliefs (rather than absolute certainty) that the error of the certified values does not exceed a certain upper bound.

## REFERENCES

1. *American Society for Testing Materials Manual on Quality Control of Materials*, Special Technical Publication 15-C, American Society for Testing Materials, Philadelphia, 1951.
2. Florence, T. M., and Y. J. Farrar, "Photometric Titration of Beryllium," *Analytical Chemistry*, **35**, 712–719 (1963).
3. Kanagy, J. R., C. W. Mann, and J. Mandel, "Study of the Variation of the Physical and Chemical Properties of Chrome-Tanned Leather and the Selections of a Sampling Location," *Journal of the Society of Leather Trades' Chemists*, **36**, 231–253 (1952).
4. Neyman, J., and E. S. Pearson, "On the Problem of the Most Efficient Tests of Statistical Hypotheses," *Phil. Trans. Roy. Soc. London*, **231**, A, 289–337 (1933).
5. Shields, W. R., D. N. Craig, and V. H. Dibeler, "Absolute Isotopic Abundance Ratio and the Atomic Weight of Silver," *J. Am. Chem. Soc.*, **82**, 5033–5036 (1960). (Some of the more detailed data did not appear in the published paper; the authors kindly supplied them.)
6. *Symbols, Definitions, and Tables, Industrial Statistics and Quality Control*, Rochester Institute of Technology, Rochester, New York, 1958.
7. Youden, W. J., "Misuse of the Average Deviation," *Technical News Bulletin of the National Bureau of Standards* (Jan. 1950).

## chapter 7

# THE METHOD OF LEAST SQUARES

### 7.1 INTRODUCTION

A simple way of improving the precision of an experimentally determined quantity is to take the average of several replicate measurements. If $\sigma$ represents the standard deviation of the population of replicate measurements, the standard error of the average of $N$ independent measurements is reduced to $\sigma/\sqrt{N}$.

The replication of measurements by the same measuring process provides perhaps the simplest case of the general problem of the "adjustment of observations." A slightly more complicated case arises in the problem of "derived" measurements, such as the determination of the area of a circle from measurements of its radius. The examples cited in Section 2.4 provide instances of still more complex situations. What is common to all these problems is the fact that from a collection of experimental values a set of well-defined quantities are to be deduced in some optimum fashion. Having once obtained values for these quantities, a further problem consists in assessing their reliability.

We will first discuss a simple problem which will allow us to discern the type of reasoning with which problems of the "adjustment of observations" are treated by the methods of mathematical statistics.

## 7.2  WEIGHTED AVERAGES

Suppose that we wish to determine the circumference of a circular disc. Three simple methods are available: (a) direct measurement with a tape; (b) calculation from a measurement of the diameter; and (c) calculation from measurements of the weight and the thickness of the disc, using the value, assumed known, of the specific gravity of its material.

Let us suppose that all three methods have been used. Clearly, the simple arithmetic average of the three numbers so obtained is not necessarily the best value that can be derived from the three measurements. Actually, two distinct problems arise in situations of this type: (1) the problem of *consistency* of the various measuring processes, and (2) the problem of combining the values obtained by them into a single "best" value. The problem of consistency is concerned with the possible presence of *systematic errors* in one or more of the measuring techniques. The second problem involves the *random errors* that affect values obtained by the methods employed. In our present discussion we deal only with the second problem, assuming that all three methods are free of systematic errors.

Let $\sigma_1$, $\sigma_2$, and $\sigma_3$ represent the standard deviations of the three methods. Consider a *linear* combination of the three measurements $p_1$, $p_2$, and $p_3$:

$$\hat{p} = a_1 p_1 + a_2 p_2 + a_3 p_3 \qquad (7.1)$$

Our problem is to select values for the coefficients $a_1$, $a_2$, and $a_3$ that will yield the highest precision for $\hat{p}$.* The variance of $\hat{p}$ is given by

$$V(\hat{p}) = a_1{}^2 \sigma_1{}^2 + a_2{}^2 \sigma_2{}^2 + a_3{}^2 \sigma_3{}^2 \qquad (7.2)$$

and this quantity must be made a minimum. On the other hand, in order that $\hat{p}$ be a value free of systematic error, its expected value must equal the true value, say $P$, of the circumference. Thus:

$$E(\hat{p}) = P$$

Since the three measuring processes were assumed to be free of systematic errors, the expected value of the data yielded by each of them is also equal to $P$. We thus have:

$$E(p_1) = E(p_2) = E(p_3) = P$$

From Eq. 7.1 it follows that

$$E(\hat{p}) = a_1 E(p_1) + a_2 E(p_2) + a_3 E(p_3)$$

Hence

$$P = a_1 P + a_2 P + a_3 P$$

* The derivation that follows could be considerably shortened through the use of *Lagrange multipliers* (3). In order to maintain an elementary level in the presentation this method has not been used.

from which we derive through division by $P$:

$$a_1 + a_2 + a_3 = 1 \qquad (7.3)$$

Using this relation we can "eliminate" one of the coefficients, say $a_3$; we have:

$$a_3 = 1 - a_1 - a_2 \qquad (7.4)$$

Introducing this value into Eq. 7.2 we obtain:

$$V(\hat{p}) = a_1{}^2\sigma_1{}^2 + a_2{}^2\sigma_2{}^2 + (1 - a_1 - a_2)^2\sigma_3{}^2$$

or

$$V(\hat{p}) = a_1{}^2(\sigma_1{}^2 + \sigma_3{}^2) + a_2{}^2(\sigma_2{}^2 + \sigma_3{}^2) - 2a_1\sigma_3{}^2 - 2a_2\sigma_3{}^2$$
$$+ 2a_1a_2\sigma_3{}^2 + \sigma_3{}^2$$

The minimum value of this quantity is obtained by equating to zero its derivatives with respect to $a_1$ and $a_2$:

$$2a_1(\sigma_1{}^2 + \sigma_3{}^2) - 2\sigma_3{}^2 + 2a_2\sigma_3{}^2 = 0$$
$$2a_2(\sigma_2{}^2 + \sigma_3{}^2) - 2\sigma_3{}^2 + 2a_1\sigma_3{}^2 = 0$$

or

$$a_1(\sigma_1{}^2 + \sigma_3{}^2) + a_2\sigma_3{}^2 = \sigma_3{}^2$$
$$a_1\sigma_3{}^2 + a_2(\sigma_2{}^2 + \sigma_3{}^2) = \sigma_3{}^2$$

Solving these equations for $a_1$ and $a_2$, we obtain:

$$a_1 = \frac{\sigma_2{}^2\sigma_3{}^2}{\sigma_1{}^2\sigma_2{}^2 + \sigma_1{}^2\sigma_3{}^2 + \sigma_2{}^2\sigma_3{}^2}$$

$$a_2 = \frac{\sigma_1{}^2\sigma_3{}^2}{\sigma_1{}^2\sigma_2{}^2 + \sigma_1{}^2\sigma_3{}^2 + \sigma_2{}^2\sigma_3{}^2}$$

Relation 7.4 then yields

$$a_3 = \frac{\sigma_1{}^2\sigma_2{}^2}{\sigma_1{}^2\sigma_2{}^2 + \sigma_1{}^2\sigma_3{}^2 + \sigma_2{}^2\sigma_3{}^2}$$

These expressions are greatly simplified by dividing both the numerator and denominator of each of the right-hand members by $\sigma_1{}^2\sigma_2{}^2\sigma_3{}^2$:

$$a_1 = \frac{1/\sigma_1{}^2}{1/\sigma_1{}^2 + 1/\sigma_2{}^2 + 1/\sigma_3{}^2}$$

$$a_2 = \frac{1/\sigma_2{}^2}{1/\sigma_1{}^2 + 1/\sigma_2{}^2 + 1/\sigma_3{}^2}$$

$$a_3 = \frac{1/\sigma_3{}^2}{1/\sigma_1{}^2 + 1/\sigma_2{}^2 + 1/\sigma_3{}^2} \qquad (7.5)$$

These relations suggest the introduction of three new quantities, defined as the reciprocals of the variances. Let us denote the reciprocal of a variance by the letter $w$ and call it a *weight factor* or more simply, *weight*. Equations 7.5 may then be written:

$$a_1 = \frac{w_1}{w_1 + w_2 + w_3}$$

$$a_2 = \frac{w_2}{w_1 + w_2 + w_3}$$

$$a_3 = \frac{w_3}{w_1 + w_2 + w_3} \tag{7.6}$$

The derivation just given is readily generalized to the case of more than three measurements. Equations 7.6 enjoy a remarkable property: if the "weights" $w_1$, $w_2$, and $w_3$ are replaced by a set of proportional quantities, say $kw_1$, $kw_2$, and $kw_3$, no change results in the coefficients $a_1$, $a_2$, and $a_3$. Thus, the weights need not be known individually. It is sufficient to know their ratios. In practical terms this means that the combination of several observations of unequal precision into a single best value does not require that the precision of each observation be known, only that the *relative* precisions of all observations with respect to each other be known. The *relative weight* of an observation is accordingly defined as its weight multiplied by any fixed positive value $k$.

Introducing the values found for the coefficients $a_1$, $a_2$, and $a_3$ into the expression $\hat{p} = a_1 p_1 + a_2 p_2 + a_3 p_3$, this quantity becomes:

$$\hat{p} = \frac{w_1 p_1 + w_2 p_2 + w_3 p_3}{w_1 + w_2 + w_3} \tag{7.7}$$

The estimate $\hat{p}$ thus obtained is referred to as a *weighted average* of the three measurements $p_1$, $p_2$, and $p_3$. Reviewing the method by which we have derived our result, we notice that it involves three conditions:

1. The "best value" is assumed to be a *linear combination* of all three measurements.

2. The "best value" is taken to be free of systematic errors; thus the best value is an *unbiased estimate* of the true value (which can also be expressed by stating that its *expected value* equals the true value).

3. The "best value" has *minimum variance* (with respect to all values satisfying requirements 1 and 2). The weighted average obtained by this procedure is therefore called a *linear unbiased minimum-variance estimate*.

Note that the weighted average becomes a simple arithmetic average in the case of equal precision in all measurements. For if $\sigma_1 = \sigma_2 = \sigma_3$ it follows that $w_1 = w_2 = w_3$, which in turn yields $a_1 = a_2 = a_3 = 1/3$ and

$\hat{p} = (p_1 + p_2 + p_3)/3$. Thus, the arithmetic average of measurements of equal precision is a special case of linear unbiased minimum-variance estimation.

By definition the weighted average has minimum variance among all linear unbiased estimates. What is the value of this minimum variance? We have:

$$V(\hat{p}) = V \left[ \left( \frac{1}{w_1 + w_2 + w_3} \right)(w_1 p_1 + w_2 p_2 + w_3 p_3) \right]$$

$$= \frac{1}{(w_1 + w_2 + w_3)^2} [w_1^2 V(p_1) + w_2^2 V(p_2) + w_3^2 V(p_3)]$$

By definition, $w_1 = 1/V(p_1)$. Hence $V(p_1) = 1/w_1$ and similarly for $w_2$ and $w_3$. Thus, we obtain:

$$V(\hat{p}) = \frac{1}{(w_1 + w_2 + w_3)^2} \left[ w_1^2 \frac{1}{w_1} + w_2^2 \frac{1}{w_2} + w_3^2 \frac{1}{w_3} \right]$$

$$= \frac{1}{w_1 + w_2 + w_3}$$

Defining the weight of $\hat{p}$, denoted by $w(\hat{p})$, as the reciprocal of its variance we find the interesting result:

$$w(\hat{p}) = w_1 + w_2 + w_3 \tag{7.8}$$

The weight of a weighted average of measurements is simply the sum of the weights of the individual measurements.

## 7.3  THE METHOD OF LEAST SQUARES

The solution we have just given for the simple problem of weighted averages is based on what we may call the principle of *linear unbiased minimum-variance estimation*. We will now show that the same result may be obtained in a simpler, but logically perhaps less satisfying, way through the use of what has become known as the method of *least squares*.

As before we denote the true value of the measured quantity by $P$ and the measured values by $p_1$, $p_2$, and $p_3$; we consider the *errors*:

$$e_1 = p_1 - P$$

$$e_2 = p_2 - P$$

$$e_3 = p_3 - P \tag{7.9}$$

The variances of $e_1$, $e_2$, $e_3$ are $\sigma_1^2$, $\sigma_2^2$, and $\sigma_3^2$ respectively, and the corresponding weights, defined as the reciprocals of the variances, are $w_1$, $w_2$, and $w_3$.

Let us now consider an estimate $\hat{p}$ based on the three measured values $p_1$, $p_2$, and $p_3$. Without specifying at this time how $\hat{p}$ is to be obtained, we can consider the "distances" of the observed values from this estimate, i.e., the quantities

$$d_1 = p_1 - \hat{p}$$

$$d_2 = p_2 - \hat{p}$$

$$d_3 = p_3 - \hat{p} \qquad (7.10)$$

These quantities, denoted as *residuals*, are in effect estimates of the errors $e_1$, $e_2$, and $e_3$. The method of least squares may now be formulated by stating that the *weighted sum of squares of the residuals must be a minimum*. Thus, we form the quantity

$$S = w_1 d_1{}^2 + w_2 d_2{}^2 + w_3 d_3{}^2 \qquad (7.11)$$

and find that value of $\hat{p}$ for which $S$ is the least (hence the term "least squares"). We have:

$$S = w_1(p_1 - \hat{p})^2 + w_2(p_2 - \hat{p})^2 + w_3(p_3 - \hat{p})^2$$

The value of $\hat{p}$ for which $S$ is minimum is one for which the derivative of $S$ with respect to $\hat{p}$ is zero. Differentiating $S$ with respect to $\hat{p}$, we obtain

$$\frac{dS}{d\hat{p}} = -[2w_1(p_1 - \hat{p}) + 2w_2(p_2 - \hat{p}) + 2w_3(p_3 - \hat{p})]$$

$$= -2[w_1 p_1 + w_2 p_2 + w_3 p_3 - (w_1 + w_2 + w_3)\hat{p}]$$

Equating this quantity to zero we obtain:

$$\hat{p} = \frac{w_1 p_1 + w_2 p_2 + w_3 p_3}{w_1 + w_2 + w_3} \qquad (7.12)$$

which agrees with the solution obtained by the method of linear unbiased minimum-variance estimation (Eq. 7.7).

Historically, the method of least squares was first introduced as an empirical rule along the lines of the preceding discussion, i.e., through minimization of the sum of squares of deviations (Legendre, 1806). It is now known that Gauss had used the method at an earlier date. He did, however, not publish his results until 1809. At that time, Gauss gave a justification of the least squares rule in terms of the normal curve of errors. Later (1821) he proposed an entirely new approach, which he himself considered far more satisfactory because of its independence from assumptions of normality. It is this second approach which we have presented, in a simple form, in Section 7.2. Essentially it is the principle of linear unbiased minimum-variance estimation. Gauss also extended the method to functions that are non-linear in the parameters. We will deal with this problem in Section 7.7.

## 7.4 THE SIMULTANEOUS DETERMINATION OF SEVERAL QUANTITIES

So far we have been concerned with the measurement of a single quantity, either by repeated determinations by the same measuring process, or by combining measurements made by different methods. Frequently, two or more quantities are determined simultaneously by measuring a number of properties that depend in a known way on these quantities. In analytical chemistry, a system composed of several constituents may be analyzed by spectrophotometric measurements made at different wavelengths. In physics, the exact determination of basic constants is made by measuring as many pertinent quantities as possible, each one involving one or several of the constants. Thus, the velocity of light $c$, Sommerfeld's fine-structure constant $\alpha$, Avogadro's number $N$, and the charge of the electron $e$ occur, either alone or in combination, in the five measurable quantities $Y_1$, $Y_2$, $Y_3$, $Y_4$, $Y_5$ given by Eqs. 2.2. Suppose that numerical values have been obtained for these five quantities. Can we then "solve" Eqs. 2.2 for the four fundamental constants? The solution presents two difficulties: we do not know the values of the errors $\varepsilon_1$, $\varepsilon_2$, $\varepsilon_3$, $\varepsilon_4$, and $\varepsilon_5$, and we have five equations for four unknowns; i.e., one more than is theoretically required. If we ignore the errors, for example by assuming them to be all zero, we will in all likelihood be faced with a set of *incompatible* equations, for the values of the constants derived from four of the equations may not satisfy the remaining equation.

In principle, this problem of "overdetermination" is already present in the case treated in the preceding section: a single quantity is measured more than once, giving rise to a set of incompatible equations; the solution is given by the weighted average, and it is based on considerations involving the experimental errors of the various measurements.

## 7.5 LINEAR RELATIONS

Suppose that in the study of a volumetric analytical procedure, titrations have been made on four solutions containing 10, 20, 30, and 40 per cent of the unknown. Suppose furthermore that a solution containing zero per cent of the unknown, when analyzed by this procedure, still requires a certain amount of the titrating agent. Let $\alpha$ represent this amount of reagent (the so-called "blank-titration") and let $\beta$ represent the quantity of reagent per per cent of the unknown. Then, apart from experimental error, the quantities of reagent for the four titrations are:

$$Y_1 = \alpha + 10\beta$$

$$Y_2 = \alpha + 20\beta$$

$$Y_3 = \alpha + 30\beta$$

$$Y_4 = \alpha + 40\beta \tag{7.13}$$

These equations may be used to determine the two quantities of interest $\alpha$ and $\beta$. Actually, the four measurements are subject to experimental error. Denoting the errors by $\varepsilon_1$, $\varepsilon_2$, $\varepsilon_3$, $\varepsilon_4$, and the measured quantities (amounts of reagent) by $y_1$, $y_2$, $y_3$, and $y_4$, we have

$$y_1 = \alpha + 10\beta + \varepsilon_1$$

$$y_2 = \alpha + 20\beta + \varepsilon_2$$

$$y_3 = \alpha + 30\beta + \varepsilon_3$$

$$y_4 = \alpha + 40\beta + \varepsilon_4 \tag{7.14}$$

A set of equations such as 7.14, which express the relations that exist between the measured ("observed") values and the unknown parameters (in this case $\alpha$ and $\beta$) is known as a set of *observational equations*.

We now make the assumption that the four errors, $\varepsilon$, belong to statistical populations of zero mean and with standard deviations equal to $\sigma_1$, $\sigma_2$, $\sigma_3$, and $\sigma_4$, respectively.

Following a well-established method in mathematics, let us assume for the moment that we have succeeded in deriving estimates for the unknown parameters $\alpha$ and $\beta$. Let us represent these estimates by $\hat{\alpha}$ and $\hat{\beta}$. Then we have a third set of quantities:

$$\hat{y}_1 = \hat{\alpha} + 10\hat{\beta}$$

$$\hat{y}_2 = \hat{\alpha} + 20\hat{\beta}$$

$$\hat{y}_3 = \hat{\alpha} + 30\hat{\beta}$$

$$\hat{y}_4 = \hat{\alpha} + 40\hat{\beta} \tag{7.15}$$

The values defined by these equations may be considered as the "theoretical estimates," corresponding to the observed values, $y_1$, $y_2$, $y_3$, $y_4$. Consider, as in the case of a single unknown, the weighted sum of squares of residuals

$$S = w_1 d_1{}^2 + w_2 d_2{}^2 + w_3 d_3{}^2 + w_4 d_4{}^2 \tag{7.16}$$

where each weight, $w$, is defined as the reciprocal of the corresponding variance: $w = 1/\sigma^2$; and the residuals, $d$, are the differences between the observed and the estimated measurements:

$$d_1 = y_1 - \hat{y}_1 = y_1 - (\hat{\alpha} + 10\hat{\beta})$$
$$d_2 = y_2 - \hat{y}_2 = y_2 - (\hat{\alpha} + 20\hat{\beta})$$
$$d_3 = y_3 - \hat{y}_3 = y_3 - (\hat{\alpha} + 30\hat{\beta})$$
$$d_4 = y_4 - \hat{y}_4 = y_4 - (\hat{\alpha} + 40\hat{\beta}) \qquad (7.17)$$

Thus, the sum of squares becomes

$$S = w_1(y_1 - \hat{\alpha} - 10\hat{\beta})^2 + w_2(y_2 - \hat{\alpha} - 20\hat{\beta})^2$$
$$+ w_3(y_3 - \hat{\alpha} - 30\hat{\beta})^2 + w_4(y_4 - \hat{\alpha} - 40\hat{\beta})^2$$

Proceeding exactly as in Section 7.3 we now require that $S$ be a minimum. In other words, the principle by which we obtain the estimates $\hat{\alpha}$ and $\hat{\beta}$ is that of minimizing the weighted sum of squares of residuals.

In Section 7.3 the quantity $S$ depended on a single variable, $\hat{p}$. The minimum was obtained by setting equal to zero the derivative of $S$ with respect to $\hat{p}$. In the present case, $S$ is a function of two variables, $\hat{\alpha}$ and $\hat{\beta}$. Accordingly, we obtain the desired minimum by setting equal to zero the *partial derivative* of $S$ with respect to each variable, $\hat{\alpha}$ and $\hat{\beta}$. This yields *two* equations, the solution of which yields the two unknowns $\hat{\alpha}$ and $\hat{\beta}$:

$$\frac{\partial S}{\partial \hat{\alpha}} = -2w_1(y_1 - \hat{\alpha} - 10\hat{\beta}) - 2w_2(y_2 - \hat{\alpha} - 20\hat{\beta})$$
$$- 2w_3(y_3 - \hat{\alpha} - 30\hat{\beta}) - 2w_4(y_4 - \hat{\alpha} - 40\hat{\beta}) = 0$$

$$\frac{\partial S}{\partial \hat{\beta}} = -20w_1(y_1 - \hat{\alpha} - 10\hat{\beta}) - 40w_2(y_2 - \hat{\alpha} - 20\hat{\beta})$$
$$- 60w_3(y_3 - \hat{\alpha} - 30\hat{\beta}) - 80w_4(y_4 - \hat{\alpha} - 40\hat{\beta}) = 0 \quad (7.18)$$

Equations 7.18 are known as the *normal equations*. In general, the normal equations are the equations obtained by equating to zero the partial derivative of the weighted sum of squares of residuals with respect to each of the unknown parameters. Thus, there will be as many normal equations as unknown parameters.

In the present case, the solution of the normal equations is simple: Eqs. 7.18 may be written:

$$w_1y_1 + w_2y_2 + w_3y_3 + w_4y_4$$
$$= (w_1 + w_2 + w_3 + w_4)\hat{\alpha} + (10w_1 + 20w_2 + 30w_3 + 40w_4)\hat{\beta}$$
$$10w_1y_1 + 20w_2y_2 + 30w_3y_3 + 40w_4y_4$$
$$= (10w_1 + 20w_2 + 30w_3 + 40w_4)\hat{\alpha}$$
$$+ (100w_1 + 400w_2 + 900w_3 + 1600w_4)\hat{\beta} \quad (7.19)$$

A numerical illustration of the least squares treatment of a problem of this type is given in Section 7.7.   In the present case, little more is involved than the elementary problem of solving two linear equations in two unknowns.   When more than two parameters are involved, the problem requires the solution of $m$ linear equations in $m$ unknowns, with $m > 2$. The computational aspects of this problem are discussed in text books on algebra and need not detain us here.   It is true that the normal equations have a symmetry that makes it possible to use special methods of solution, but the widespread use of high-speed computers has tended to relegate this problem to the field of programming.   We prefer to devote this book to the discussion of statistical principles, and refer the reader who is interested in computational details for the general case to references 1, 2, 4, and 6.

Let us, however, make two important observations.   In the first place we note that if the quantities $w_1$, $w_2$, $w_3$, and $w_4$ are replaced by a proportional set

$$w_1' = kw_1$$

$$w_2' = kw_2$$

$$w_3' = kw_3$$

$$w_4' = kw_4 \qquad (7.20)$$

the substitution of the $w'$ for the $w$ in the normal equations will lead to exactly the same solutions for $\hat{\alpha}$ and $\hat{\beta}$ as those obtained with the original $w$ values.   Thus, here again, the solution of the least squares problem does not require that we know the actual weights but only their ratios with respect to each other.

Secondly, we observe that merely obtaining numerical values for the parameters $\alpha$ and $\beta$ does not constitute a complete solution of the problem. The estimates $\hat{\alpha}$ and $\hat{\beta}$, while derived through the use of a sound principle, cannot be expected to be free of error.   Being based on observed quantities that are subject to experimental error, they are themselves subject to error. In other words, $\hat{\alpha}$ and $\hat{\beta}$ are random variables.   This is more readily visualized by considering a hypothetical sequence of experiments, each of which is a repetition of the one actually run.   More specifically, each experiment would consist of four titrations, performed on solutions containing 10, 20, 30, and 40 per cent of the unknown.   Thus, each experiment would yield a set of four values, such as in Eq. 7.14.   The true values $\alpha$ and $\beta$ would be the same for all experiments, but the errors $\varepsilon_1$, $\varepsilon_2$, $\varepsilon_3$, and $\varepsilon_4$ would differ from experiment to experiment in accordance with the laws of chance expressed by their distribution functions.   This would cause each experiment to yield different $y$ values and consequently a different solution $(\hat{\alpha}, \hat{\beta})$. In other words, the successive experiments will yield a sequence of estimates

$$\hat{\alpha}_1 \quad \hat{\alpha}_2 \quad \hat{\alpha}_3 \quad \ldots$$

$$\hat{\beta}_1 \quad \hat{\beta}_2 \quad \hat{\beta}_3 \quad \ldots$$

Each of these two sequences, when continued indefinitely, constitutes a statistical population.   The principle of least squares insures that: (a) the mean of the $\hat{\alpha}$ sequence is $\alpha$, and similarly for $\beta$; (b) the variance of $\hat{\alpha}$ is smaller (or at any rate, not greater) than that provided by any other linear estimation method, and similarly for the variance of $\hat{\beta}$.

It is now clear why, in addition to the estimates $\hat{\alpha}$ and $\hat{\beta}$, we are also interested in their variances.   It is also evident that the fluctuations measured by these variances are caused by the errors $\varepsilon$ and are, therefore, functions of these errors.   An important aspect of least squares theory is the evaluation of the variance (or of the standard deviation) of each estimated parameter.   In the example to be presented in the following section we will include in the analysis the evaluation of the precision of the estimates.

## 7.6   A NUMERICAL EXAMPLE IN LINEAR LEAST SQUARES

We refer to the analytical problem of the preceding section, with the following detailed information: (a) The numerical values for $y_1$, $y_2$, $y_3$, and $y_4$ are those given in Table 7.1.   (b) We assume that $y_1$, $y_3$, and $y_4$ are each the result of a single titration, but that $y_2$ is the average of four titrations. (c) The variance of the error in a single titration is $\sigma^2$ (not known).

Under these assumptions, the variance of $\varepsilon_1$, $\varepsilon_3$, and $\varepsilon_4$ is $\sigma^2$, but the variance of $\varepsilon_2$ is $\sigma^2/4$.   The data and the assumptions are given in Table 7.1.

**TABLE 7.1**

| Measurement | Observational equation | Variance of error |
|---|---|---|
| $y_1 = 33.8$ | $y_1 = \alpha + 10\beta + \varepsilon_1$ | $\sigma^2$ |
| $y_2 = 62.2$ | $y_2 = \alpha + 20\beta + \varepsilon_2$ | $(1/4)\sigma^2$ |
| $y_3 = 92.0$ | $y_3 = \alpha + 30\beta + \varepsilon_3$ | $\sigma^2$ |
| $y_4 = 122.4$ | $y_4 = \alpha + 40\beta + \varepsilon_4$ | $\sigma^2$ |

The weights of the four observations are:

$$w_1 = 1/\sigma^2, \quad w_2 = 4/\sigma^2, \quad w_3 = 1/\sigma^2, \quad w_4 = 1/\sigma^2$$

We know that it is permissible to replace these absolute weights by a proportional set of relative weights; such a set is given by the four numbers

1, 4, 1, 1. Thus, we obtain the set of observational equations and relative weights shown in Table 7.2.

**TABLE 7.2**

| Measurement | Coefficient of Unknown | | Relative weight, $w'$ |
|---|---|---|---|
| | $\alpha$ | $\beta$ | |
| $y_1 = 33.8$ | 1 | 10 | 1 |
| $y_2 = 62.2$ | 1 | 20 | 4 |
| $y_3 = 92.0$ | 1 | 30 | 1 |
| $y_4 = 122.4$ | 1 | 40 | 1 |

We first derive the two normal equations. A simple procedure for obtaining the normal equation *corresponding to a given parameter* (e.g., $\alpha$), is as follows:

*a.* For each row, form a "multiplier," obtained as the product of the weight $w'$ by the coefficient of the parameter. Thus, the "multipliers" for $\alpha$ are $1 \times 1, 1 \times 4, 1 \times 1, 1 \times 1$, or 1, 4, 1, 1.

*b.* Multiply the elements in each row by its "multiplier"; add the products in each column. The calculations for $\alpha$ are shown in Table 7.3.

**TABLE 7.3**

| $y$ | $\alpha$ | $\beta$ |
|---|---|---|
| $1 \times 33.8$ | $1 \times 1$ | $1 \times 10$ |
| $4 \times 62.2$ | $4 \times 1$ | $4 \times 20$ |
| $1 \times 92.0$ | $1 \times 1$ | $1 \times 30$ |
| $1 \times 122.4$ | $1 \times 1$ | $1 \times 40$ |
| 497.0 | 7 | 160 |

*c.* Using these sums, form an equation exactly analogous to the observational equations; thus, for the example above

$$497.0 = 7\hat{\alpha} + 160\hat{\beta}$$

This is the first normal equation. The reader is urged at this point to turn back to Eqs. 7.19 and to satisfy himself that the calculations we have just made are an exact description of the first of these two equations.

The second normal equation, the one corresponding to $\beta$, is obtained by the same procedure. The "multipliers" are $10 \times 1, 20 \times 4, 30 \times 1, 40 \times 1$, or 10, 80, 30, and 40.

**TABLE 7.4**

| $y$ | $\alpha$ | $\beta$ |
|---|---|---|
| $10 \times 33.8$ | $10 \times 1$ | $10 \times 10$ |
| $80 \times 62.2$ | $80 \times 1$ | $80 \times 20$ |
| $30 \times 92.0$ | $30 \times 1$ | $30 \times 30$ |
| $40 \times 122.4$ | $40 \times 1$ | $40 \times 40$ |
| 12,970 | 160 | 4200 |

Using the sums of Table 7.4, the second normal equation is

$$12{,}970 = 160\hat{\alpha} + 4200\hat{\beta}$$

Again the reader should verify that this procedure exactly describes the second equation in 7.19.

Solving the set of normal equations

$$497 = 7\hat{\alpha} + 160\hat{\beta}$$
$$12{,}970 = 160\hat{\alpha} + 4200\hat{\beta} \tag{7.21}$$

we obtain

$$\hat{\alpha} = 3.21$$
$$\hat{\beta} = 2.966 \tag{7.22}$$

We now turn to the estimation of the precision of these estimates. In line with the general nature of this book we will employ an elementary approach, using the numerical example to illustrate general principles, rather than computational details.

The key to the estimation of the standard errors of the estimates is the following observation. A glance at the derivation of the normal equations, either in the form of Eqs. 7.19, or in terms of the rules leading to the numerical Eqs. 7.21, shows that in each normal equation, only one member contains quantities subject to random errors (namely the measurements $y$). The other member contains, in addition to the parameters to be calculated, a set of coefficients depending on known and fixed quantities. Keeping this in mind, let us rewrite Eqs. 7.21 in the following form:

$$M_1 = 7\hat{\alpha} + 160\hat{\beta}$$
$$M_2 = 160\hat{\alpha} + 4200\hat{\beta} \tag{7.23}$$

where only $M_1$ and $M_2$ are functions of random variables.

Solving these equations for $\hat{\alpha}$ and $\hat{\beta}$, we obtain:

$$\hat{\alpha} = \frac{1}{3800}(4200M_1 - 160M_2)$$
$$\hat{\beta} = \frac{1}{3800}(-160M_1 + 7M_2) \tag{7.24}$$

Now, retracing our steps (first column of Tables 7.3 and 7.4), we find that

$$M_1 = y_1 + 4y_2 + y_3 + y_4$$

$$M_2 = 10y_1 + 80y_2 + 30y_3 + 40y_4 \qquad (7.25)$$

Introducing these expressions in Eq. 7.24 we obtain:

$$\hat{\alpha} = \frac{1}{3800} [(4200 - 1600)y_1 + (16,800 - 12,800)y_2$$
$$+ (4200 - 4800)y_3 + (4200 - 6400)y_4]$$

$$\hat{\beta} = \frac{1}{3800} [(-160 + 70)y_1 + (-640 + 560)y_2$$
$$+ (-160 + 210)y_3 + (-160 + 280)y_3]$$

or

$$\hat{\alpha} = \frac{1}{3800} [2600y_1 + 4000y_2 - 600y_3 - 2200y_4]$$

$$\hat{\beta} = \frac{1}{3800} [-90y_1 - 80y_2 + 50y_3 + 120y_4]$$

Now, the error variances of the $y$'s are $\sigma^2$, $\sigma^2/4$, $\sigma^2$, and $\sigma^2$ respectively. Using the law of propagation of errors (which for linear functions we know to be exact) we obtain:

$$V(\hat{\alpha}) = \left[\frac{1}{3800}\right]^2 \left[(2600)^2 + \frac{(4000)^2}{4} + (600)^2 + (2200)^2\right]\sigma^2$$

$$V(\hat{\beta}) = \left[\frac{1}{3800}\right]^2 \left[(90)^2 + \frac{(80)^2}{4} + (50)^2 + (120)^2\right]\sigma^2$$

which, after simplification, becomes:

$$V(\hat{\alpha}) = 1.105\sigma^2$$

$$V(\hat{\beta}) = 0.00184\sigma^2 \qquad (7.26)$$

The problem of determining the precision of the least squares estimates $\hat{\alpha}$ and $\hat{\beta}$ is solved by Eq. 7.26, provided that $\sigma^2$ is known. We have, however, assumed that $\sigma^2$ is not known. It is therefore necessary to carry the analysis one step further by attempting to obtain an estimate of $\sigma^2$ from the data themselves. It may perhaps seem surprising that this is at all possible. The argument goes as follows.

The observational equations of Table 7.1 show that the errors of measurement and their variances are those given in Table 7.5:

**TABLE 7.5**

| Error | Variance |
|-------|----------|
| $\varepsilon_1 = y_1 - \alpha - 10\beta$ | $\sigma^2$ |
| $\varepsilon_2 = y_2 - \alpha - 20\beta$ | $(1/4)\sigma^2$ |
| $\varepsilon_3 = y_3 - \alpha - 30\beta$ | $\sigma^2$ |
| $\varepsilon_4 = y_4 - \alpha - 40\beta$ | $\sigma^2$ |

We also know that the expected value of each of the $\varepsilon$ values is zero. We conclude that the 4 quantities $\varepsilon_1$, $(\sqrt{4})\varepsilon_2$, $\varepsilon_3$, and $\varepsilon_4$, are random drawings from a single population of mean zero and of variance equal to $\sigma^2$. Consequently, an estimate of $\sigma^2$ is obtained from the relation:

$$\hat{\sigma}^2 = \tfrac{1}{4}[\varepsilon_1^2 + (\sqrt{4}\,\varepsilon_2)^2 + \varepsilon_3^2 + \varepsilon_4^2] = \tfrac{1}{4}(\varepsilon_1^2 + 4\varepsilon_2^2 + \varepsilon_3^2 + \varepsilon_4^2) \tag{7.27}$$

The $\varepsilon$ quantities are not known because the exact values of $\alpha$ and $\beta$ are not known. We can, however, obtain estimates for the $\varepsilon$ values by substituting in the equations of Table 7.5 the estimates $\hat{\alpha}$ and $\hat{\beta}$ for the true values $\alpha$ and $\beta$. For $y_1$, $y_2$, $y_3$, and $y_4$ we use, of course, the observed values. Denoting the estimates of $\varepsilon$ by the symbol $d$, we obtain:

$$d_1 = y_1 - \hat{\alpha} - 10\hat{\beta} = 33.8 - 3.21 - 10(2.966) = 0.93$$

$$d_2 = y_2 - \hat{\alpha} - 20\hat{\beta} = 62.2 - 3.21 - 20(2.966) = -0.33$$

$$d_3 = y_3 - \hat{\alpha} - 30\hat{\beta} = 92.0 - 3.21 - 30(2.966) = -0.19$$

$$d_4 = y_4 - \hat{\alpha} - 40\hat{\beta} = 122.4 - 3.21 - 40(2.966) = 0.55 \tag{7.28}$$

These quantities are precisely the *residuals* defined in Section 7.5.

At this point one might be tempted to substitute the residuals for the $\varepsilon$ quantities in Eq. 7.27 in order to obtain an estimate for $\sigma^2$. This would not be correct because, unlike the errors $\varepsilon$, the residuals $d$ are not statistically independent; they all depend on the two random variables $\hat{\alpha}$ and $\hat{\beta}$. It can be shown that the only modification required as a result of this lack of statistical independence is a reduction of the degrees of freedom by a number equal to the number of estimated parameters occurring in the residuals. In the present case the original number of degrees of freedom is 4, corresponding to the 4 errors, $\varepsilon_1$, $\varepsilon_2$, $\varepsilon_3$, and $\varepsilon_4$. The number of estimated parameters is 2 ($\hat{\alpha}$ and $\hat{\beta}$). Consequently, the number of degrees of freedom properly allocated to the *residuals* (in contradistinction to the *errors*) is $4 - 2 = 2$. We thus obtain the correct estimate

$$\hat{\sigma}^2 = \frac{d_1^2 + 4d_2^2 + d_3^2 + d_4^2}{4 - 2} = \frac{d_1^2 + 4d_2^2 + d_3^2 + d_4^2}{2} \tag{7.29}$$

More generally, the equation for the estimate of $\sigma^2$ is:

$$\hat{\sigma}^2 = \frac{\sum\limits_i w_i d_i^2}{N - p} \tag{7.30}$$

where $d_i$ is the $i$th residual, $w_i$ the corresponding weight, $N$ the total number of observations, and $p$ the number of estimated parameters. The quantity $N - p$ represents the number of degrees of freedom available for the estimation of $\sigma^2$.

It is important to understand the true meaning of the variance $\sigma^2$: this quantity is the *variance of an observation whose weight factor $w$ is equal to unity*. In general, the weight factors are only *proportional* (rather than *equal*) to the reciprocals of the variances. We therefore have $w_i = k/V_i$; for $w_i = 1$, we obtain $k = V_i$. The quantity $\sigma^2$ estimated by Eq. 7.30 is none other than the positive factor of proportionality $k$ (i.e., $\sigma^2 = k$). In the special case in which the weighting factors are exactly equal to the reciprocals of the variances, we have $k = 1$ and consequently $\sigma^2 = 1$.

Substituting in Eq. 7.29 numerical values for the residuals we obtain:

$$\hat{\sigma}^2 = \frac{1.6391}{2} = 0.82 \tag{7.31}$$

Finally, by using this estimate in Eqs. 7.26, we obtain the desired variance estimates for $\hat{\alpha}$ and $\hat{\beta}$. To indicate that these quantities are estimates rather than true variances, we place a caret on the variance symbol:

$$\hat{V}(\hat{\alpha}) = 1.105\hat{\sigma}^2 = 0.906$$

$$\hat{V}(\hat{\beta}) = 0.00184\hat{\sigma}^2 = 0.001508 \tag{7.32}$$

The uncertainty in these values is due solely to that of $\hat{\sigma}^2$. Therefore, these variance estimates, as well as their square roots, the estimated standard errors of $\hat{\alpha}$ and $\hat{\beta}$, are also said to be estimated with 2 degrees of freedom (the degrees of freedom corresponding to $\hat{\sigma}^2$).

Note the significance of the double caret in each of these expressions: The caret on $\alpha$ indicates that $\hat{\alpha}$ is but an estimate of the true $\alpha$, whereas the caret on the $V$ indicates that the variance of the random variable $\hat{\alpha}$ is itself but an estimate of the true variance $V(\hat{\alpha})$.

Thus the values 0.906 and 0.001508, which measure the uncertainty of the estimates $\hat{\alpha} = 3.21$ and $\hat{\beta} = 2.966$, are themselves not entirely trustworthy. In the present example, in which the variance estimates are based on only 2 degrees of freedom, they are particularly untrustworthy.

Even with more numerous degrees of freedom, there generally is appreciable uncertainty in the estimated variances of the parameter estimates.

Account can be taken of both uncertainties—of the estimate and of its variance—by constructing confidence intervals for the parameters. Following the general procedure of Section 6.8, we write

$$\hat{\alpha} - t\hat{\sigma}_{\hat{\alpha}} < \alpha < \hat{\alpha} + t\hat{\sigma}_{\hat{\alpha}}$$

and

$$\hat{\beta} - t\hat{\sigma}_{\hat{\beta}} < \beta < \hat{\beta} + t\hat{\sigma}_{\hat{\beta}}$$

Using the estimates of Eqs. 7.22 and 7.32 and Student's $t$ for 2 degrees of freedom and 95 per cent confidence (0.05 level of significance), we obtain the confidence intervals:

$$3.21 - (4.30)(\sqrt{0.906}) < \alpha < 3.21 + (4.30)(\sqrt{0.906})$$

$$2.966 - (4.30)(\sqrt{0.001508}) < \beta < 2.966 + (4.30)(\sqrt{0.001508})$$

or

$$-0.88 < \alpha < 7.30$$

$$2.80 < \beta < 3.13$$

These intervals are probably quite unsatisfactory. In particular, the interval for $\alpha$ extends from a negative to a positive value, telling us little about a possible blank titration. The reasons for this state of affairs are to be found largely in the small number of degrees of freedom available for the estimation of the error-variance. The confidence interval reflects, in this case, perhaps not so much the error of the estimated value ($\hat{\alpha}$ or $\hat{\beta}$) as our ignorance of this error. Rather than being blamed as inadequate, the statistical analysis in a situation of this type, should be given credit for giving us a realistic picture of our state of knowledge or lack of it; the analysis is a diagnostic tool, not a remedy for poor data.

## 7.7 NON-LINEAR RELATIONS

The method of the weighted average, as well as the more general procedure explained in Sections 7.3, 7.4, and 7.5 have an important requirement in common: the relationships between the measured quantities and the unknown parameters are all linear. Thus, in the case of several measurements, $p_i$, of an unknown quantity, $P$, we had relationships of the type $p_i = P + \varepsilon_i$, where $\varepsilon_i$ represents the error of the measurement $p_i$. This is the simplest form of a linear relation. In the more general case of several measurements involving several unknowns, the relationships were of the type represented by Eqs. 7.14, which are again linear. In all these cases we were therefore justified in postulating that the best estimates of the unknown parameters be linear functions of the measurements. This no longer holds for the frequently occurring situations in which the basic equations relating the measurements to the unknown parameters are non-linear. Equations 2.2 are non-linear and provide good examples of this more general situation.

The least squares treatment for non-linear relations was also developed by Gauss, as a natural generalization of the procedure for the linear case. It is based on the process of linearization, which we have already encountered in Section 4.7.

Consider again the non-linear relations 2.2 expressing 5 measured quantities as functions of 4 fundamental physical constants; $c$, $\alpha$, $N$, and $e$. Suppose that approximate values are known for these 4 constants: $c_0$, $\alpha_0$, $N_0$, and $e_0$. Our aim is to improve these approximations by adding to them the "corrections" $\Delta c$, $\Delta \alpha$, $\Delta N$, and $\Delta e$. The basis for the calculations of the corrections is provided by the measured values.

To illustrate the process of linearization, we consider the third measurement in the set 2.2:

$$y_3 = Y_3 + \varepsilon_3 = \frac{Ne^2}{\alpha^3 c^2} + \varepsilon_3$$

Expanding the function $Y = Ne^2/\alpha^3 c^2$ in a Taylor's series around the approximate values $c_0$, $\alpha_0$, $N_0$, and $e_0$, we obtain

$$Y_3 - \frac{N_0 e_0{}^2}{\alpha_0{}^3 c_0{}^2} = \frac{\partial Y_3}{\partial N}(N - N_0) + \frac{\partial Y_3}{\partial e}(e - e_0)$$
$$+ \frac{\partial Y_3}{\partial \alpha}(\alpha - \alpha_0) + \frac{\partial Y_3}{\partial c}(c - c_0) + \text{terms of higher order}$$

From this relation we derive:

$$y_3 - \frac{N_0 e_0{}^2}{\alpha_0{}^3 c_0{}^2} = \frac{\partial Y_3}{\partial N}(N - N_0) + \frac{\partial Y_3}{\partial e}(e - e_0) + \frac{\partial Y_3}{\partial \alpha}(\alpha - \alpha_0)$$
$$+ \frac{\partial Y_3}{\partial c}(c - c_0) + \text{(terms of higher order)} + \varepsilon$$

In this relation, first of all, we neglect the terms of higher order. Next we observe that the differences $N - N_0$, $e - e_0$, $\alpha - \alpha_0$, and $c - c_0$, may be considered as the corrections $\Delta N$, $\Delta e$, $\Delta \alpha$, and $\Delta c$, provided that the measured value $y_3$ is considered as a better value than the first approximation $N_0 e_0{}^2/\alpha_0{}^3 c_0{}^2$. Thus, writing

$$y_3 - \frac{N_0 e_0{}^2}{\alpha_0{}^3 c_0{}^2} = z$$

$$N - N_0 = w$$

$$e - e_0 = v$$

$$\alpha - \alpha_0 = u$$

$$c - c_0 = x \tag{7.33}$$

we obtain the relation

$$z = \frac{\partial Y_3}{\partial N} w + \frac{\partial Y_3}{\partial e} v + \frac{\partial Y_3}{\partial \alpha} u + \frac{\partial Y_3}{\partial c} x + \varepsilon \qquad (7.34)$$

in which $z$ is a "measurement" and $w$, $v$, $u$, and $x$ the unknown parameters. The error $\varepsilon$ is a random variable with zero mean; its standard deviation is of course that of the measuring process $Y_3$.

Strictly speaking, the partial derivatives should be taken at the correct values $N$, $e$, $\alpha$, and $c$. Little error will result, however, if we take them at the neighboring values $N_0$, $e_0$, $\alpha_0$, and $c_0$. But then, these partial derivatives are known numbers and Eq. 7.34 becomes, apart from the presence of the error $\varepsilon$, a *linear relation* between the measured value $z$ and the unknown corrections $x$, $u$, $v$, and $w$.

The introduction of the approximate set of values for the unknown constants has allowed us to replace the third non-linear equation of Eqs. 2.2 by the linear relation (Eq. 7.34). A similar process is carried out for each of the remaining relations of 2.2. The unknown parameters are the 4 corrections to the approximate values of the constants rather than the constants themselves. From this point on, the solution proceeds exactly as explained in Sections 7.4 and 7.5, using the method of least squares for linear relationships.

The linearization process may fail if the initial values are poor approximations. For then the higher order terms may be too large to be neglected. It is also futile to attempt a "correction" of a good set of initial values by a least squares solution based on very poor measurements. On the other hand, if the errors $\varepsilon$ of the measured quantities are small, it is always possible to obtain a good set of initial values by selecting from the complete set of measurements a subset equal in number to the number of the unknowns. For example, 4 of the 5 equations in 2.2 would provide, by regular algebraic solution, a set of values for the four constants, $c$, $\alpha$, $N$, and $e$. These solutions can then be used as an initial set for the linearization of all 5 equations. The adequacy of the corrections finally obtained can be tested by repeating the entire process, using the corrected values as a new initial set: if the second set of corrections is very small in comparison with the first, no further calculations are required; if not, the process can be repeated once more, or as many times as required. In practice it is seldom necessary to resort to more than two iterations.

It is not sufficiently realized that the process of least squares "adjustment"—a common terminology for the procedure just described—cannot possibly provide reliable values for the unknown parameters unless the measurements are free of large systematic errors. Its success depends also, as we have seen, on a reasonable degree of precision (small $\varepsilon$ values) of the measured quantities.  What is commonly known as the *statistical design of experiments* is, when properly used, an application of least squares employing a correct set of relations between the measurements and the unknown parameters.  Applications have been made in which it is based on a set of relations that are merely *assumed* to be correct, very often without an adequate basis for this assumption.  It should be clear from what precedes that statistical procedures based on least squares theory, including the statistical design of experiments, provide no magical formula for obtaining usable "adjusted" values from poor measurements.  The notion that statistical procedures are especially appropriate for measurements with large errors is a deplorable fallacy.   On the contrary, the practical usefulness, and to some extent even the validity, of many statistical procedures is vastly increased when the measurements themselves are of high precision and accuracy.

## 7.8  AN IMPORTANT SPECIAL CASE

Equations 2.2, which relate a set of fundamental physical constants to measured quantities are, apart from experimental error, of the purely multiplicative type; they involve no operations other than multiplication or division.   Such equations can be linearized in a particularly simple way.   As an example, consider once more the relations

$$Y = \frac{Ne^2}{\alpha^3 c^2}$$

and

$$y = Y + \varepsilon \tag{7.35}$$

Thus, $Y$ is the true value of the measurement $y$, which differs from it by the error $\varepsilon$.   Taking natural logarithms of the first of these relations, we obtain

$$\ln Y = \ln N + 2 \ln e - 3 \ln \alpha - 2 \ln c \tag{7.36}$$

If $\Delta N$, $\Delta e$, $\Delta \alpha$, and $\Delta c$ are corrections to a set of initial values $N_0$, $e_0$, $\alpha_0$, and $c_0$, the corresponding change in $Y$, $\Delta Y$, is obtained as before by expansion into a Taylor's series and retention of the linear terms only. The differential of a natural logarithm being given by the formula $d(\ln x) = dx/x$, we derive from Eq. 7.36:

$$\frac{dY}{Y} = \frac{dN}{N} + 2\frac{de}{e} - 3\frac{d\alpha}{\alpha} - 2\frac{dc}{c}$$

and consequently, by neglecting non-linear terms in the expansion:

$$\frac{\Delta Y}{Y} = \frac{\Delta N}{N} + 2\frac{\Delta e}{e} - 3\frac{\Delta \alpha}{\alpha} - 2\frac{\Delta c}{c}$$

The initial values being $N_0$, $e_0$, $\alpha_0$, $c_0$, with the corresponding $Y_0$, we can write this equation as:

$$\frac{Y - Y_0}{Y_0} = \frac{N - N_0}{N_0} + 2\frac{e - e_0}{e_0} - 3\frac{\alpha - \alpha_0}{\alpha_0} - 2\frac{c - c_0}{c_0}$$

In view of Eq. 7.35, the first member can also be written:

$$\frac{Y - Y_0}{Y_0} = \frac{y - \varepsilon - Y_0}{Y_0} = \frac{y - Y_0}{Y_0} - \frac{\varepsilon}{Y_0}$$

Therefore we have:

$$\frac{y - Y_0}{Y_0} = \left(\frac{N - N_0}{N_0} + 2\frac{e - e_0}{e_0} - 3\frac{\alpha - \alpha_0}{\alpha_0} - 2\frac{c - c_0}{c_0}\right) + \frac{\varepsilon}{Y_0} \quad (7.37)$$

We now define a new set of "unknowns":

$$\frac{N - N_0}{N_0} = \frac{\Delta N}{N_0} = w$$

$$\frac{e - e_0}{e_0} = \frac{\Delta e}{e_0} = v$$

$$\frac{\alpha - \alpha_0}{\alpha_0} = \frac{\Delta \alpha}{\alpha_0} = u$$

$$\frac{c - c_0}{c_0} = \frac{\Delta c}{c_0} = x \quad (7.38)$$

The difference between these quantities and the corresponding ones in Eqs. 7.33 should be noted. The latter are the corrections themselves whereas the quantities now defined are *relative corrections*.

Similarly we define the *relative error*

$$\delta = \frac{\varepsilon}{Y_0} \quad (7.39)$$

and the *relative measurement change*:

$$z = \frac{y - Y_0}{Y_0} \quad (7.40)$$

Equation 7.37 can now be written:

$$z = w + 2v - 3u - 2x + \delta \quad (7.41)$$

This linear relation is considerably simpler than Eq. 7.34. The other equations in set 2.2 are treated in a similar manner. The coefficients in Eq. 7.41 are, of course, simply the exponents of the unknowns in the original relation of Eqs. 2.2: $+1$ for $N$, $+2$ for $e$, $-3$ for $\alpha$, and $-2$ for $c$.

Estimates of $w$, $v$, $u$, and $x$ are obtained by the method of least squares, as developed for the linear case, by observing that the variance of the "error" $\delta$ is given by:

$$V(\delta) = \frac{1}{Y_0^2} V(\varepsilon) \qquad (7.42)$$

Thus, if $w_\varepsilon$ is the weight of $\varepsilon$, the weight of $\delta$ is given by

$$w_\delta = \frac{1}{V(\delta)} = \frac{Y_0^2}{V(\varepsilon)} = Y_0^2 w_\varepsilon \qquad (7.43)$$

In the case exemplified by Eqs. 2.2 there is no valid reason for believing that the relative errors $\delta$, corresponding to the five equations of the set, all belong to the same population. There are, however, many situations in which such an assumption is justified. In those cases one can dispense with weight factors in carrying out the analysis. In our example Eqs. 2.2, the weight factors $w_\delta$ are derived from extensive knowledge of the five measuring processes involved (5).

## 7.9 GENERALIZATION OF THE LEAST SQUARES MODEL

In Section 7.4 we have discussed a situation in which each measured quantity is a known function of a number of unknown parameters. While each such function contains, in general, several parameters, it does not involve more than a single measurement. Thus, each observational equation in Table 7.1 involves two parameters, $\alpha$ and $\beta$, but only one measurement $y_i$. The reason for this is that the quantities 10, 20, 30, and 40, which represent the various concentrations of the constituent in which we are interested, were assumed to be known without error, and are therefore known numerical constants, rather than measurements. Had this not been the case, then each observational equation would involve *two* measurements, the volume of reagent $y$, and the concentration of the constituent in the solution, say $x$. The value used for this concentration, $x$, might have been obtained by a different method of analysis; it would therefore not be a "true" value but rather another measurement, subject to experimental error. We will deal with this particular problem in Chapter 12. However, this viewpoint suggests a more general approach to the least squares problem, which can be formulated as follows:

Let there be a number of measured quantities

$$x_1, x_2, \ldots, x_1, \ldots, x_k$$

whose weights* are

$$w_1, w_2, \ldots, w_i, \ldots, w_k$$

* The weight is the reciprocal of the error-variance.

and whose true values (unknown) are

$$X_1, X_2, \ldots, X_i, \ldots, X_k$$

Now, let there be $N$ relations each of which may involve one, several, or all of the measurements (or rather their true values) as well as some unknown parameters $\alpha, \beta, \gamma$, etc.   These relations may be represented by the equations

$$F_1(X_1, X_2, \ldots, X_k; \alpha, \beta, \gamma, \ldots) = 0$$

$$F_2(X_1, X_2, \ldots, X_k; \alpha, \beta, \gamma, \ldots) = 0$$

$$F_N(X_1, X_2, \ldots, X_k; \alpha, \beta, \gamma, \ldots) = 0 \qquad (7.44)$$

For example, in the case represented by Table 7.1, the measured quantities are

$$x_1, y_1, x_2, y_2, x_3, y_3, x_4, y_4$$

where the $x$ stand for the measurements replacing the quantities 10, 20, 30, and 40.

Thus, $k = 8$ in this example.   The relations 7.44 for this case are

$$Y_1 - \alpha - \beta X_1 = 0$$

$$Y_2 - \alpha - \beta X_2 = 0$$

$$Y_3 - \alpha - \beta X_3 = 0$$

$$Y_4 - \alpha - \beta X_4 = 0 \qquad (7.45)$$

and are 4 in number: $N = 4$.

Now we note that the quantities $X_i$ occurring in Eqs. 7.44 will forever be unknown to us.   Therefore we will never be able to make certain that our estimates for the parameters $\alpha, \beta, \gamma$, etc., will actually satisfy these equations.   One might be tempted to try to satisfy the equations in terms of the *measured* quantities $x_i$, but a glance at Eqs. 7.45 will show that this will, in general, not be possible.   Suppose indeed that estimates $\hat{\alpha}$ and $\hat{\beta}$ have been found for which the first two relations, in terms of the measured quantities $x$ and $y$, are satisfied:

$$y_1 - \hat{\alpha} - \hat{\beta} x_1 = 0$$

$$y_2 - \hat{\alpha} - \hat{\beta} x_2 = 0$$

Then it is unlikely that these same estimates $\hat{\alpha}$ and $\hat{\beta}$ will also satisfy the remaining two relations

$$y_3 - \hat{\alpha} - \hat{\beta} x_3 = 0$$

$$y_4 - \hat{\alpha} - \hat{\beta} x_4 = 0$$

It follows from this line of reasoning that the measured values must themselves be "adjusted" to satisfy the set of Eqs. 7.44. We consequently consider a set of "condition equations":

$$F_1(\hat{x}_1, \hat{x}_2, \ldots, \hat{x}_k; \hat{\alpha}, \hat{\beta}, \hat{\gamma}, \ldots) = 0$$

$$F_2(\hat{x}_1, \hat{x}_2, \ldots, \hat{x}_k; \hat{\alpha}, \hat{\beta}, \hat{\gamma}, \ldots) = 0$$

$$F_N(\hat{x}_1, \hat{x}_2, \ldots, \hat{x}_k; \hat{\alpha}, \hat{\beta}, \hat{\gamma}, \ldots) = 0 \qquad (7.46)$$

These equations must be *exactly* satisfied for the estimates $\hat{\alpha}$, $\hat{\beta}$, $\hat{\gamma}$, etc.; and in order to make this possible the quantities $X_1, X_2, \ldots, X_k$ have been replaced, *not* by the corresponding measurements $x_1, x_2, \ldots, x_k$, but rather by the "adjusted" values

$$\hat{x}_1, \hat{x}_2, \ldots, \hat{x}_k$$

The difference between a measurement and its adjusted value is called a *residual*:

$$d_i = x_i - \hat{x}_i \qquad (7.47)$$

Deming (4), who formulated the preceding approach, then defines the method of least squares as the following solution to this problem.

To obtain both the adjusted values $\hat{x}_i$ and the parameter estimates $\hat{\alpha}$, $\hat{\beta}$, $\hat{\gamma}, \ldots$, it is sufficient to minimize the sum of squares of weighted residuals

$$U = \sum d_i^2 w_i \qquad (7.48)$$

subject to the set of conditions 7.46.

The minimization is carried out with respect to both the adjusted values $\hat{x}_i$ *and* the parameters estimates $\hat{\alpha}$, $\hat{\beta}$, $\hat{\gamma}$, etc. For details of the procedure, which involves a linearization process similar to that discussed in Section 7.7, the reader is referred to Deming's book, which appears to be the only source in which the problem is discussed with this degree of generality.*

Deming shows that this approach to the method of least squares covers not only the curve-fitting case, exemplified by the data of Table 7.1, but also the case of "parameter-free" least squares adjustments. To illustrate the latter we will apply Deming's procedure to the following simple problem.

Suppose that an alloy of copper, tin, and antimony is analyzed for all three of these components. Then, if it is known that no other components are contained in the alloy, the sum of the per cent compositions of these three components should add up to 100 exactly. Suppose that the experimental results are 92.6 per cent copper, 5.9 per cent tin, and 1.7 per cent

---

* A briefer account, along the same lines, is given in Wilson (7).

antimony, and that the standard errors of these results are 0.11 for copper, 0.05 for tin, and 0.04 for antimony.

Since the sum of the experimental results is not equal to 100 exactly, "adjustments" are required.   Let $A$, $B$, and $C$ represent the adjusted values for the copper, tin, and antimony contents, respectively.   According to the principle of least squares, we are to minimize the quantity

$$U = (92.6 - A)^2 \frac{1}{(0.11)^2} + (5.9 - B)^2 \frac{1}{(0.05)^2} + (1.7 - C)^2 \frac{1}{(0.04)^2}$$

$$(7.49)$$

But we must also satisfy the condition

$$A + B + C = 100 \tag{7.50}$$

From Eq. 7.50 we obtain:

$$C = 100 - A - B \tag{7.51}$$

Substituting this value in Eq. 7.49 we have:

$$U = (92.6 - A)^2 \frac{1}{(0.11)^2} + (5.9 - B)^2 \frac{1}{(0.05)^2}$$

$$+ (1.7 - 100 + A + B)^2 \frac{1}{(0.04)^2}$$

The minimum of $U$ is obtained by equating to zero its partial derivatives with respect to $A$ and $B$:

$$\frac{\partial U}{\partial A} = \frac{-2}{(0.11)^2} (92.6 - A) + \frac{2}{(0.04)^2} (1.7 - 100 + A + B) = 0 \quad (7.52)$$

$$\frac{\partial U}{\partial B} = \frac{-2}{(0.05)^2} (5.9 - B) + \frac{2}{(0.04)^2} (1.7 - 100 + A + B) = 0 \quad (7.53)$$

Subtracting Eq. 7.53 from Eq. 7.52 we have:

$$\frac{5.9 - B}{25 \times 10^{-4}} - \frac{92.6 - A}{121 \times 10^{-4}} = 0$$

hence

$$B = \frac{25A - 1601.1}{121} \tag{7.54}$$

Substituting this value in Eq. 7.53 we then have,

$$\frac{92.6 - A}{121} = \frac{-98.3 + A + (25A - 1601.1)/121}{16}$$

which yields:

$$A = 92.45$$

From Eq. 7.54 we then obtain

$$B = 5.87$$

and from Eq. 7.51:

$$C = 100 - 92.45 - 5.87 = 1.68$$

Thus, the adjusted values are:

$$A = 92.45 \quad B = 5.87 \quad C = 1.68$$

These results could have been obtained in a far simpler way. Our reason for discussing this example, however, was to demonstrate the applicability of the general method of least squares to problems in which no unknown parameters are involved, and in which the main objective is the adjustment of the observations.

## 7.10 CONCLUDING REMARKS

In the preceding pages we have attempted to describe the concepts involved in least squares rather than details pertaining to calculations or to the application of the method to special cases. When it is realized that a considerable portion of statistical practice is based on the method of least squares, it becomes clear that an inadequate understanding of the underlying reasoning can lead to serious errors. Indeed, one frequently encounters misapplications, many of which can be traced to the following points:

1. The assumption of an erroneous mathematical model. This can happen either because an erroneous theory is advanced for the physical or chemical processes involved, or because of an exaggerated desire to simplify matters, for example by assuming linearity for non-linear processes. Further discussion of this point occurs in the following chapter.

2. Correlated errors. If the errors pertaining to the various measurements are not statistically independent, the method as developed here and as usually applied is not valid. Known correlations between errors can generally be handled through modified formulas. We will discuss some aspects of this question in greater detail in Chapter 12.

3. Failure to apply proper weighting factors. It should be remembered that the method is based on *known* values for the weights or more precisely, on known values of their ratios to each other. This point is too often neglected and the method is used without weighting. Wherever the error-variances are all of the same order of magnitude, this procedure will lead to no serious misapplications but quite often the variances of errors are functions of the magnitude of the measurements. In such cases, the

parameter estimates derived from the unweighted analysis may be seriously in error, and the evaluation of their reliability thoroughly misleading. This matter will receive further discussion in Chapter 13.

Finally, the following point deserves the most careful consideration. By its very nature, the method of least squares estimates parameters in such a way as to maximize the agreement between the data actually obtained and those derived from these parameter estimates. Since most situations involve several parameters, it is, generally speaking, always possible to obtain reasonable agreement (i.e., reasonably small residuals) as a result of adjusting simultaneously the values of all the parameters. For example, if a given straight line is a poor fit for a given set of fairly linear data, one can shift the line, by changing both its slope and its intercept, until it fits the data in the best possible manner. But this does not insure that the value of the intercept for this "best fitting" line is necessarily close to the true intercept. The values obtained for the intercept and the slope are really compromises aimed at minimizing the departures between the observed and "fitted" values. The method ignores any difference that may exist in the mind of the experimenter regarding the relative importance of getting a very good value for the slope versus one for the intercept. If such differences do exist, they must be given due attention in the planning of the investigation. Once the experimental values have been obtained, little more can be done than fitting the data in the least squares sense. This matter will be made clearer in our discussion of curve fitting and surface fitting.

## 7.11 SUMMARY

When a given physical quantity is measured by different methods, the combination of the various test results into a single over-all measure requires the consideration of the relative precisions of all individual values. This is accomplished by assigning weights to the individual measurements and then computing their weighted average. Each weight is obtained as the reciprocal of the variance of the corresponding measuring process. This procedure provides an application of the principle of linear, unbiased, minimum-variance estimation. The weighted average is indeed a linear combination of the individual test results; it is unbiased in the sense that if the individual test results contain no systematic error, it will also be free of systematic errors; and the use of the reciprocal variances as weighting factors ensures that the variance of the combined value is the smallest that can be achieved for any linear, unbiased combination of the individual test results.

The method of least squares may be considered as a generalization of the weighted average procedure. Its need arises when several quantities are simultaneously determined, not by direct measurement, but rather through the measurement of a number of other quantities that are functionally related to them. For example, the basic physical constants are determined by measuring a number of related quantities, more readily amenable to direct measurement and each of which can be expressed as a mathematical function of one or more of the physical constants. Generally the number of measured quantities exceeds the number of constants to be determined, and leads thereby to an over-determined set of equations. The method of least squares was originally developed for the case in which the equations of this set are all linear in the unknown constants. For that case it provides a computational procedure for deriving the "best" solutions of the over-determined set of equations. By "best" solution is meant a set of values that are linear functions of the measured values, that are free of systematic errors, and that have a minimum variance. However, the method can also be applied to non-linear equations by using the "linearization" device of expanding the non-linear expression into Taylor's series about a set of initial values for the unknown constants. In practice, such a set of initial values is always available. The least squares procedure then provides a set of positive or negative corrections to be added to the initial values. When the equations relating the measured quantities to the unknown constants, the so-called observational equations, are of the multiplicative type, i.e., if they contain only multiplications and divisions (and raising to powers), but no additions or subtractions, the linearization process can be considerably simplified.

The arithmetic process of least squares computations for linear observational equations consists of two steps. First, a set of so-called "normal equations" is derived. This is a set of linear equations equal in number to the number of unknown constants. Then, this set is solved by appropriate algebraic procedures. If the number of unknowns is larger than 4, the computations involved in the second step may become time-consuming, and are best carried out on a high-speed computer.

An important aspect of the method of least squares is that it also provides, in addition to estimates for the unknown constants, measures for their precision. Since the estimates of the constants are functions of the measured values, they are subject to uncertainties induced by the experimental errors of these values. By applying the law of propagation of errors one obtains estimates for the uncertainty of the estimates of the constants. It is also possible to derive confidence intervals for these constants, through application of Student's $t$ distribution.

The method of least squares, as described so far, covers those situations only in which each observational equation contains a single measured value though it may contain several of the unknown constants. The method has been extended to the more general case in which each observational equation may contain more than one measured value, each of these having its own precision. In this formulation, the method also covers the adjustment of observations, i.e., the calculation of adjusted values for a number of measured quantities, the adjustment being necessitated by additional mathematical conditions that the values must satisfy. For example, if the three angles of a triangle are independently measured, experimental errors will cause their sum to differ from the theoretical value of 180 degrees. The adjustment procedure consists in correcting each measured angle by an appropriate amount, based on the precision of the measurement, such that the sum of the corrected values will indeed be 180 degrees. This problem is but a simple case of a class of problems that occur frequently in astronomy, geodesy, and other fields. As formulated here it contains no unknown quantities beyond the measured values. In the broader formulation of the generalized method of least squares, unknown parameters may occur in addition to the values that are to be adjusted.

Basic to the method of least squares is the requirement that the relationships expressed by the observational equations be completely known, except for the values of the unknown parameters. Thus, the application of the method assumes that a model be given prior to the analysis; it does not allow for a change of the model as a result of an examination of the data. In this sense the method of least squares is not powerful as a diagnostic tool. In Chapters 8 and 9 we will deal with statistical methods that are more particularly designed to throw light on the model underlying the data.

## REFERENCES

1. Anderson, R. L., and T. A. Bancroft, *Statistical Theory in Research*, McGraw-Hill, New York, 1952.
2. Bennett, C. A., and N. L. Franklin, *Statistical Analysis in Chemistry and the Chemical Industry*, Wiley, New York, 1954.
3. Courant, R., *Differential and Integral Calculus*, Nordeman, New York, 1945.
4. Deming, W. E., *Statistical Adjustment of Data*, Wiley, New York, 1943.
5. DuMond, J. W. M., and E. R. Cohen, "Least Squares Adjustment of the Atomic Constants 1952," *Revs. Modern Phys.*, **25**, 691–708 (July 1953).
6. Hald, A., *Statistical Theory with Engineering Applications*, Wiley, New York, 1952.
7. Wilson, E. B., *Introduction to Scientific Research*, McGraw-Hill, New York, 1952.

# chapter 8

# TESTING THE STATISTICAL MODEL

## 8.1 FURTHER REMARKS ON THE METHOD OF LEAST SQUARES

We have stated at some length the conditions upon which the method of least squares is based. As long as these conditions are fulfilled, the method will yield the desired results. It is, however, a pertinent question whether the method is still satisfactory under conditions that depart in one or more ways from the ideal state of affairs. A method whose validity is not seriously impaired by departures from the conditions of its highest degree of validity is said to be *robust*. Our first question concerns the robustness of the method of least squares.

Taken literally, the method of least squares is a mere computational device. We have seen, however, that it can be justified in a more fundamental way by the principle of linear unbiased minimum-variance estimation.

Consider the purely computational aspect: it consists in minimizing the sum of squares of residuals. All residuals are treated equally provided that the observations have equal weight. The simplest case of the application of the principle of least squares is the estimation of the mean of a population on the basis of $N$ randomly selected observations. The least squares solution is then the sample mean. Indeed, the sum

$$\sum_i (x_i - \mu)^2$$

160

is minimum when $\mu$ is made equal to $\bar{x}$, $\mu = \bar{x}$. Denoting the residuals by $d_1, d_2, \ldots, d_N$, the sum of squares that is made a minimum is

$$d_1{}^2 + d_2{}^2 + \cdots + d_N{}^2$$

It is clear that the minimum is achieved by reducing more particularly the *larger* residuals. But if one of the measurements happened to be affected by a substantial error, due to a blunder or to a momentary change in the environmental conditions, then this measurement would be likely to depart radically from the other (unbiased) measurements, and would tend to give a large residual. The method of least squares will then necessarily operate principally on this large residual for the purpose of achieving the minimum sum of squares. This can only be done by moving the estimated average closer to the abnormal value. Thus the presence of biased measurements may seriously impair the validity of the results of least squares estimation. The method is far from robust with respect to biased observations. Since so much of classical statistical methodology is based on least squares, the vulnerability of this technique is transmitted to many of these methods. In practical applications, where strict conformance with the assumptions underlying the technique can never be taken for granted, the only safeguard against erroneous conclusions is a careful examination of the residuals as well as of the final estimates. Nothing can be more harmful to the evaluation of data than the blind application of so-called "optimum" techniques (such as least squares) without regard to the possible anomalies that might vitiate the method. We will have several opportunities to discuss this point in detail.

A second characteristic of the method of least squares that may lead to error is its limited ability to act as a diagnostic tool. We have introduced the method as a means of estimating the *unknown* parameters of a *known* functional form. The method is much less suited to a diagnosis as to which functional form is most appropriate. Should an inappropriate form be selected, the application of the method of least squares for the estimation of parameters would not be too likely to reveal the wrong selection. It is therefore necessary, in most problems of statistical inference, to introduce further means for the verification of the assumed model, or for the selection of a suitable model from among a set of candidates.

The problem we have just raised has, so far, not received a general solution. Suppose, for example, that $y$ is known to depend on $x$ and we measure $y$ at certain fixed values of $x$. A plot of $y$ versus $x$ might suggest a quadratic relationship, but an exponential relation would perhaps provide an equally good fit. There exists no general theory that will allow us to pick the most appropriate type of function in such situations.

However, if we were in doubt between a quadratic and a cubic function, we would receive considerable help from existing statistical theory. What distinguishes this latter case from the former?

Consider the two alternatives in the second example:

$$y = a_1 + b_1 x + c_1 x^2$$
$$y = a_2 + b_2 x + c_2 x^2 + d_3 x^3$$

Both equations contain three sets of elements: (1) the measured value $y$; (2) the independent variable $x$ and its powers, $x^2$ and $x^3$; and (3) the unknown parameters $a, b, c, \ldots$. Now, even though $y$ is not a linear function of $x$, it *is* a linear function of the parameters $(a, b, c, \ldots)$ both for the quadratic and the cubic case. Similarly, the functions

$$y = a + bx + ce^x$$
$$y = a + \frac{b}{\sin x} + c \log x$$

are also linear in the parameters even though they involve transcendental functions of $x$. In all cases in which the measured value is a strictly linear function of the *parameters*, we can make use of a general statistical technique known as the *general linear hypothesis*. On the other hand, the function

$$y = ae^{bx}$$

is not linear in the parameter $b$. It is therefore not *directly* tractable by the method of the general linear hypothesis. Of course, this does not mean that a function of this type is beyond the resources of statistical analysis. In fact, there exist many special devices for dealing with non-linear functions, but so far they have not been integrated into a theory of the same degree of generality as that which applies to the linear case. In a subsequent chapter we will examine some special non-linear cases and develop techniques for dealing with them.

## 8.2    A NUMERICAL EXAMPLE

The data in Table 8.1 were obtained in the study of a method for determining the natural rubber content in mixtures of natural and synthetic rubber by infrared spectroscopy (7). A plot of the measured value $y$ versus the known concentration of natural rubber $x$ is almost a straight line. We ask, "Is a quadratic fit significantly better than a linear one?"

**TABLE 8.1**

| $x$ | 0 | 20 | 40 | 60 | 80 | 100 |
|---|---|---|---|---|---|---|
| $y$ | 0.734 | 0.885 | 1.050 | 1.191 | 1.314 | 1.432 |

The linear model is represented by the relation

$$y = a_1 + b_1 x + \varepsilon \qquad (8.1)$$

and the quadratic model by:

$$y = a_2 + b_2 x + c_2 x^2 + \varepsilon \qquad (8.2)$$

where $\varepsilon$ represents random experimental error.  Of course, $\varepsilon$ is a function of the experiment, not of the model.

Both equations can be fitted by the method of least squares, by minimizing the sum of squares of residuals.  The estimates of the coefficients, as well as the calculated values of $y$, and the residuals, $d$, are listed in Table 8.2 for both the linear and the quadratic models.

**TABLE 8.2**

| $x$ | $y$ | Linear model $\hat{y}$ | Linear model $d_L$ | Quadratic model $\hat{y}$ | Quadratic model $d_Q$ |
|---|---|---|---|---|---|
| 0 | 0.734 | 0.7497 | $-0.0157$ | 0.7299 | $+0.0041$ |
| 20 | 0.885 | 0.8902 | $-0.0052$ | 0.8942 | $-0.0092$ |
| 40 | 1.050 | 1.0307 | $+0.0193$ | 1.0466 | $+0.0034$ |
| 60 | 1.191 | 1.1713 | $+0.0197$ | 1.1871 | $+0.0039$ |
| 80 | 1.314 | 1.3118 | $+0.0022$ | 1.3157 | $-0.0017$ |
| 100 | 1.432 | 1.4523 | $-0.0203$ | 1.4325 | $-0.0005$ |
|  |  | $\sum d_L{}^2 = 0.001451$ | | $\sum d_Q{}^2 = 0.000131$ | |
|  |  | $\hat{a}_1 = 0.7497 \quad \hat{b}_1 = 0.007026$ | | $\hat{a}_2 = 0.7299 \quad \hat{b}_2 = 0.008512$ $\hat{c}_2 = -0.00001487$ | |

The sum of squares of residuals is smaller for the quadratic model than for the linear model.  In the case of the linear model, the estimate of the error-variance is given by

$$\hat{V}_1(\varepsilon) = \frac{\sum d_L{}^2}{6-2} = \frac{0.001451}{4} = 0.000363 \qquad (8.3)$$

For the quadratic model, the estimate of the error-variance is

$$\hat{V}_2(\varepsilon) = \frac{\sum d_Q{}^2}{6-3} = \frac{0.000131}{3} = 0.000044 \qquad (8.4)$$

The denominator is $6 - 2$, or 4, for the linear fit, because two parameters, $a_1$ and $b_1$, are estimated from the data; and $6 - 3$, or 3, for the quadratic fit, because in this case three parameters, $a_2$, $b_2$, and $c_2$, are estimated from the data. All this is in accordance with Eq. 7.30.

It is seen that $\hat{V}_2(\varepsilon)$ is smaller than $\hat{V}_1(\varepsilon)$, which would appear to indicate conclusively the superiority of the quadratic fit. However, we must remember that both $\hat{V}_1(\varepsilon)$ and $\hat{V}_2(\varepsilon)$ are only *estimates* of variances. As such—and especially in view of the fact that the number of degrees of freedom is small for both—these estimates may differ considerably from the true variance they represent. The question therefore is not whether $\hat{V}_2(\varepsilon)$ is numerically smaller than $\hat{V}_1(\varepsilon)$, but rather whether it is *sufficiently* smaller, so that even after allowing for the uncertainty of both estimates, we can still state with confidence that the quadratic model is a superior representation of the data. The posing of such a question and its answer are known as a *test of significance*.

### 8.3 TESTS OF SIGNIFICANCE

For clarity of presentation we will first present the basic mechanics of the test, in terms of the example of the preceding section, and then discuss its meaning.

Table 8.3, which is usually referred to as an *analysis of variance*, presents all necessary quantities and computations. The table is self-explanatory, except for the last column: the "mean square" is by definition the sum of squares divided by the corresponding number of degrees of freedom.

**TABLE 8.3**   Analysis of Variance

| Number of parameters in model | Remaining degrees of freedom | Sum of squares of residuals | Mean square |
|---|---|---|---|
| 2 | $6 - 2 = 4$ | $SS_L = 0.001451$ | $M_L = 0.000363$ |
| 3 | $6 - 3 = 3$ | $SS_Q = 0.000131$ | $M_Q = 0.000044$ |
| Difference  1 | 1 | $SS_\Delta = 0.001320$ | $M_\Delta = 0.001320$ |
| | $F = \dfrac{M_\Delta}{M_Q} = \dfrac{0.001320}{0.000044} = 30.0$ | | |

Thus, in the last row, the mean square, $M_\Delta$, is not the difference of the two numbers above it but rather the quotient of the sum of squares, 0.001320, by the degrees of freedom, one in this case. The test is carried out by considering the ratio of the two mean squares $M_\Delta$ and $M_Q$

$$F = \frac{M_\Delta}{M_Q} = \frac{0.001320}{0.000044} = 30.0 \tag{8.5}$$

This ratio of mean squares, which is commonly denoted by the letter $F$, is said, in this example, to have 1 degree of freedom in the numerator and 3 degrees of freedom in the denominator. The numerical value $F = 30.0$ is now compared with a tabulated value to assess its "significance," but before doing this, let us first examine the meaning of these computations.

The denominator of $F$ is, of course, the variance estimate $\hat{V}_2(\varepsilon)$. If the quadratic model is the correct one, this estimate is evidently a valid estimate of the error variance. It can be shown that it is also a valid, though not the best, estimate of the error-variance, in case the linear model is the correct one. This may appear surprising, but is due to the fact that fitting a quadratic to linear data is tantamount to wasting one degree of freedom on an unnecessary coefficient $b_2$. As a consequence, the residuals become smaller but so does the number of degrees of freedom corresponding to these residuals (see Table 6.3). On the average there is exact compensation so that the ratio of the sum of squares of residuals to the degrees of freedom (the mean square) is still an unbiased estimate of the true error-variance.

Let us now examine the numerator, $M_\Delta$, of $F$. If the linear model is correct, this quantity corresponds precisely to the one wasted degree of freedom. It is therefore itself an unbiased estimate of the true error-variance. If, on the other hand, the quadratic model is correct, the degree of freedom corresponding to $M_\Delta$ should actually have been used for the estimation of the quadratic coefficient $b_2$. In that case $M_\Delta$ is therefore not a measure of pure error, and will, in fact, tend to be larger than the variance of the experimental error.

Summarizing the above considerations we obtain the following situation:

1. If the true model is linear, both $M_Q$ and $M_\Delta$ are unbiased estimates of the same quantity, namely the true error-variance.

2. If the true model is quadratic, $M_Q$ is an unbiased estimate of the true error-variance, but $M_\Delta$ will tend to be larger than the error-variance.

The ratio $F$ is thus a usable criterion for discriminating between the two possibilities: if the true model is linear, $M_Q$ and $M_\Delta$ will tend to be alike and $F$ will tend to be close to unity; if, on the other hand, the true model is the quadratic one, then $M_\Delta$ will tend to be larger than $M_Q$ and $F$ will tend to exceed unity.

The problem has now been reduced to deciding when $F$ is sufficiently greater than unity to warrant a preference for the quadratic model.

Actually, in problems of this type, the alternatives are generally not formulated as linear versus quadratic, but rather as linear versus any polynomial of degree higher than one. Thus the second alternative includes the possibility that the true model be quadratic, cubic, and so forth. This gives the first alternative a preferential status as being clearly and completely defined whereas the second alternative includes several possibilities. The well-defined alternative is called the *null-hypothesis*. From a mathematical point of view, there is a definite advantage in having a null-hypothesis of this type. For if we accept, provisionally, the null-hypothesis, we can derive *exactly* the probability distribution of the quantity $F$, provided that we may assume the errors $\varepsilon$ to be normally distributed. That $F$ is a random variable is obvious, since both its numerator and its denominator are sample estimates and therefore random variables. The fact that, *under the null-hypothesis*, they are sample-estimates of the same quantity, is particularly felicitous, and allows for a relatively simple derivation of the frequency-distribution of their ratio, $F$.

So far the reasoning has been purely *deductive*. We have omitted mathematical details but the assertions made can be substantiated by purely mathematical proofs. At this point, however, an entirely new principle must be invoked. This principle belongs to the domain of *inductive* reasoning, which is at the root of the entire field of statistical inference.

Essentially, the principle states that if the numerical value of $F$ actually obtained has a small probability of occurrence, then the null-hypothesis is considered unacceptable. Actually, since $F$ is a continuous variable the probability of obtaining *any particular* value is infinitesimally small. This can be made plausible by an analogy: if an alloy contains exactly 70 per cent copper, an analysis of the alloy might give the value 69.83 per cent. To obtain exactly the same value (69.83) in a second determination is very unlikely, but not impossible. However, the value 69.83 is only a rounded approximation of the copper content actually yielded by the analysis. If the error is a continuous variable, the actual value yielded by the analysis might be 69.832792..., with an infinite number of decimal places. To obtain *exactly* the same infinite succession of decimal figures in a second analysis is an event of probability zero. Returning now to the $F$ test, a more correct form of our principle of inductive reasoning states that the null-hypothesis is unacceptable whenever the value of $F$ belongs to a certain *set* of possible values, the total probability of which is small.

To clarify the last part of this statement, consider the probability distribution of $F$ in Fig. 8.1. The total area under the curve is unity. Compare the areas denoted A and B.

Both areas correspond to $F$ intervals of the same width, yet area $A$ is considerably larger than area $B$. In other words, the probability of obtaining values near $F_1$ is considerably larger than that of obtaining values near $F_2$, if the word "near" is interpreted in identical manner for both values. Thus the right "tail" of the curve constitutes a region of low probabilities of occurrence. If we now select a value $F_c$ (see Fig. 8.2)

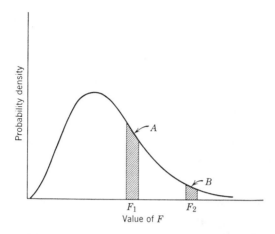

**Fig. 8.1** Probability distribution of $F$; the shaded areas have equal width, but correspond to regions of different probability density.

such that the area to its right is equal to 0.05, or 5 per cent, then we have isolated a region of low probability values, such that the total probability of falling *anywhere* in this region is only 5 per cent. Such a region is called a 5 per cent *critical region*. By selecting a value $F_c'$ appropriately

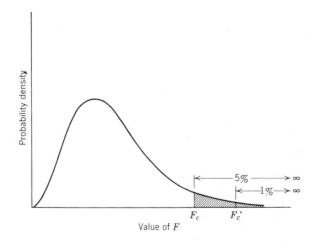

**Fig. 8.2** Critical areas in the $F$ distribution; the portion under the curve to the right of the critical value $F_c$ has a preselected area (5 per cent and 1 per cent in the figure).

to the right of $F_c$ we can, of course, construct a 1 per cent critical region, or, more generally, a critical region of any desired "size."

The principle of inductive inference stated above can now be formulated more simply by stipulating that if the $F$ value actually obtained falls in the critical region, the null-hypothesis is considered unacceptable. It is at once clear that this statement is incomplete unless the size of the critical region is also stipulated. This stipulated value is known as the *level of significance* of the test.

The function of the data on which a test of significance is based is called the *test-statistic*. In our example, the test-statistic is the quantity $F$. If the test-statistic yields a value falling in the critical region of a certain size, the result is said to be *statistically significant* (or simply *significant*) at the level corresponding to that size. In the contrary case the result is termed "non-significant" at that level. In our example, the value $F = 30.0$ is significant at the 5 per cent level of significance, but not quite significant at the 1 per cent level. This is seen by comparing this value with the tabulated "critical" $F$ values for 1 and 3 degrees of freedom* in Tables III and IV of the Appendix: the 5 per cent critical value is 10.13 and the 1 per cent critical value is 34.12.

It may appear disturbing that the test of significance, thus formulated, involves a completely arbitrary element: the level of significance. A null-hypothesis that is considered unacceptable on the 5 per cent level of significance may be accepted when a 1 per cent level is chosen. Is this not a very unsatisfactory state of affairs?

The answer requires a deeper insight into the nature of inductive reasoning. We have already stated that the principle under discussion is not of a deductive mathematical nature. It is not even of a logical nature, if by logic we understand the classical field of deductive reasoning. We were faced with the problem of deciding whether a particular set of data could be considered linear. If the data had been free of error the decision would have been simple and final: any departure, no matter how small, from a straight line would unequivocally lead to the decision that the data are not linear. But the actual data are known to be subject to experimental errors. Departures from a straight line are therefore susceptible of two alternative interpretations: either the particular pattern exhibited by these departures is merely the result of chance and the line is really straight, or the pattern is actually a reflection of real curvature in the true model. It is impossible to choose between these alternative interpretations without risk of error. The reason for this is that the interplay of chance

---

* The $F$ distribution depends on two parameters: the number of degrees of freedom in the numerator, $\nu_1$, and the number of degrees of freedom in the denominator, $\nu_2$. In the present case $\nu_1 = 1$ and $\nu_2 = 3$.

effects occasionally produces "improbable" situations.    What we have to decide is how large a risk we are willing to take in mistaking chance for reality, and this decision is made by selecting the level of significance.    A choice of a 5 per cent level simply means that we have decided to become suspicious of the straight line hypothesis (the null-hypothesis) every time the departures from this line have caused the $F$ value to be in the low probability 5 per cent tail.    If, upon observing an $F$ value falling in this region, we *reject* the null-hypothesis, we may well have erred, for the low probability may merely be a "trick" played on us by chance.    Our consolation is that the chance of making such erroneous rejections of a valid null-hypothesis is only 5 in 100.    This does *not* mean that the chance of a correct decision is 95 per cent.    For the rejection of a valid hypothesis is only one way of erring.    Another one is to accept an erroneous null-hypothesis.    This becomes clear when one considers the logical alternatives in Table 8.4.

**TABLE 8.4**

| Choice | Null-hypothesis | |
|---|---|---|
| | True | Not true |
| Acceptance of Null-hypothesis | Correct | Error II |
| Rejection of Null-hypothesis | Error I | Correct |

The null-hypothesis is either true or not true; and our choice, after performing the experiment, is either to accept or to reject it.    There are thus four possible situations, represented by the four entries in the table. Of these four, only two are correct, namely to accept the null-hypothesis when it is actually true and to reject it when it is actually false.    The two other entries represent erroneous choices denoted in the table as "Error I" and "Error II."    These errors are generally referred to as the *error of the first type* and the *error of the second type*.    The *level of significance* can now be defined as *the probability of committing an error of the first type.* Any probability statement requires a frame of reference.    In the present instance, this frame of reference consists of all situations in which the null-hypothesis happens to be true.    Thus, the level of significance pertains only to those situations in which the null-hypothesis is true and tells us nothing about the situations in which the null-hypothesis is false.

To make the point more specific, suppose that in 100 problems subjected to tests of significance, 40 involved a true null-hypothesis and the remaining 60 a false null-hypothesis.    If a level of significance of 5 per cent was used, we might expect an error of the first type to have occurred in 5 per cent of the situations in which the null-hypothesis is true. In the present case, this is 5 per cent of 40, or in 2 cases. The remaining 38 problems would have led to a correct choice. However, this accounts only for 40 out of 100 cases. What about the remaining 60 problems? Here, the level of significance is of no help to us. It simply does not pertain to those cases. Consequently, the total expected percentage of erroneous choices could vary all the way from 2 to a maximum of 62, despite the adoption of a 5 per cent level of significance.

Looking at the matter in a slightly different way, we can state it in the following manner. Given a set of data and an hypothesis to be tested, the level of significance provides some assurance against an unwarranted rejection of the hypothesis. Where it fails, however, is in the assessment of the *sufficiency* of the data for the purpose of testing the hypothesis. This problem of sufficiency deserves as much attention in experimentation and in the analysis of data as that of extracting all the information from a given set of data. In subsequent chapters, the statistical techniques available for that purpose will be discussed in some detail. At this point we merely point out that by rejecting the hypothesis that the data in Table 8.1 are adequately represented by a straight line, we have not learned what the true model is; it could be quadratic, but it also could be exponential, or of a still more complex nature.

In the preceding discussion we have assumed that a level of significance has been chosen prior to the examination of the data and that significance or non-significance is assessed in terms of that level. An alternative approach is to await the outcome of the experiment without prior commitment to a fixed level of significance. The probability associated with the value yielded by the experiment is then taken as an objective measure of the degree of support that the experiment lends to the null-hypothesis: for example, a value just significant at the 3 per cent level lends little support to the null-hypothesis, while a value just significant at the 60 per cent level shows no conflict between the data and the null-hypothesis. Under this approach, the test of significance does not necessarily result in an act of "acceptance" or "rejection" of the null-hypothesis. It merely measures a strength of belief. For the interpretation of scientific data, this approach appears to be more cogent than that consisting in automatic acceptance or rejection "decisions."

One final point requires clarification. In Fig. 8.1, we have defined the critical region as a region of appropriate area in the *right* tail of the curve. It is, however, apparent that the left tail also constitutes a region of low probability density values. Why do we not use the left tail area in defining the critical region? The reason is that a critical region must be composed of those values of the test-statistic that tend to *contradict* the null-hypothesis.   In our case, a value of $F$ in the left tail of the curve, i.e., a low $F$ value, would indicate that $M_\Delta$ is small with respect to $M_Q$ (see Table 8.3).   But then $SS_Q$ would be little different from $SS_L$ and there would be no valid reason to doubt the appropriateness of the linear model (the null-hypothesis), since adoption of the quadratic model would show little reduction in the size of the residuals.   It is thus seen that only large values of $F$ tend to contradict the null-hypothesis, and therefore the critical region for this test must be entirely in the right tail of the curve.

It is evident from the preceding discussion that the reasoning underlying tests of significance is of a subtle nature.   This is because significance testing is one form of inductive reasoning, and inductive reasoning is a difficult and controversial subject.   Nevertheless, the physical scientist who uses statistics as a research tool must have an understanding of the nature of inductive inference.   Only thus can he avoid substituting mechanical manipulations, that may or may not be appropriate, for his own expert understanding of the subject of his investigations.

## 8.4   THE GENERAL LINEAR HYPOTHESIS

The problem discussed in the preceding two sections is a typical example of a general class of situations in which the statistical test is referred to as the *general linear hypothesis*.   The word "linear" in this expression does not refer to a straight line situation but rather to the fact that the statistical models considered are all linear in terms of all parameters.   Thus, both Eqs. 8.1 and 8.2 are linear models in this sense, since both equations are linear in the parameters $a$, $b$, and $c$.

In the general linear hypothesis situation it is assumed that the measured quantity $y$ is known to depend on at most $p + q$ variables: $x_1, x_2, \ldots, x_{p+q}$, thus

$$y = \beta_1 x_1 + \beta_2 x_2 + \cdots + \beta_{p+q} x_{p+q} + \varepsilon \qquad (8.6)$$

where $\varepsilon$ is an experimental error.

The null-hypothesis to be tested is that some of these variables, namely $q$ of them, are superfluous; in other words that the coefficients $\beta$ corresponding to these variables are zero (for $y$ does, of course, no longer depend on an $x$, when this $x$ is preceded by a zero-coefficient).

Since the $x$ variables are given numerical quantities, and not random variables, they can be defined in any desired way, and need not be independent of each other.   For example, we may have

$$x_1 = 1; \qquad x_2 = x; \qquad x_3 = x^2$$

and consider a measurement $y$ obeying the relation:

$$y = \beta_0 x_1 + \beta_1 x_2 + \beta_2 x_3 + \varepsilon$$

or

$$y = \beta_0 + \beta_1 x + \beta_2 x^2 + \varepsilon \tag{8.7}$$

In this example, therefore, we suppose that $y$ is at most a quadratic function of $x$. The null-hypothesis is to the effect that the function is linear in $x$. This is equivalent to the statement that $\beta_2 = 0$. Denoting the null-hypothesis symbolically by $H_0$, we write

$$H_0: \qquad \beta_2 = 0 \tag{8.8}$$

Another example is provided by a situation in which the measurement $y$ is known to depend linearly on concentration, $c$ and on temperature, $T$. It is also suspected to depend on $R$, the relative humidity, but this is not definitely known. Then the general model is

$$y = \beta_1 c + \beta_2 T + \beta_3 R + \varepsilon \tag{8.9}$$

The null-hypothesis is

$$H_0: \qquad \beta_3 = 0 \tag{8.10}$$

More generally, the model is

$$y = (\beta_1 x_1 + \beta_2 x_2 + \cdots + \beta_p x_p) + (\gamma_1 z_1 + \gamma_2 z_2 + \cdots + \gamma_q z_q) + \varepsilon \tag{8.11}$$

and the null-hypothesis is that the variables $z_1, z_2, \ldots, z_q$ are really without effect on $y$, so that

$$H_0: \qquad \gamma_1 = \gamma_2 = \cdots = \gamma_q = 0 \tag{8.12}$$

This hypothesis is equivalent to assuming that the terms $\gamma_1 z_1, \gamma_2 z_2, \ldots,$ $\gamma_q z_q$ do not occur at all in model 8.11.

We assume that the total *number of measurements* is $N$. To test hypothesis 8.12 we proceed as in the example of Section 8.2. Using the method of least squares we fit two models.

### Fit of Model I

The model is

$$y = \beta_1 x_1 + \beta_2 x_2 + \cdots + \beta_p x_p + \varepsilon \tag{8.13}$$

The estimates are $\hat{\beta}_1, \ldots, \hat{\beta}_p$, and the estimate of the error variance is:

$$V_1(\varepsilon) = \frac{\sum d_I^2}{N - p} \tag{8.14}$$

where $d_I$ denotes a residual, the subscript I merely denoting the model fitted.

**Fit of Model II**

The model is

$$y = (\beta_1 x_1 + \beta_2 x_2 + \cdots + \beta_p x_p) + (\gamma_1 z_1 + \gamma_2 z_2 + \cdots + \gamma_q z_q) + \varepsilon \tag{8.15}$$

The estimates are

$$\beta_1', \beta_2', \ldots, \beta_p'; \hat{\gamma}_1, \hat{\gamma}_2, \ldots, \hat{\gamma}_q$$

where the primes of the $\beta$ estimates serve to remind us that the inclusion of $q$ more variables will in all probability also change the estimates of the coefficients of the first $p$ variables.

The estimate of the error-variance is

$$V_2(\varepsilon) = \frac{\sum d_{\text{II}}^2}{N - (p + q)} \tag{8.16}$$

We now construct the analysis of variance, Table 8.5, exactly analogous to Table 8.3.

**TABLE 8.5**  Analysis of Variance

| Number of parameters in model | Remaining degrees of freedom | Sum of squares of residuals | Mean square[a] |
|:---:|:---:|:---:|:---:|
| $p$ | $N - p$ | $\text{SS}_{\text{I}} = \sum d_{\text{I}}^2$ | $M_{\text{I}} = \dfrac{\text{SS}_{\text{I}}}{N - p}$ |
| $p + q$ | $N - (p + q)$ | $\text{SS}_{\text{II}} = \sum d_{\text{II}}^2$ | $M_{\text{II}} = \dfrac{\text{SS}_{\text{II}}}{N - (p + q)}$ |
| Difference   $q$ | $q$ | $\text{SS}_{\Delta} = \text{SS}_{\text{I}} - \text{SS}_{\text{II}}$ | $M_{\Delta} = \dfrac{\text{SS}_{\Delta}}{q}$ |

[a] The reader is reminded that $M_{\Delta}$ is not a difference, but rather a quotient of two differences.

The test-statistic is $F$, defined as for Table 8.3 by

$$F = \frac{M_{\Delta}}{M_{\text{II}}} \tag{8.17}$$

The ratio $F$ has $q$ degrees of freedom in the numerator and $N - (p + q)$ in the denominator.

Reasoning as in the example of Section 8.2, we agree to suspect the null-hypothesis whenever the numerical value of $F$ falls in the critical region for $F$ with $q$ and $N - p - q$ degrees of freedom, for the level of significance chosen.

The quantity denoted $SS_A$ is often referred to as the *reduction in the sum of squares* associated with the fitting of the additional $q$ parameters. This terminology reflects the fact that the larger the number of parameters in the model, the smaller will be the sum of squares of residuals, i.e., the sum of squares used for the estimation of error. Thus each additional parameter in the model has the effect of reducing the residual sum of squares. It is important to note that the order in which the parameters are introduced in the model is, in general, not immaterial. Thus the reduction in the sum of squares, $SS_A$, due to the $q$ additional parameters in Table 8.5, would generally not have been the same if it had followed a fit different from that involving the $p$ parameters in the first row; for example, if this initial fit had involved fewer or more than $p$ parameters, or a different set of variables. We will return to this matter in Section 8.5.

The statement we made in Section 8.1 concerning the difference between a comparison of two linear models and one between a linear and a non-linear model (or between two non-linear models) is now more easily understood. For the theory of the general linear hypothesis does not help us in the choice, say, between the models

$$y = a + bx + cx^2 + \varepsilon$$

and

$$y = y_0 + Ke^{bx} + \varepsilon$$

The reasons for this are twofold. In the first place, the second model is not linear in the parameter $b$, and, secondly, neither hypothesis can be "embedded" in the other (as can be done in the comparison of a quadratic and a cubic fit).

In spite of this limitation, the general linear hypothesis is one of the most powerful tools in applied statistics. Indeed, the class of problems that can be treated by this method is appreciably larger than appears from the formulation we have given above.

Suppose that the hypothesis to be tested is not whether $y$ depends linearly on a certain variable $z$ but rather whether the coefficient relating $y$ to $z$ is equal to a certain value $\gamma_0$. That is: in the general model (8.11) we wish to test the hypothesis, for example, that

$$H_0: \qquad \gamma_2 = 5 \qquad (8.18)$$

We do this simply by writing:

$$\gamma_2 = 5 + \gamma_2' \qquad (8.19)$$

from which we derive

$$\gamma_2 z_2 = 5z_2 + \gamma_2' z_2$$

and

$$y = [\beta_1 x_1 + \cdots + \beta_p x_p] + [\gamma_1 z_1 + (5z_2 + \gamma_2' z_2) + \cdots + \gamma_q z_q] + \varepsilon$$

Subtracting $5z_2$ from both members, we obtain:

$$y - 5z_2 = [\beta_1 x_1 + \cdots + \beta_p x_p] + [\gamma_1 z_1 + \gamma_2' z_2 + \cdots + \gamma_q z_q] + \varepsilon \tag{8.20}$$

Since $5z_2$ is a known constant for any given value of $z_2$ we can consider a new variable

$$u = y - 5z_2 \tag{8.21}$$

the error-variance of which is identical to that of $y$.

Now, in view of Eq. 8.19 we may write the null-hypothesis of Eq. 8.18 as:

$$H_0: \qquad \gamma_2' = 0 \tag{8.22}$$

Thus we have reduced the problem under discussion to the first formulation we have given of it, by replacing the measured variable $y$ by $y - 5z_2$.

With this generalization, the general linear hypothesis allows us to test the values of any number of coefficients in the model expressing $y$ as a linear function of independent variables.

## 8.5  AN APPLICATION IN CHEMICAL ANALYSIS

Consider the data in Table 8.6, representing a study of a new analytical procedure for the determination of fatty acid in rubber.  The column labeled $x$ lists the true values for ten samples in solution.  The column headed $y$ contains the values, obtained by the procedure under study, corrected for "blank."  The question we wish to answer concerns the accuracy of this method.  More specifically, are we justified in stating

**TABLE 8.6**  Fatty Acid in Rubber

| Sample | Milligrams of fatty acid | |
| :---: | :---: | :---: |
| | Added ($x$) | Found ($y$) |
| 1 | 20.0 | 20.6 |
| 2 | 20.0 | 17.1 |
| 3 | 50.0 | 51.1 |
| 4 | 50.0 | 50.4 |
| 5 | 150.0 | 150.4 |
| 6 | 153.7 | 155.8 |
| 7 | 250.0 | 250.4 |
| 8 | 250.0 | 251.9 |
| 9 | 500.0 | 505.0 |
| 10 | 500.0 | 501.8 |

that, *apart from random experimental errors*, the value $y$ obtained for any sample will be equal to $x$, the true fatty acid content of that sample?

Suppose that it is known that the variance of the random error $\varepsilon$ affecting the measurement $y$ is essentially the same for all samples. Then the null-hypothesis to be tested is

$$H_0: \qquad y = x + \varepsilon \qquad (8.23)$$

where $\varepsilon$ is a random error of constant, though unknown variance.

In order to apply the general linear hypothesis we must "embed" this hypothesis within a model of greater generality, just as Eq. 8.13 (the equation expressing the null-hypothesis) is embedded within Eq. 8.15. The ways in which this can be done are unlimited in number. However, a simple and plausible general model is given by the linear relationship

$$y = \alpha + \beta x + \varepsilon \qquad (8.24)$$

Then the null-hypothesis asserts that

$$H_0: \qquad \alpha = 0 \quad \text{and} \quad \beta = 1 \qquad (8.25)$$

Applying the method of least squares both to model 8.23 and to model 8.24, we obtain the calculated values and residuals shown in Table 8.7. The analysis of variance is given in Table 8.8. The $F$ value is found to be significant at the 5 per cent level. If we adopt a critical region of size 0.05,

**TABLE 8.7**

| | | Model I: $y = x + \varepsilon$ | | Model II: $y = \alpha + \beta x + \varepsilon$ | |
|---|---|---|---|---|---|
| $x$ | $y$ | $\hat{y}$ | $d_{\mathrm{I}}$ | $\hat{y}$ | $d_{\mathrm{II}}$ |
| 20.0 | 20.6 | 20.0 | +0.6 | 19.746 | +0.854 |
| 20.0 | 17.1 | 20.0 | −2.9 | 19.746 | −2.646 |
| 50.0 | 51.1 | 50.0 | +1.1 | 49.976 | +1.124 |
| 50.0 | 50.4 | 50.0 | +0.4 | 49.976 | +0.424 |
| 150.0 | 150.4 | 150.0 | +0.4 | 150.741 | −0.341 |
| 153.7 | 155.8 | 153.7 | +2.1 | 154.469 | +1.331 |
| 250.0 | 250.4 | 250.0 | +0.4 | 251.506 | −1.106 |
| 250.0 | 251.9 | 250.0 | +1.9 | 251.506 | +0.394 |
| 500.0 | 505.0 | 500.0 | +5.0 | 503.418 | +1.582 |
| 500.0 | 501.8 | 500.0 | +1.8 | 503.418 | −1.618 |
| | | $\sum d_{\mathrm{I}}^2 = 46.72$ | | $\sum d_{\mathrm{II}}^2 = 17.56$ | |
| | | $\alpha = 0; \quad \beta = 1$ | | $\hat{\alpha} = -0.407;$ | |
| | | | | $\hat{\beta} = 1.00765$ | |

**TABLE 8.8**    Analysis of Variance

| Number of parameters in model | Remaining degrees of freedom | Sum of squares of residuals | Mean square |
|:---:|:---:|:---:|:---:|
| 0 | $10 - 0 = 10$ | 46.72 | 4.672 |
| 2 | $10 - 2 = 8$ | 17.56 | 2.195 |
| Difference = 2 | 2 | 29.16 | 14.58 |

$$F_{2,8} = \frac{14.58}{2.195} = 6.64$$

we will therefore look with suspicion upon the null-hypothesis $y = x + \varepsilon$. In chemical terms, we would consider the method to be affected by systematic errors.

At this point a second question arises: is the null-hypothesis in error in asserting that $\alpha = 0$ or in stating that $\beta = 1$, or are both parts of the hypothesis wrong? Questions of this type always arise when the null-hypothesis involves more than one parameter. The analysis can readily be extended to provide answers to these questions. For example, in order to test the null-hypothesis

$$H_0': \quad \alpha = 0 \tag{8.26}$$

without making any assumptions regarding the value of $\beta$, we embed this

**TABLE 8.9**

| | | Model | | | |
|:---:|:---:|:---:|:---:|:---:|:---:|
| | | I': $y = \beta x + \varepsilon$ | | II: $y = \alpha + \beta x + \varepsilon$ | |
| $x$ | $y$ | $\hat{y}$ | $d_{I'}$ | $\hat{y}$ | $d_{II}$ |
| 20.0 | 20.6 | 20.130 | $+0.470$ | 19.746 | $+0.854$ |
| 20.0 | 17.1 | 20.130 | $-3.030$ | 19.746 | $-2.646$ |
| 50.0 | 51.1 | 50.324 | $+0.776$ | 49.976 | $+1.124$ |
| 50.0 | 50.4 | 50.324 | $+0.076$ | 49.976 | $+0.424$ |
| 150.0 | 150.4 | 150.972 | $-0.572$ | 150.741 | $-0.341$ |
| 153.7 | 155.8 | 154.696 | $+1.104$ | 154.469 | $+1.331$ |
| 250.0 | 250.4 | 251.620 | $-1.220$ | 251.506 | $-1.106$ |
| 250.0 | 251.9 | 251.620 | $+0.280$ | 251.506 | $+0.394$ |
| 500.0 | 505.0 | 503.240 | $+1.760$ | 503.418 | $+1.582$ |
| 500.0 | 501.8 | 503.240 | $-1.440$ | 503.418 | $-1.618$ |

$$\sum d_{I'}^2 = 18.29 \qquad \sum d_{II}^2 = 17.56$$
$$\alpha = 0; \quad \hat{\beta} = 1.00648 \qquad \hat{\alpha} = -0.407;$$
$$\hat{\beta} = 1.00765$$

hypothesis in the more general assumption expressed by Eq. 8.24. Thus, we construct Table 8.9, in which Model II is the same as in Table 8.7, but Model I is replaced by Model I':

$$y = \beta x + \varepsilon \tag{8.27}$$

The corresponding analysis of variance, shown in Table 8.10, retains the second line of Table 8.8, pertaining to Model II, but has a different first line, because of the changed null-hypothesis.

Similarly we test the hypothesis

$$H_0'': \qquad \beta = 1 \tag{8.28}$$

by constructing Table 8.11 and the corresponding analysis of variance in Table 8.12.

**TABLE 8.10**   Analysis of Variance

| Number of parameters in model | Remaining degrees of freedom | Sum of squares of residuals | Mean square |
|---|---|---|---|
| 1 | $10 - 1 = 9$ | 18.29 | 2.032 |
| 2 | $10 - 2 = 8$ | 17.56 | 2.195 |
| Difference = 1 | 1 | 0.73 | 0.73 |

$$F_{1,8} = \frac{0.73}{2.195} = 0.33$$

**TABLE 8.11**

| | | Model | | | |
|---|---|---|---|---|---|
| | | I'': $y = \alpha + x + \varepsilon$ | | II: $y = \alpha + \beta x + \varepsilon$ | |
| $x$ | $y$ | $\hat{y}$ | $d_{\mathrm{I''}}$ | $\hat{y}$ | $d_{\mathrm{II}}$ |
| 20.0 | 20.6 | 21.08 | $-0.48$ | 19.746 | $+0.854$ |
| 20.0 | 17.1 | 21.08 | $-3.98$ | 19.746 | $-2.646$ |
| 50.0 | 51.1 | 51.08 | $+0.02$ | 49.976 | $+1.124$ |
| 50.0 | 50.4 | 51.08 | $-0.68$ | 49.976 | $+0.424$ |
| 150.0 | 150.4 | 151.08 | $-0.68$ | 150.741 | $-0.341$ |
| 153.7 | 155.8 | 154.78 | $+1.02$ | 154.469 | $+1.331$ |
| 250.0 | 250.4 | 251.08 | $-0.68$ | 251.506 | $-1.106$ |
| 250.0 | 251.9 | 251.08 | $+0.82$ | 251.506 | $+0.394$ |
| 500.0 | 505.0 | 501.08 | $+3.92$ | 503.418 | $+1.582$ |
| 500.0 | 501.8 | 501.08 | $+0.72$ | 503.418 | $-1.618$ |

$$\sum d_{\mathrm{I''}}^2 = 35.06 \qquad\qquad \sum d_{\mathrm{II}}^2 = 17.56$$

$$\hat{\alpha} = 1.08; \ \beta = 1 \qquad\qquad \hat{\alpha} = -0.407;$$
$$\hat{\beta} = 1.00765$$

**TABLE 8.12**    Analysis of Variance

| Number of parameters in model | Remaining degrees of freedom | Sum of squares of residuals | Mean square |
|---|---|---|---|
| 1 | $10 - 1 = 9$ | 35.06 | 3.896 |
| 2 | $10 - 2 = 8$ | 17.56 | 2.195 |
| Difference $= 1$ | 1 | 17.50 | 17.50 |

$$F_{1,8} = \frac{17.50}{2.195} = 7.97$$

The results of these analyses are quite informative: the data are seen to be in no conflict with the hypothesis $\alpha = 0$, whereas the test for $\beta = 1$ shows significance at the 5 per cent level.    We have learned in this way that the analytical method under study is probably subject to a systematic error of a relative type: the ratio of the measured value to the true value is 1.0065.    In other words the method yields values that are high by about 0.65 per cent, regardless of the amount of fatty acid present in the solution.

**TABLE 8.13**    Analysis of Variance

| Null-hypothesis | Model | Remaining degrees of freedom | Sum of squares of residuals | Mean square |
|---|---|---|---|---|
| $\alpha = 0; \beta = 1$ | I: $y = x + \varepsilon$ | 10 | 46.72 | 4.672 |
| $\alpha = 0$ | I′: $y = \beta x + \varepsilon$ | 9 | 18.29 | 2.032 |
| $\beta = 1$ | I″: $y = \alpha + x + \varepsilon$ | 9 | 35.06 | 3.896 |
| — | II: $y = \alpha + \beta x + \varepsilon$ | 8 | 17.56 | 2.195 |

| Null-hypothesis | Model | Reduction in DF | Reduction in SS | MS |
|---|---|---|---|---|
| $\alpha = 0; \beta = 1$ | I | 2 | 29.16 | 14.58 |
| $\alpha = 0$ | I′ | 1 | 0.73 | 0.73 |
| $\beta = 1$ | I″ | 1 | 17.50 | 17.50 |

Tables 8.7, 8.9, and 8.11 can be combined, to obviate recopying the common part pertaining to model 8.24.    Similarly, the analysis of variance tables can be combined as indicated in Table 8.13.    Examination of this table reveals an important fact: the reduction in the sum of squares due to the combined hypothesis $\alpha = 0$ *and* $\beta = 1$ is *not* (as might have been

expected) equal to the sum of the reductions due to these two hypotheses taken one at a time.   The reason for this is briefly as follows: if a set of points tend to fall along a straight line for which the intercept is different from zero, and the slope different from unity, then if we fit to these points a straight line through the origin, we can only do this by changing the slope from its correct value.   Similarly, a line of slope "one" can be fitted to these data only by changing the intercept from its correct value. In other words, an error in the slope is compensated by an error in the intercept, and vice-versa.   Consequently the slope estimate and the intercept estimate are not statistically independent.   The reduction in the sum of squares due to assuming $\beta = 1$ is therefore dependent on what has been assumed regarding $\alpha$, and the reduction in the sum of squares due to assuming $\alpha = 0$ depends on what has been assumed regarding $\beta$.   Such a situation is denoted as *non-orthogonality*.   We will see later that there exist many situations in which the reductions in the sum of squares due to various hypothesis are strictly *additive*.   One then refers to the sample estimates of the parameters involved in these hypotheses as *orthogonal functions*.

The non-orthogonality of the functions $\hat{\alpha}$ and $\hat{\beta}$ can be ascertained directly from Tables 8.9 and 8.11: in the former, a change in $\hat{\alpha}$ from $-0.407$ to the hypothesized value $\alpha = 0$ results in a change in $\hat{\beta}$ from 1.00765 to 1.00648.   Had $\hat{\alpha}$ and $\hat{\beta}$ been orthogonal, then a change in $\hat{\alpha}$ would have had no effect whatsoever on $\hat{\beta}$.   A similar statement holds for Table 8.11.

It should be noted that for each hypothesis we have considered, we have computed all individual residuals.   It is in this fashion that we obtained the sums of squares for the analyses of variance.   There exist short-cut methods for calculating these sums of squares, without computing each individual residual.   The practice of using these short-cut methods, although extremely widespread, is highly deplorable.   The analysis of variance technique is highly sensitive to "outlying" observations; thus, a single bad value in a set of straight line data may cause a highly significant quadratic term, leading to the erroneous conclusion that the quadratic fit is more appropriate than the linear fit.   If in such a case, the residuals are individually computed, the discrepant observation is generally revealed by a conspicuously large residual.

In examining a set of residuals one is guided by the fact that in a good fit the residuals tend to show a pattern of randomness.   For example, in Table 8.7, the residuals for Model I are, with a single exception, all positive. Since the residuals are the sample-image of random errors, such a pattern is highly unlikely.   This in itself would lead us to suspect the appropriateness of Model I.   Similarly, in the example discussed in Section 8.2, the linear model (Table 8.2) exhibits residuals with a definite non-random pattern: negative, positive, negative. This pattern, together with the actual values of the residuals, strongly suggests curvature. The analysis of variance may often be considered as a formal confirmation of the conclusions drawn from a careful examination of the residuals. A note of warning is necessary, however. The residuals resulting from a least-squares fit are not statistically independent. Failure to take their inter-correlations into account may lead to erroneous conclusions. The fewer the data, the greater is the correlation among residuals. The correlation also increases when the number of parameters in the model increases. All this leads to the fairly obvious conclusion that the greater the complexity of the model, the larger a sample will be required if valid inferences are to be made.

Before turning to a different application of the general linear hypothesis, we make the following observation. In the analysis of the data of Table 8.6, the initial null-hypothesis is a natural expression of the chemical problem: "Is the new method for determining fatty acid in rubber accurate?" Less obvious is the more general hypothesis, Eq. 8.24, in which the null-hypothesis model is embedded. The choice of the general hypothesis is determined, in the final analysis, by knowledge of the subject matter underlying the data as well as by good intuitive judgment. That the statistical treatment of data should thus involve a subjective element is neither surprising nor distressing: it is merely a consequence of the fact, already mentioned, that statistical inference is a particular form of inductive reasoning. Since an inductive inference depends on *all* the pertinent facts, rather than on a limited set of facts, it is bound to vary with the state of knowledge and background information available. This implies that inductive inferences involve subjective as well as objective elements, since no two persons have an absolutely identical background, even in regard to scientific matters.

## 8.6 APPLICATION TO THE LORENTZ–LORENZ EQUATION

According to the Lorentz–Lorenz formula, the index of refraction, $n$, is related to the density, $D$, by means of the equation

$$\frac{n^2 - 1}{n^2 + 2} \frac{1}{D} = K \tag{8.29}$$

where $K$ is a constant independent of the pressure and the temperature. The data in Table 8.14 are part of a study in which the value $K$ was measured at various pressures and temperatures. For the five values in this table, obtained for methanol, the temperature was held constant at 25 °C, but the pressure, $P$, was varied over a wide range. The question to be answered is whether $K$ is really independent of the pressure $P$.

The values of $K$ calculated separately for the various values of $P$ appear to decrease slightly as $P$ increases. Since the Lorentz–Lorenz equation has theoretical justification, it is important to make certain that this apparent decrease is not just a result of chance fluctuations in the data.

**TABLE 8.14** Specific Refractivity of Methanol at 25°C

| Pressure (bars), $P$ | Refractive Index, $n$ | Density, $D$ | Specific Refractivity, $K$ |
|---|---|---|---|
| 1 | 1.32770 | 0.786553 | 0.257731 |
| 259 | 1.33668 | 0.809088 | 0.256778 |
| 492 | 1.34459 | 0.825832 | 0.256915 |
| 739 | 1.35098 | 0.841029 | 0.256491 |
| 1023 | 1.35764 | 0.856201 | 0.256246 |

Let us first rewrite Eq. 8.29 in the form:

$$\frac{n^2 - 1}{n^2 + 2} = KD \qquad (8.30)$$

We will assume, without justification at this point, that the experimental error in $D$ is small with respect to that of the quantity $(n^2 - 1)/(n^2 + 2)$. The exact sense in which such a statement may be made will be discussed in Chapter 14. Denoting $D$ by $x$ and $(n^2 - 1)/(n^2 + 2)$ by $y$, and allowing for an error $\varepsilon$ in the measurement of $y$, we obtain from Eq. 8.30 the relation:

$$y = Kx + \varepsilon \qquad (8.31)$$

where

$$y = \frac{n^2 - 1}{n^2 + 2}$$

$$x = D \qquad (8.32)$$

Table 8.15 lists $x$ and $y$ for the five values of $P$. The question of interest may now be reformulated statistically by considering Eq. 8.31 as the null-hypothesis and embedding it in a more general model that reflects the possible dependence of $K$ on $P$. The simplest way of doing this is to consider $K$ as a linear function of $P$:

$$K = K_0 + \beta P \qquad (8.33)$$

**TABLE 8.15**

| $P$ | $x$ | $y$ |
|---|---|---|
| 1 | 0.786553 | 0.202719 |
| 259 | 0.809088 | 0.207756 |
| 492 | 0.825832 | 0.212169 |
| 739 | 0.841029 | 0.215716 |
| 1023 | 0.856201 | 0.219398 |

With this expression for $K$, the general model becomes

$$y = K_0(x) + \beta(xP) + \varepsilon \tag{8.34}$$

with the null-hypothesis

$$H_0: \qquad \beta = 0 \tag{8.35}$$

In Eq. 8.34 the independent variables are $x$ and $xP$; the coefficients to be estimated are $K_0$ and $\beta$. Table 8.16 shows the least-squares fits for both Eq. 8.31 and Eq. 8.34. The analysis of variance is given in Table 8.17.

**TABLE 8.16**

| $x$ | $xP$ | $y$ | I: $y = Kx + \varepsilon$ $\hat{y}$ | $d_{\mathrm{I}} \times 10^6$ | II: $y = K_0x + \beta(xP) + \varepsilon$ $\hat{y}$ | $d_{\mathrm{II}} \times 10^6$ |
|---|---|---|---|---|---|---|
| 0.786553 | 0.786 | 0.202719 | 0.201990 | +729 | 0.202514 | +205 |
| 0.809088 | 209.554 | 0.207756 | 0.207777 | −21 | 0.208050 | −294 |
| 0.825832 | 406.309 | 0.212169 | 0.212077 | +92 | 0.212111 | +58 |
| 0.841029 | 621.520 | 0.215716 | 0.215980 | −264 | 0.215749 | −33 |
| 0.856201 | 875.894 | 0.219398 | 0.219876 | −478 | 0.219331 | +67 |

$\sum d_{\mathrm{I}}^2 = 838{,}526 \times 10^{-12}$
$\hat{K} = 0.256804$

$\sum d_{\mathrm{II}}^2 = 137{,}403 \times 10^{-12}$
$\hat{K}_0 = 0.257472$;
$\hat{\beta} = -1.275 \times 10^{-6}$

**TABLE 8.17**  Analysis of Variance

| Model | Residual DF | Sum of squares of residuals | Mean square |
|---|---|---|---|
| $y = Kx + \varepsilon$ | 4 | $838{,}526 \times 10^{-12}$ | $209{,}632 \times 10^{-12}$ |
| $y = K_0x + \beta(xP) + \varepsilon$ | 3 | $137{,}403 \times 10^{-12}$ | $45{,}801 \times 10^{-12}$ |
| Difference | 1 | $701{,}123 \times 10^{-12}$ | $701{,}123 \times 10^{-12}$ |

$$F_{1,3} = \frac{701{,}123 \times 10^{-12}}{45{,}801 \times 10^{-12}} = 15.31$$

The $F$ value is significant at the 5 per cent level, indicating reasonably strong evidence against the null-hypothesis. We are therefore inclined to consider the specific refractivity, $K$, as not strictly independent of the pressure. The data suggest that the dependence of $K$ on $P$ for methanol at 25 °C may be represented by the relation

$$K = 0.257472 - 1.275 \times 10^{-6}P \tag{8.36}$$

This relation is of course merely an empirical representation, valid only over the range of $P$ values of the experiment. The relation has neither theoretical meaning, nor should it be used for extrapolation purposes. The important inference from the data is not Eq. 8.36, but rather the fact that the data provide evidence against the hypothesis that $K$ is independent of $P$.

A conclusion of this type is often the point of departure for further research and experimentation that eventually leads to a better understanding of the physical phenomena involved. It is worth noting that the difficulty in this problem is that the dependence of $K$ on $P$ is essentially of the same order of magnitude as the experimental measurement errors. The point here is that statistical analysis is especially useful in problems involving high-precision data. For increased precision generally leads to the detection of effects that would have been completely obscured by larger experimental errors, and are just barely apparent with the improved methods of measurement. In such situations the distinction between real effects and the result of chance fluctuations is of special pertinence.

The example we have just discussed provides a further illustration of the observation made at the end of the preceding section. The general model 8.34 is dictated primarily by the nature of the problem and the available information. The problem is the possible effect of pressure on the specific refractivity $K$, and the information consists in numerical values of the pressure $P$. Numerous other models would have been possible. For example, one might attempt to explain the trend of the $K$ values in Table 8.14 in terms of models such as

$$\frac{n^2 - 1}{n^2 + 2} = A + KD$$

or

$$\frac{n^2 - 1}{n^2 + 2} = KD + BD^2$$

but such models would ignore the nature of the problem as well as the information provided by the $P$ values. An analysis that merely seeks a "good fit" may not be of great value as a scientific tool. The example provides a further illustration of the basic fact that statistical inferences depend not only on the data themselves, but also, and to a very marked extent, on the background information available.

## 8.7   SIGNIFICANCE TESTS USING STUDENT'S *t*

In many situations involving tests of significance, the test can be carried out by a far simpler procedure than that based on reductions of sums of squares.   The basis for this simpler treatment is the fact that an *F* value *with one degree of freedom in the numerator* (and any number of degrees of freedom in the denominator) can be expressed as the square of Student's *t* statistic.   The number of degrees of freedom of this *t* value is precisely that of the *denominator* of *F*.   The use of *t* instead of *F* in such cases is not only computationally simpler but has the further advantage of simpler conceptual interpretation.

We will first illustrate the procedure by means of two examples and then discuss its use in a more general way.

The data in Table 8.18 represent the determination of bitumen in expansion joint fillers, an important material used in highway construction (3).   The specified procedure called for a 4 inch × 4 inch test specimen, but a different method of analysis was proposed and it was believed that the more economical 2 inch × 2 inch specimen would give identical results. The reduction in size of the test specimen would constitute a great advantage because the method involves an extraction procedure, the duration of which increases with the size of the specimen.

**TABLE 8.18**   Bitumen Content of Expansion Joint Fillers

| Specimen-pair | 2″ × 2″ specimen | 4″ × 4″ specimen | Difference |
|:---:|:---:|:---:|:---:|
| 1 | 64.3 | 65.4 | −1.1 |
| 2 | 65.9 | 66.1 | −0.2 |
| 3 | 66.4 | 66.7 | −0.3 |
| 4 | 65.1 | 66.2 | −1.1 |
| 5 | 66.3 | 66.4 | −0.1 |
| 6 | 66.7 | 66.5 | +0.2 |
| 7 | 66.7 | 66.3 | +0.4 |
| 8 | 64.6 | 66.1 | −1.5 |
| 9 | 64.5 | 64.5 | 0 |
| Average | 65.61 | 66.02 | −0.41 |

In this example the values are "paired," i.e., the experiment was performed in such a way that to each result obtained on a 2 inch × 2 inch specimen there corresponds a *particular* result obtained on a 4 inch × 4 inch specimen and vice versa.    Pairing can be accomplished in many ways.    In the present case it was done by taking two *adjacent* specimens, one of each size, for each pair, so that the two specimens of a pair had approximately the same location on the sheet of material from which the test specimens were drawn.

Let us number the pairs from 1 to 9 and refer to the two sizes as $y$ (for the 2 inch × 2 inch specimen) and $z$ (for the 4 inch × 4 inch specimen). Then each measurement may be represented by a symbol of the type $y_i$ or $z_i$, where $i$ denotes the number of the pair.    Allowing for errors of measurement we consider also the expected values $\mu_i$ and $\nu_i$ corresponding to $y_i$ and $z_i$.    Thus, representing the error of measurement for a 2 inch × 2 inch specimen by $\varepsilon$, and that for a 4 inch × 4 inch specimen by $\delta$, we have:

$$y_i = \mu_i + \varepsilon_i$$
$$z_i = \nu_i + \delta_i \tag{8.37}$$

The hypothesis we wish to test is expressed by the model:

Model I (null-hypothesis):    $\mu_i = \nu_i,$    for all $i$

As always we must embed this hypothesis into a more general one.    A natural choice for the latter is the assumption that there exists a *systematic, constant difference* between the results obtained on specimens of the two sizes.    Representing this difference by $K$, we have, for the general model:

Model II (alternative hypothesis):    $\mu_i = \nu_i + K$

Thus the null-hypothesis becomes:

$$H_0: \quad K = 0$$

It is advantageous, in this type of problem, to consider a new variable, say $v_i$, defined as the difference between the two values in a pair:

$$v_i = y_i - z_i \tag{8.38}$$

Let us represent the expected value of $v_i$ by $\tau_i$; thus:

$$E(v_i) = \tau_i \tag{8.39}$$

We then have:

| | | | |
|---|---|---|---|
| Model I: | $\tau_i = 0,$ | for all $i$ | (8.40) |
| Model II: | $\tau_i = K,$ | for all $i$ | (8.41) |

and the null-hypothesis is:

$$H_0: \quad K = 0 \tag{8.42}$$

In terms of the $v_i$, there are 9 observations, but whereas Model II involves the estimation of one parameter, $K$, Model I involves no parameters at all. According to Model I, the $i$th residual is $v_i - 0 = v_i$; according to Model II, it is $v_i - \hat{K}$ and it is readily seen that $\hat{K} = \bar{v}$, the average of the $v_i$. Consequently, the analysis is as shown in Table 8.19, from which we see at once that

$$SS_\Delta = \sum v_i^2 - \sum (v_i - \bar{v})^2 = 9\bar{v}^2 \tag{8.43}$$

and hence:

$$F = \frac{9\bar{v}^2}{\sum (v_i - \bar{v})^2/(9-1)} = \left[ \frac{\bar{v}}{\sqrt{\sum (v_i - \bar{v})^2/9(9-1)}} \right]^2 \tag{8.44}$$

**TABLE 8.19**  Analysis of Bitumen Data

| Model | Residual DF | Sum of squares of residuals | Mean square |
|---|---|---|---|
| $v \equiv y - z = 0 + (\varepsilon - \delta)$ | 9 | $SS_I = \sum v^2 = 5.01$ | 0.55 |
| $v \equiv y - z = K + (\varepsilon - \delta)$ | 8 | $SS_{II} = \sum (v - \bar{v})^2 = 3.49$ | 0.44 |
| Difference | 1 | $SS_\Delta = 1.52$ | 1.52 |

$$F = \frac{1.52}{0.44} = 3.45$$

$$t = \frac{\bar{v}}{\sqrt{\dfrac{\sum (v - \bar{v})^2}{9(9-1)}}} = \frac{-0.41}{0.22} = -1.86$$

$$t^2 = (-1.86)^2 = 3.45 = F$$

The quantity inside the brackets is simply the ratio of the average $\bar{v}$ to its estimated standard error. Furthermore, according to the null-hypothesis, $E(v_i) = 0$, hence $E(\bar{v}) = 0$. Therefore, the quantity inside the brackets can be written

$$\frac{\bar{v} - E(\bar{v})}{\hat{\sigma}_{\bar{v}}} \tag{8.45}$$

According to Eq. 6.20, a ratio of this kind is distributed in accordance with Student's $t$ (provided that the $v_i$ are independently and normally distributed).

Writing the null-hypothesis in the form: $E(\bar{v}) = 0$, we see therefore that a simpler approach to the problem is to test *directly*, by means of Student's $t$, the hypothesis that the population mean corresponding to a given homogeneous sample is zero.

If Student's $t$ is used for the construction of a confidence interval for $K$, it will be found that every time the confidence interval includes the value zero, the least squares estimate for $K$ is not significantly different from zero, provided that the confidence coefficient, $\beta$, is equal to $1 - \alpha$, where $\alpha$ is the level of significance. This is consistent with our previous remark (Section 6.11) that any value inside the confidence interval is a *possibility* for the parameter; therefore the latter is "not significantly different" from any value inside the interval.

Let us now turn to an application of the $t$ test in which the observations are not paired. We use once more the data of Shields *et al.* (6), on the isotopic abundance ratio of silver. This time we wish to compare the results of Table 6.1, obtained on a commercial sample of silver nitrate, with those obtained on a sample of natural silver from Norway. For convenience both sets of data are given in Table 8.20. The question we wish to answer is whether the isotopic ratio is the same for both samples.

**TABLE 8.20**    Isotopic Abundance Ratio of Natural Silver, Comparison of Two Sources

| Commercial Ag NO$_3$ | | | Norway sample |
|---|---|---|---|
| 1.0851 | 1.0842 | 1.0803 | 1.0772 |
| 834 | 786 | 811 | 824 |
| 782 | 812 | 789 | 788 |
| 818 | 784 | 831 | 830 |
| 810 | 768 | 829 | |
| 837 | 842 | 825 | |
| 857 | 811 | 796 | |
| 768 | 829 | 841 | |
| Average | | 1.081483 | 1.080350 |
| Sum of Squares of Deviations | | $1.5351 \times 10^{-4}$ | $0.2355 \times 10^{-4}$ |
| Degrees of freedom | 23 | | 3 |

$$t = \frac{1.081483 - 1.080350}{\left[\dfrac{1.7706 \times 10^{-4}}{26}\left(\dfrac{1}{24} + \dfrac{1}{4}\right)\right]^{1/2}} = \frac{0.00113}{0.00141} = 0.91$$

In the discussion of this problem we will omit the treatment by the $F$ test, and use the $t$ test from the start. The crucial quantity is, of course, the difference between the two averages. Using a line of reasoning already explained in Section 6.5 (Eq. 6.1), we compute the standard error of the difference of means. If $\sigma_x$ is the standard deviation corresponding to the silver nitrate sample and $\sigma_y$ that for the Norway sample, and if $\bar{x}$ and $\bar{y}$ represent the sample averages for these two materials, we have, for the variance of $\bar{x} - \bar{y}$,

$$V(\bar{x} - \bar{y}) = \sigma_{\bar{x}-\bar{y}}^2 = \sigma_{\bar{x}}^2 + \sigma_{\bar{y}}^2 = \frac{\sigma_x^2}{24} + \frac{\sigma_y^2}{4} \tag{8.46}$$

Now both $\sigma_x$ and $\sigma_y$ are unknown but can be estimated from the data. This would give us an estimate of $\sigma_{\bar{x}-\bar{y}}$ involving *two* distinct estimates of standard deviations. The *t*-statistic, on the other hand, involves only a *single* estimate of a standard deviation and is therefore not directly applicable to this problem. Fortunately we can, in the present case, assume with good justification that $\sigma_x$ and $\sigma_y$ are equal to each other, since they depend on the same measuring process applied to essentially the same basic system. Equation 8.46 can therefore be written

$$\sigma_{\bar{x}-\bar{y}}^2 = \sigma^2\left(\frac{1}{24} + \frac{1}{4}\right) \tag{8.47}$$

where $\sigma^2$ is the common variance. An estimate of this common variance is obtained by the usual pooling procedure (see Section 4.4, Eq. 4.17):

$$\hat{\sigma}^2 = \frac{\sum(x - \bar{x})^2 + \sum(y - \bar{y})^2}{(24 - 1) + (4 - 1)} \tag{8.48}$$

From Eq. 8.47 we then derive the estimate

$$\hat{\sigma}_{\bar{x}-\bar{y}}^2 = \frac{\sum(x - \bar{x})^2 + \sum(y - \bar{y})^2}{(24 - 1) + (4 - 1)}\left[\frac{1}{24} + \frac{1}{4}\right] \tag{8.49}$$

The numerical value for this quantity is

$$\hat{\sigma}_{\bar{x}-\bar{y}}^2 = \frac{15{,}351 + 2355}{23 + 3}\left(\frac{1}{24} + \frac{1}{4}\right) \times 10^{-8} = 199 \times 10^{-8} \tag{8.50}$$

and, consequently $\hat{\sigma}_{\bar{x}-\bar{y}} = 14.1 \times 10^{-4}$. Consequently, the *t* statistic is given by

$$\begin{aligned}
t &= \frac{(\bar{x} - \bar{y}) - E(\bar{x} - \bar{y})}{\sigma_{\bar{x}-\bar{y}}} \\
&= \frac{(1.08148 - 1.08035) - E(\bar{x} - \bar{y})}{0.00141}
\end{aligned} \tag{8.51}$$

Now the hypothesis we wish to test is precisely that $E(\bar{x} - \bar{y}) = 0$. Under this hypothesis, we have

$$t = \frac{0.00113}{0.00141} = 0.91 \qquad DF = 26 \tag{8.52}$$

This value is not significant, even at the 10 per cent level of significance. The data therefore show no evidence of a systematic difference in the isotopic abundance ratio for the two samples.

## 8.8 NOTES ON THE USE OF THE *t*-STATISTIC
## AND ON SIGNIFICANCE TESTING

Student's *t* test is perhaps the most popular of all statistical tests of significance. It is used essentially for two types of problems: to test the mean of a single population or to test the difference of the means of two populations. In the latter case, as we have just seen, the samples may be paired or nonpaired.

In the many applications that have been made of the *t* test, the assumptions underlying it have not always been sufficiently considered. We shall mention only a few important points.

In the first place, the test is subject to the limitations inherent in all tests of significance. Such tests depend not only on the choice of the null hypothesis but also, implicitly or explicitly, on the particular alternative hypotheses considered in the study. In the terminology we have used in this chapter for the general linear hypothesis we would say that the model representing the null-hypothesis can be embedded in a more general model in a large number of ways. Which of these is selected depends on the general background knowledge available to the experimenter on the subject of his experimentation. For example, in the paired-comparisons case, Student's *t* tests the hypothesis of *no* difference versus that of a *constant* difference. An often more meaningful alternative is that the difference depends on the level of the value, in which case regression methods are more appropriate than a *t* test on paired values (see Chapters 11, 12, and 13).

The pooling of variance estimates in the non-paired case should never be made without reasonable justification. Modifications of the test are available when the pooling is not justified (1, p. 235ff).

Finally, in many cases, the over-all problem involves more than two populations. The individual application of a test of significance to a selected pair of these populations then involves some basic difficulties in the interpretation of the test. We will not discuss this matter in this book but refer the reader to discussions on multiple comparisons (1, pp. 252–4) and joint confidence regions (1, pp. 314–7; 5).

Recently the entire subject of significance testing has been examined from a novel viewpoint, involving considerations of the risks associated with the making of wrong decisions. This new field, known as "Decision Theory," has received strong support in some quarters but has also been severely criticized, at least in terms of its applicability to scientific research. Fairly elementary discussions will be found in references 2 and 4.

It is evident from this chapter that significance testing is an important, but conceptually difficult subject. It is being vigorously discussed by statisticians as well as by philosophers of science. Perhaps a more coherent picture will soon emerge from these discussions.

## 8.9  SUMMARY

The method of least squares, in its classical form, was not intended as a diagnostic tool. To make inferences from a set of data about the mathematical model that most likely underlies them, it is necessary to introduce other statistical aids. Among those, the theory of testing hypothesis constitutes an important tool.

The theory may be illustrated in terms of the choice between a linear and a quadratic relation for the representation of a given set of points. The more general relation, i.e., the quadratic, is tentatively adopted as the model, and the coefficients are estimated by least squares. If the true model is linear, one should obtain an equally good fit by fitting a straight line to the data. Since both fits are subject to statistical uncertainty, the question as to whether the straight line fit is as good as the quadratic must be reformulated as follows: could the difference between the two fits be attributed to random error? The statistical procedure that poses a question of this type and provides an answer to it is called a test of significance. However, the answer given by a test of significance is correct only in a probability sense. Each test of significance involves a null-hypothesis. In our example, this is the assumption that the coefficient of the quadratic term is zero, which is equivalent to the statement that the linear model is the correct one. If the null-hypothesis represents the true state of affairs, the probability that the test of significance will provide the correct answer can be exactly computed. The probability of the opposite event, i.e., that the test of significance will provide the wrong answer, is called the level of significance. If the null-hypothesis is actually untrue, the probability that the test of significance will provide the correct answer is no longer a single quantity, since it depends on the numerical extent by which the null-hypothesis is in error.

The "general linear hypothesis" may be considered as a generalization of the problem of deciding between a linear and a quadratic fit. All problems covered by this general theory can be treated in terms of a particular type of probability distribution function, known as the $F$-distribution, and the statistical technique used in the solution of these problems is called the analysis of variance.

In some cases, a problem tractable by the general linear hypothesis can be treated more simply, but in a mathematically equivalent way, by the use of Student's $t$ distribution.

Statisticians are not agreed on what constitutes the best approach to problems of statistical inference of the type discussed in this chapter. The subject is under active consideration by statisticians as well as by philosophers of science. At the time of this writing, no single approach has emerged with sufficient clarity to be regarded as definitive.

## REFERENCES

1. Brownlee, K. A., *Statistical Theory and Methodology in Science and Engineering*, Wiley, New York, 1960.
2. Chernoff, H., and L. E. Moses, *Elementary Decision Theory*, Wiley, New York, 1960.
3. Horowitz, E., and J. Mandel, "Determination of Bitumen Content in Expansion Joint Fillers," *Materials Research and Standards*, 3, No. 9, 723–725 (Sept. 1963).
4. Lindgren, B. W., *Statistical Theory*, MacMillan, New York, 1962.
5. Mandel, J., and F. J. Linnig, "Study of Accuracy in Chemical Analysis Using Linear Calibration Curves," *Analytical Chemistry*, 29, 743–749 (May 1957).
6. Shields, W. R., D. N. Craig, and V. H. Dibeler, "Absolute Isotopic Abundance Ratio and the Atomic Weight of Silver," *J. Am. Chem. Soc.*, 82, 5033–5036 (1960). (The detailed data, which do not appear in this paper, were kindly supplied by the authors.)
7. Tryon, M., E. Horowitz, and J. Mandel, "Determination of Natural Rubber in GR-S-Natural Rubber Vulcanizates by Infrared Spectroscopy," *J. Research Natl. Bur. Standards*, 55, 219–222 (Oct. 1955), Research Paper 2623.

simply the average of all elev

sis, differences between groups a

esis of Eq. 9.2, the estimate of $\mu_i$

s in group $i$.

| | Model | | |
|---|---|---|---|
| $+ \varepsilon_{ij}$ | II:   $y_{ij} = \mu_i + \varepsilon_{ij}$ | | |
| $d_{ij}$ | | $\hat{y}_{ij}$ | $d_{ij}$ |
| $+672$ | | 1106 | $+5$ |
| $+662$ | | 1106 | $-5$ |
| $+756$ | | 1195 | $0$ |
| $-167$ | | 237 | $+35$ |
| $-237$ | | 237 | $-35$ |
| $-301$ | | 139 | $-1$ |
| $-299$ | | 139 | $+1$ |
| $-287$ | | 167 | $-15$ |
| $-308$ | | 167 | $-36$ |
| $-306$ | | 167 | $-34$ |
| $-188$ | | 167 | $+84$ |
| $^2 = 2{,}031{,}637$ | | $\sum d_{II}^2 = 12{,}235$ | |

ded as shown in Table 9.1.

umn labeled "Conventional Terminology" in
oy the following considerations. The degrees of
the reduction in the sum of squares due to the
os is $5 - 1$, because the introduction of groups
5 parameters whereas the null-hypothesis requires
gle parameter. For this reason, a reduction of sum
e is often referred to as the *sum of squares between*
to have $k - 1$ degrees of freedom, where $k$ is the
We will see that this interpretation of the analysis of
to a certain extent but presents some danger of
he true nature of a mean square.

res of residuals for the general hypothesis, i.e., the value
3 is seen from Table 9.2 to be a pooled value of the sum
iduals obtained separately for each group. The corre-
quare, 2039, is therefore a valid measure of within-group
rdless of the truth of the null-hypothesis, and it repre-
d estimate of the unknown parameter $\sigma_\varepsilon^2$. It is of interest
t the manner in which this pooled estimate is obtained is
explained in Section 4.4. The term "total" for the model

---

*chapter 9*

# THE ANALYSIS OF STRUCTURED DATA

## 9.1   INTRODUCTION

It frequently happens that a set of data can naturally be partitioned into two or more classes, the question of interest being whether the classification is of any relevance regarding the measured value. For example, human longevity data can be classified according to sex, to observe whether, on the average, males live longer than females, or vice versa. The grades of students in a mathematics class can be classified according to the field in which the students major. The question here is whether, for example, students majoring in physics tend to obtain better grades in mathematics than students majoring in chemistry, or vice versa. In either of these examples there is only one *criterion of classification*: sex in the first example, and major field in the second.

It is clear that the classification may involve several criteria. Thus, tread wear data of automobile tires can be classified according to manufacturer but they can also be classified according to the speed at which the tires are tested. Both classifications can be carried out simultaneously by arranging the data in a rectangular array, in which each row represents a manufacturer and each column a speed value.

The general problem of classification data and its statistical treatment is best explained in terms of specific examples. We will show that, statistically, these problems may often be considered as applications of the general linear hypothesis.

## 9.2   BETWEEN—WITHIN CLASSIFICATIONS

The data in Table 9.1 are taken from a study dealing with the determination of the faraday (1).   In this part of the study, 11 individual determinations were made of the electrochemical equivalent of silver, and expressed in milligrams of silver per coulomb.   The results are shown in the column labeled "Value."   The purpose was to elucidate the effect of oxygen that might be present in the purified silver used for the determinations.   The first 2 determinations were made on untreated specimens of the silver, while the remaining 9 determinations were made on silver specimens treated in various ways, as indicated.

**TABLE 9.1**   Electrochemical Equivalent of Silver

| Determination number | Sample of silver | Treatment | Value | Coded Value |
|---|---|---|---|---|
| 1 | II | None | 1.119111 | 1111 |
| 2 | II | None | 1.119101 | 1101 |
| 3 | II | Etched only | 1.119195 | 1195 |
| 4 | II | Annealed only | 1.118272 | 272 |
| 5 | II | Annealed only | 1.118202 | 202 |
| 6 | II | Annealed and Etched | 1.118138 | 138 |
| 7 | II | Annealed and Etched | 1.118140 | 140 |
| 8 | I | Annealed and Etched | 1.118152 | 152 |
| 9 | I | Annealed and Etched | 1.118131 | 131 |
| 10 | I | Annealed and Etched | 1.118133 | 133 |
| 11 | I | Annealed and Etched | 1.118251 | 251 |

The differences between the various determinations become more apparent, and the computations are greatly simplified, if the data are "coded" by subtracting the constant quantity 1.118 from each value and multiplying the difference by $10^6$.   These coded values are listed in the last column of Table 9.1.

The question that is to be resolved concerns the differences between values obtained by different treatments.

A classification of this type is often called a *between–within classification*, because different values belong either within the same group, or to different groups; in the latter case they may reflect between-group differences.

In the language of the general linear hypothesis, we are to test the null-hypothesis that all groups represent samples from one single population, versus the alternative hypothesis that there are 5 populations, not all identical.

**TABLE 9.3**   Analysis of Variance[a]

| Model | Conventional terminology | DF | SS | MS |
|-------|--------------------------|----|----|----|
| I (null) | Total | 10 | $SS_I = 2,031,637$ | $MS_I = 203,164$ |
| II | Within groups | 6 | $SS_{II} = 12,235$ | $MS_{II} = 2039$ |
| Difference | Between groups | 4 | $SS_{III} = 2,019,402$ | $MS_{III} = 504,851$ |

$$F_{4,6} = \frac{504,851}{2039} = 248$$

[a] All data are coded as shown in Table 9.1.

representing the null-hypothesis is explained by the fact that its sum of squares is equal to the sum of the other two sums of squares.

Further insight into the analysis is gained by considering the algebraic quantities for the three sums of squares.

Let us represent the average of the observations in the $i$th group by $\bar{y}_i$, and the average of all observations by $\bar{\bar{y}}$. Then the sum of squares of residuals under the null-hypothesis is

$$SS_I = \sum_i \sum_j (y_{ij} - \bar{\bar{y}})^2 \tag{9.4}$$

The sum of squares of residuals within-groups is

$$SS_{II} = \sum_i \left[ \sum_j (y_{ij} - \bar{y}_i)^2 \right] \tag{9.5}$$

It can readily be proved by expanding the square and summing all terms that

$$SS_I = \sum_i \sum_j y_{ij}^2 - N\bar{\bar{y}}^2 \tag{9.6}$$

where $N$ is the total number of observations.   (In our example, $N = 11$.) Similarly, representing by $n_i$ the number of observations in the $i$th group, it can be shown that

$$SS_{II} = \sum_i \sum_j y_{ij}^2 - \sum_i n_i \bar{y}_i^2 \tag{9.7}$$

The sum of squares *between groups* is the difference $SS_I - SS_{II}$, i.e.,

$$SS_{III} = SS_I - SS_{II} = \sum_i n_i \bar{y}_i^2 - N\bar{\bar{y}}^2 \tag{9.8}$$

Thus the mean square *between groups* is equal to

$$MS_{III} = \frac{\sum_i n_i \bar{y}_i^2 - N\bar{\bar{y}}^2}{k - 1} \tag{9.9}$$

where $k$ represents the number of groups.   (In our example $k = 5$.)

To understand the meaning of $MS_{III}$, let us consider a case in which each group contains the same number $n$ of observations. Then all $n_i$ are equal to $n$, and $N = kn$. Under those conditions we obtain

$$MS_{III} = \frac{n \sum_i \bar{y}_i^2 - kn\bar{\bar{y}}^2}{k - 1} = n \frac{\sum_i \bar{y}_i^2 - k\bar{\bar{y}}^2}{k - 1}$$

But the quantity $\sum_i \bar{y}_i^2 - k\bar{\bar{y}}^2$ is merely another way of writing $\sum_i (\bar{y}_i - \bar{\bar{y}})^2$ and therefore

$$MS_{III} = n \frac{\sum_i (\bar{y}_i - \bar{\bar{y}})^2}{k - 1} \tag{9.10}$$

Thus $MS_{III}$ is *not* an expression for the variance of group averages; it represents in fact $n$ times this quantity.

From the viewpoint of the test of significance, this multiplication by $n$ presents no difficulty. Indeed, the null hypothesis to be tested asserts that all groups belong to a single population. In that case, the quantity $\sum_i (\bar{y}_i - \bar{\bar{y}})^2/(k - 1)$ measures the variance of averages of $n$ observations *from a single population*. We know that such a variance equals $1/n$ times the variance of individual determinations. Hence, in that case, the numerator of $F$, $MS_{III}$, is an estimate of

$$n\left[\frac{1}{n} V(\varepsilon)\right] = V(\varepsilon)$$

We have already seen that the denominator of $F$ is an estimate of $V(\varepsilon)$. The ratio $F$ is therefore, under the null-hypothesis, the ratio of two estimates of the same variance, as it should be.

However, if the null-hypothesis is false, the so-called *mean square between groups* is not an estimate of the true variance between group means. In a later chapter we will show how this true variance between group means can be estimated from the mean squares of the analysis of variance.

Returning to our numerical example, we find an $F$ value with 4 and 6 degrees of freedom equal to 248. In view of this result, we would be very reluctant to accept the null-hypothesis. In fact, the authors of this study concluded from the evidence resulting from these data and from other experiments that the silver samples that had not been heated to dull redness (annealed) contained dissolved oxygen and therefore yielded systematically high results for the electrochemical equivalent. Using the within-group mean square as a measure of $\sigma^2$, we obtain, after decoding:

$$\hat{\sigma} = 10^{-6}\sqrt{2039} = 0.000045$$

With an average value of the measured quantity equal to 1.118439 this represents a replication error characterized by a per cent coefficient of variation approximately equal to

$$100 \times \frac{0.000045}{1.118} = 0.0040\%$$

Apart from the estimate of replication error, the statistical analysis up to this point seems to have contributed very little in this example; a mere inspection of Table 9.1 would also have revealed the striking effect of annealing on the results. Actually, Table 9.1 suggests four specific questions: (*a*) What is the effect of annealing? (*b*) What is the effect of etching? (*c*) What is the *joint* effect of annealing *and* etching, *beyond* the sum of their individual effects? (*d*) Is there a difference between samples I and II.

The experiment was not specifically designed for the study of these questions, and the data are therefore not sufficient to obtain precise answers to them. We will nevertheless carry the analysis further in order to illustrate some of the statistical principles that are involved in situations of this type.

In Table 9.4 the 8 combinations of the 3 factors, annealing, etching, and sample, are represented by the letters *A* to *H* and displayed in diagrammatic form. A comparison of this table with Table 9.1 shows that for the conditions represented by the letters *A*, *C*, and *E* no data are available; *B*, *F*, and *H* are represented by two measurements each, *D* by one measurement, and *G* by four measurements.

**TABLE 9.4**  Effects of Three Factors

|  | Sample I | Sample II |
|---|---|---|
| Not annealed | | |
| Not etched | *A* | *B* |
| Etched | *C* | *D* |
| Annealed | | |
| Not etched | *E* | *F* |
| Etched | *G* | *H* |

Let us imagine for the moment that data in equal number were available for all *cells** in Table 9.4 and explore ways of answering questions (*a*) through (*d*). A difficulty arises from the fact that the effect of each

---

* A *cell* is the intersection of a particular row with a particular column of the table (see Section 9.3).

factor may be influenced by one or more of the other factors. Thus, the effect of etching may be different according as it is studied for annealed or unannealed samples. The way in which this problem is usually approached is to distinguish between *main effects* and *interactions*. The main effect of a factor is a measure of its effect averaged over all levels of the other factors, while the interaction between two factors is a measure of the differences observed in the effect of one of the factors when the level of the second factor is varied. These concepts may be illustrated by means of Table 9.4. The main effect of "etching" is the average difference between the results for all etched specimens and those for all non-etched specimens, i.e.,

$$\tfrac{1}{4}(C + D + G + H) - \tfrac{1}{4}(A + B + E + F)$$

The interaction between annealing and etching, which is designated as "annealing $\times$ etching" is the difference between the average effect of annealing on etched specimens and its average effect on non-etched specimens:

$$[\tfrac{1}{2}(G + H) - \tfrac{1}{2}(C + D)] - [\tfrac{1}{2}(E + F) - \tfrac{1}{2}(A + B)]$$

or

$$\tfrac{1}{2}(A + B + G + H) - \tfrac{1}{2}(C + D + E + F)$$

It is readily verified that the interactions *annealing $\times$ etching* and *etching $\times$ annealing* are conceptually identical, and therefore measured by the same quantity.

An interaction, such as between annealing and etching, can be interpreted in still a different way. It is really the *additional* effect of etching *and* annealing, *beyond* the sum of these effects taken individually in the absence of each other. This is shown as follows. The effect of etching on non-annealed specimens is:

$$\tfrac{1}{2}(C + D) - \tfrac{1}{2}(A + B)$$

the effect of annealing in the absence of etching is:

$$\tfrac{1}{2}(E + F) - \tfrac{1}{2}(A + B)$$

the effect of both annealing and etching is

$$\tfrac{1}{2}(G + H) - \tfrac{1}{2}(A + B)$$

therefore, the additional effect of both treatments, beyond the sum of their individual effect is:

$$[\tfrac{1}{2}(G + H) - \tfrac{1}{2}(A + B)]$$
$$- [\tfrac{1}{2}(C + D) - \tfrac{1}{2}(A + B) + \tfrac{1}{2}(E + F) - \tfrac{1}{2}(A + B)]$$

which is equal to

$$\tfrac{1}{2}(A + B + G + H) - \tfrac{1}{2}(C + D + E + F)$$

This last interpretation explains why interaction is taken to be a measure of "non-additivity," since it measures the amount by which the simple addition of individual effects fails to account for their joint effect.

The measures for main effects and interaction we have derived for our example are linear functions of the observations. This arises from the fact that each factor is studied at exactly two levels. Such measures are called *contrasts*. Table 9.5 lists the contrasts for four effects of interest in our problem.

**TABLE 9.5**  Some Contrasts Arising From Three Factors At Two Levels Each

| Effect studied | Contrast |
|---|---|
| Sample main effect | $\tfrac{1}{4}(A + C + E + G) - \tfrac{1}{4}(B + D + F + H)$ |
| Annealing main effect | $\tfrac{1}{4}(A + B + C + D) - \tfrac{1}{4}(E + F + G + H)$ |
| Etching main effect | $\tfrac{1}{4}(A + B + E + F) - \tfrac{1}{4}(C + D + G + H)$ |
| Annealing × etching | $\tfrac{1}{2}(A + B + G + H) - \tfrac{1}{2}(C + D + E + F)$ |

There are two additional two-factor interactions: sample × annealing and sample × etching, but these have little physical meaning. Furthermore, there exists a "three-factor interaction": "sample × annealing × etching" which we will not discuss, and which is of little interest in the problem under discussion.

We can present the information contained in a Table 9.5 in a different, and very instructive way. Let us write the symbols $A$ through $H$ as column-heads (see Table 9.6) and under each letter, the sign $(+)$ or $(-)$ according as the letter, for the effect considered, occurs with a $(+)$ or a $(-)$ sign in the corresponding formula in Table 9.5. Thus, for the Sample Main Effect in Table 9.6, $A$, $C$, $E$, and $G$ have the sign $(+)$, because they have that sign in the expression

$$\tfrac{1}{4}(A + C + E + G) - \tfrac{1}{4}(B + D + F + H)$$

while the remaining four letters have the sign $(-)$.  Tables 9.5 and 9.6

**TABLE 9.6**  Symbolic Representation of Contrasts

| Effect studied | $A$ | $B$ | $C$ | $D$ | $E$ | $F$ | $G$ | $H$ |
|---|---|---|---|---|---|---|---|---|
| Sample main effect | + | − | + | − | + | − | + | − |
| Annealing main effect | + | + | + | + | − | − | − | − |
| Etching main effect | + | + | − | − | + | + | − | − |
| Annealing × etching | + | + | − | − | − | − | + | + |

are actually identical except for omission, in the latter, of the numerical multipliers $\frac{1}{4}$ and $\frac{1}{2}$.

An interesting feature of Table 9.6 is that if the corresponding symbols in any two of the four rows are multiplied (by the usual algebraic rules) and the products added, the sum is always zero. Thus, for the two effects "sample" and "annealing × etching," we have:

$$(+1)(+1) + (-1)(+1) + (+1)(-1) + (-1)(-1) + (+1)(-1)$$
$$+ (-1)(-1) + (+1)(+1) + (-1)(+1)$$
$$= 1 - 1 - 1 + 1 - 1 + 1 + 1 - 1 = 0$$

When the sum of cross-products of corresponding elements for two effects is zero, these effects are said to be *orthogonal*. Statistically, orthogonality is a very desirable feature, because it simplifies the least squares calculations, and causes the covariance between any two contrasts to be zero. A consequence of this feature is that if the experimental errors of the observations are normally distributed and have all the same variance, the orthogonal contrasts will all be statistically independent of each other.

**TABLE 9.7**  Effect of Three Factors on Electrochemical Equivalent of Silver[a]

|  | Sample I | Sample II |
|---|---|---|
| Not annealed |  |  |
|    Not etched | — | 1, 2 |
|    Etched | — | 3 |
| Annealed |  |  |
|    Not etched | — | 4, 5 |
|    Etched | 8, 9, 10, 11 | 6, 7 |

[a] The numbers in the body of the table are the "determination numbers" of Table 9.1.

Let us now examine the contrasts that can actually be evaluated in the experiment on the electrochemical equivalent of silver. In Table 9.7, the 11 determinations, as numbered in the first column of Table 9.1, are arranged as factor-combinations.

From Table 9.7 we easily derive the contrasts shown in Table 9.8, and from these we construct the table of "multipliers" for the 11 determinations shown in Table 9.9. An examination of Table 9.9 shows that orthogonality is no longer achieved between any two rows. Furthermore, each of the four contrasts involves some but not all of the 11 determinations whereas in the orthogonal situation of Table 9.6, each contrast involved all

**TABLE 9.8**   Contrasts for Electrochemical Equivalent of Silver[a]

| Effect studied | Contrast in terms of determination numbers |
|---|---|
| Sample | $\frac{1}{4}(8 + 9 + 10 + 11) - \frac{1}{2}(6 + 7)$ |
| Annealing | $\frac{1}{4}(4 + 5 + 6 + 7) - \frac{1}{2}[\frac{1}{2}(1 + 2) + 3]$ |
| Etching | $\frac{1}{2}[3 + \frac{1}{2}(6 + 7)] - \frac{1}{4}(1 + 2 + 4 + 5)$ |
| Annealing × etching | $\frac{1}{4}(1 + 2 + 6 + 7) - \frac{1}{2}[3 + \frac{1}{2}(4 + 5)]$ |

[a] The numbers inside the parentheses represent the determination numbers of the first column of Table 9.1; they also occur in Table 9.7.

**TABLE 9.9**   Contrasts for Electrochemical Equivalent of Silver, Table of Multipliers[a]

| Effect studied | Determination number | | | | | | | | | | |
|---|---|---|---|---|---|---|---|---|---|---|---|
| | **1** | **2** | **3** | **4** | **5** | **6** | **7** | **8** | **9** | **10** | **11** |
| Sample | 0 | 0 | 0 | 0 | 0 | −2 | −2 | +1 | +1 | +1 | +1 |
| Annealing | −1 | −1 | −2 | +1 | +1 | +1 | +1 | 0 | 0 | 0 | 0 |
| Etching | −1 | −1 | +2 | −1 | −1 | +1 | +1 | 0 | 0 | 0 | 0 |
| Annealing × etching | +1 | +1 | −2 | −1 | −1 | +1 | +1 | 0 | 0 | 0 | 0 |

[a] See Table 9.7; all entries in the body of the table have been multiplied by 4, to avoid fractional numbers.

available determinations.   Consequently, the arrangement of Table 9.1 is less efficient for the study of the four effects than a completely orthogonal design of the type of Table 9.4.   Nevertheless, we can, and should, use the data for whatever they are worth, to obtain information that may be of interest.   Using the combinations listed in Table 9.8, and substituting the data for their corresponding symbols, we obtain (after proper decoding) the following numerical values for the four contrasts:

| Effect studied | Contrast |
|---|---|
| Sample | $28.75 \times 10^{-6}$ |
| Annealing | $-963. \times 10^{-6}$ |
| Etching | $-5. \times 10^{-6}$ |
| Annealing × etching | $-94. \times 10^{-6}$ |

Our next task is to derive the variance of the error for each of these contrasts.   This is easily done through application of the law of propagation of errors (Eq. 4.9).   We illustrate the calculation for the third contrast (etching). Denoting by $\sigma^2$ the error-variance for a single determination and referring to the expression for the contrast given in Table 9.8, we have for the error variance of this contrast:

$$V_e(\text{etching}) = \frac{1}{4}\left[\sigma_e^2 + \frac{2\sigma_e^2}{4}\right] + \frac{1}{16}\left[4\sigma_e^2\right] = \frac{5}{8}\sigma_e^2$$

Thus, the standard error for this contrast is $\sigma_e\sqrt{5/8}$. Now, an estimate is available for $\sigma_e$. we have found from the analysis of variance that

$$\hat{\sigma}_e = 0.000045 \quad \text{or} \quad 45 \times 10^{-6}$$

We can therefore use Student's $t$ to test the hypothesis that the effect of *etching* has any preassigned expected value. In our present problem, we are of course interested in testing whether etching has *any* effect at all; therefore the expected value to be tested is zero:

$$t = \frac{-5 \times 10^{-6} - 0}{(\sqrt{5/8})(45 \times 10^{-6})} = -0.14$$

This value fails to reach significance at any interesting level and we conclude that the data show no evidence of an effect of etching on the experimental value of the electrochemical equivalent of silver. Table 9.10 summarizes the information for all four effects.

**TABLE 9.10** Electrochemical Equivalent of Silver, Tests of Significance for Contrasts

| Effect | Contrast Observed value | Standard error | Student's $t$ |
|---|---|---|---|
| Sample | $28.75 \times 10^{-6}$ | $\sqrt{3/4}\,\sigma_e$ | $0.74$ |
| Annealing | $-963. \times 10^{-6}$ | $\sqrt{5/8}\,\sigma_e$ | $-27.06$ |
| Etching | $-5. \times 10^{-6}$ | $\sqrt{5/8}\,\sigma_e$ | $-0.14$ |
| Annealing $\times$ etching | $-94. \times 10^{-6}$ | $\sqrt{5/8}\,\sigma_e$ | $-2.64$ |

These results can be considered as a refinement of the over-all analysis of variance in Table 9.3. They show, as might have been expected, that the significant differences between groups of measurements are due to a very large extent to the effect of annealing. However, the results in Table 9.10 also suggest that the total effect of annealing and etching is greater than that due to annealing only, although etching alone seems to have little effect. This conclusion must be considered as tentative and subject to further experimental verification. A full investigation of the effects of annealing and etching would require more elaborate experimental work than is shown in Table 9.1.

## 9.3  TWO-WAY CLASSIFICATIONS

When data are classified simultaneously according to two criteria, they are exhibited most advantageously in the form of a rectangular array. The rows will represent the different categories (called *levels*) of one of the criteria of classification and the columns the levels of the other criterion. An illustration is shown in Table 9.11, representing measurements of the thickness of the wearing surface of floor coverings, in a study involving 4 materials and 7 participants (2).*   Here the two criteria of classification are materials, with 4 levels, and participants, with 7 levels.   The materials are represented by the columns, the participants by the rows.   The purpose of the study was to determine the reproducibility of the method of measurement when carried out by different operators.   We will denote the 28 places in the table as *cells*.   The $i, j$ cell is the intersection of row $i$ with column $j$, where $i$ and $j$, in our example, denote the numbers 1 through 7, and 1 through 4 respectively.   Each cell is occupied by a measured value represented by $y_{ij}$.

**TABLE 9.11**   Thickness of Wearing Surface, Mils

|  | Material | | | |
| --- | --- | --- | --- | --- |
| Participant | Z | Y | X | W |
| A | 5.50 | 13.40 | 19.50 | 20.40 |
| B | 5.03 | 11.75 | 18.90 | 24.13 |
| C | 6.75 | 13.65 | 19.65 | 23.45 |
| D | 5.48 | 11.74 | 19.82 | 22.51 |
| E | 5.30 | 10.02 | 15.12 | 19.36 |
| G | 5.85 | 13.55 | 18.65 | 23.40 |
| H | 5.26 | 12.17 | 18.57 | 21.45 |

The analysis of a set of data such as Table 9.11 consists in setting up several mathematical models for comparison and choosing the one which is most likely to represent the data.   The parameters of this model are then estimated by least squares and the error variance is estimated from the residuals.   Estimates of uncertainty for the parameters, as well as

---

* Actually, $G$ and $H$ represent the same experimenter, using two different instruments.

confidence intervals may, if desired, also be obtained by the methods explained in Chapter 7.

The models usually tried on data classified in two ways are as follows:

*Model I.* Each cell represents a sample from a different population. Symbolically:

$$y_{ij} = \mu_{ij} + \varepsilon_{ij} \tag{9.11}$$

Here $\mu_{ij}$ is the population mean corresponding to cell $(ij)$, and $\varepsilon_{ij}$ is the "error," i.e., the amount by which the measurement $y_{ij}$ differs from its mean $\mu_{ij}$.

In this general model, no assumptions are made regarding the relationships between the $\mu_{ij}$ corresponding to different cells. However, it is often assumed that the standard deviation of $\varepsilon_{ij}$ is the same for all cells. If this is the case, the populations represented by the different cells are said to be *homoscedastic*. Otherwise they are said to be *heteroscedastic*. In the latter case, it is sometimes possible, through a change in the *scale* of the measurements, to achieve homoscedasticity. A familiar example of a transformation of scale is the expression of hydrogen–ion concentrations as pH-values. Here the transformation of scale is expressed by the relation

$$\text{pH} = -\log_{10} c \tag{9.12}$$

or, if $y$ represents the original scale (concentration $c$), and $z$ the new scale (pH), by the transformation:

$$z = -\log_{10} y \tag{9.13}$$

Why a transformation of scale may achieve homoscedasticity will be discussed in a subsequent section. Until then, we will assume that the populations, as given, are homoscedastic.

*Model II.* All cells in the *same row* represent samples from the same population, but the populations corresponding to *different rows* may be different (i.e., they may have different mean values). Symbolically:

$$y_{ij} = \mu_i + \varepsilon_{ij} \tag{9.14}$$

*Model III.* All cells in the *same column* represent samples from the same population, but the populations corresponding to *different columns* may be different (i.e., they may have different mean values). Symbolically:

$$y_{ij} = \mu_j + \varepsilon_{ij} \tag{9.15}$$

*Model IV.* All cells of the entire table represent samples from a single population. Symbolically:

$$y_{ij} = \mu + \varepsilon_{ij} \tag{9.16}$$

Before proceeding with the analysis, let us examine the four models in terms of the problem exemplified in Table 9.11.   Model I states that the value obtained by a given participant for a given material bears no relation to the value obtained by the same participant for any other material, nor to that obtained by any other participant for any material. It ignores therefore the possible existence of any pattern among the populations corresponding either to the same participant or to the same material.   Clearly this model is too general to be of any practical value. To render it more useful, some assumptions must be introduced restricting its generality.

Model II states that for any given participant, the same population of values is obtained for all materials.   For our example, this is an absurd hypothesis, since the materials $Z$, $Y$, $X$, and $W$ were chosen to be different. However, if $Z$, $Y$, $X$, and $W$ had represented, for example, the same material made by four different manufacturers, Model II might be quite plausible.

Model III states that for any given material, all participants obtain results belonging to one and the same population.   This hypothesis is certainly not absurd, but experience shows it to be rather unlikely.

Finally, Model IV states that no systematic differences exist, either between participants or between materials.   The latter part of this state-ment is utterly unacceptable for our example, but, as has been pointed out in connection with Model II, a slight modification of the problem would make such a model at least conceptually possible.

It is customary to write Models I, II, and III in a slightly different form. Whichever of these three models we adopt, we can always represent the average of the population means of all cells by the symbol $\mu$, and express each individual cell population mean by its deviation from $\mu$.   Thus, we obtain:

$$\text{Model I:} \qquad y_{ij} = \mu + \nu_{ij} + \varepsilon_{ij} \qquad\qquad (9.17)$$

$$\text{Model II:} \qquad y_{ij} = \mu + \rho_i + \varepsilon_{ij} \qquad\qquad (9.18)$$

$$\text{Model III:} \qquad y_{ij} = \mu + \gamma_j + \varepsilon_{ij} \qquad\qquad (9.19)$$

$$\text{Model IV:} \qquad y_{ij} = \mu + \varepsilon_{ij} \qquad\qquad (9.20)$$

in which $\nu_{ij}$, $\rho_i$, and $\gamma_i$ are defined by:

$$\nu_{ij} = \mu_{ij} - \mu \qquad\qquad (9.21)$$

$$\rho_i = \mu_i - \mu \qquad\qquad (9.22)$$

$$\gamma_j = \mu_j - \mu \qquad\qquad (9.23)$$

It is easily verified that the definition of $\mu$ as the average of all the population means of the table implies that the averages of either all $\nu_{ij}$, all $\rho_i$, or all $\gamma_j$, must be zero. Thus:

$$\bar{\nu} = 0 \tag{9.24}$$

$$\bar{\rho} = 0 \tag{9.25}$$

$$\bar{\gamma} = 0 \tag{9.26}$$

We now note that any one of the last three models can be embedded in Model I. For example, Model II is a special case of I, if in the latter we make all $\nu_{ij}$ of the $i$th row equal to each other and represent their common value by $\rho_i$.

It would however be futile to attempt to test Models II, III, and IV against Model I since the latter involves as many parameters as there are data (one parameter for each cell) and leaves therefore no degrees of freedom for the estimation of error. We must therefore modify Model I in such a way that: (a) fewer parameters are to be estimated; and (b) the modified form still contains Models II, III, and IV as special cases. This can be done in a number of ways. The most commonly used consists in introducing the so-called *additive* model:

$$\textit{Model A}: \qquad y_{ij} = \mu + \rho_i + \gamma_j + \varepsilon_{ij} \tag{9.27}$$

Here, as before we assume $\bar{\rho} = 0$ and $\bar{\gamma} = 0$.

Let us now count the degrees of freedom available for error in each of the five models. They are given in Table 9.12 for the data exhibited in Table 9.11.

**TABLE 9.12**

| Model | Equation | Number of parameters | Residual degrees of freedom |
|-------|----------|----------------------|------------------------------|
| I | $y_{ij} = \mu + \nu_{ij} + \varepsilon_{ij}$ | 28 | 0 |
| A | $y_{ij} = \mu + \rho_i + \gamma_j + \varepsilon_{ij}$ | 10 | 18 |
| II | $y_{ij} = \mu + \rho_i + \varepsilon_{ij}$ | 7 | 21 |
| III | $y_{ij} = \mu + \gamma_j + \varepsilon_{ij}$ | 4 | 24 |
| IV | $y_{ij} = \mu + \varepsilon_{ij}$ | 1 | 27 |

This table is easily derived. For example, for Model II the parameters are $\mu$, and the 7 $\rho_i$. However, since $\bar{\rho} = 0$, only 6 of the 7 $\rho_i$ values are independent. Thus, the number of parameters is $1 + 6 = 7$. Similarly, for Model $A$, there are $1 + 6 + 3 = 10$ parameters. The residual degrees of freedom are then obtained by subtracting the number of parameters from 28.

The least squares fits are given in Table 9.13.    Model I is omitted since, for it, $\hat{y}_{ij} = y_{ij}$ and all the residuals are zero.    The least squares calculations are extremely simple, in all four cases.    In Model IV, the only parameter is $\mu$.    Its estimate, $\hat{\mu}$, is simply the average of all 28 observations (the *grand average*).    This same estimate also applies, for $\mu$, in all other models.

**TABLE 9.13**

|  | Data | A | | II | | III | | IV | |
|---|---|---|---|---|---|---|---|---|---|
|  | $y$ | $\hat{y}$ | $d$ | $\hat{y}$ | $d$ | $\hat{y}$ | $d$ | $\hat{y}$ | $d$ |
| A | 5.50 | 5.64 | −0.14 | 14.70 | −9.20 | 5.60 | −0.10 | 14.66 | −9.16 |
| B | 5.03 | 5.89 | −0.86 | 14.95 | −9.92 | 5.60 | −0.57 | 14.66 | −9.63 |
| C | 6.75 | 6.82 | −0.07 | 15.88 | −9.13 | 5.60 | +1.15 | 14.66 | −7.91 |
| D | 5.48 | 5.83 | −0.35 | 14.89 | −9.41 | 5.60 | −0.12 | 14.66 | −9.18 |
| E | 5.30 | 3.39 | +1.91 | 12.45 | −7.15 | 5.60 | −0.30 | 14.66 | −9.36 |
| G | 5.85 | 6.30 | −0.45 | 15.36 | −9.51 | 5.60 | +0.25 | 14.66 | −8.81 |
| H | 5.26 | 5.30 | −0.04 | 14.36 | −9.10 | 5.60 | −0.34 | 14.66 | −9.40 |
| A | 13.40 | 12.37 | +1.03 | 14.70 | −1.30 | 12.33 | +1.07 | 14.66 | −1.26 |
| B | 11.75 | 12.62 | −0.87 | 14.95 | −3.20 | 12.33 | −0.58 | 14.66 | −2.91 |
| C | 13.65 | 13.54 | +0.11 | 15.88 | −2.23 | 12.33 | +1.32 | 14.66 | −1.01 |
| D | 11.74 | 12.56 | −0.82 | 14.89 | −3.15 | 12.33 | −0.59 | 14.66 | −2.92 |
| E | 10.02 | 10.12 | −0.10 | 12.45 | −2.43 | 12.33 | −2.31 | 14.66 | −4.64 |
| G | 13.55 | 13.03 | +0.52 | 15.36 | −1.81 | 12.33 | +1.22 | 14.66 | −1.11 |
| H | 12.17 | 12.03 | +0.14 | 14.36 | −2.19 | 12.33 | −0.16 | 14.66 | −2.49 |
| A | 19.50 | 18.65 | +0.85 | 14.70 | +4.80 | 18.60 | +0.90 | 14.66 | +4.84 |
| B | 18.90 | 18.90 | 0 | 14.95 | +3.95 | 18.60 | +0.30 | 14.66 | +4.24 |
| C | 19.65 | 19.82 | −0.17 | 15.88 | +3.77 | 18.60 | +1.05 | 14.66 | +4.99 |
| D | 19.82 | 18.83 | +0.99 | 14.89 | +4.93 | 18.60 | +1.22 | 14.66 | +5.16 |
| E | 15.12 | 16.40 | −1.28 | 12.45 | +2.67 | 18.60 | −3.48 | 14.66 | +0.46 |
| G | 18.65 | 19.31 | −0.66 | 15.36 | +3.29 | 18.60 | +0.05 | 14.66 | +3.99 |
| H | 18.57 | 18.31 | +0.26 | 14.36 | +4.21 | 18.60 | −0.03 | 14.66 | +3.91 |
| A | 20.40 | 22.14 | −1.74 | 14.70 | +5.70 | 22.10 | −1.70 | 14.66 | +5.74 |
| B | 24.13 | 22.30 | +1.73 | 14.95 | +9.18 | 22.10 | +2.03 | 14.66 | +9.47 |
| C | 23.45 | 23.32 | +0.13 | 15.88 | +7.57 | 22.10 | +1.35 | 14.66 | +8.79 |
| D | 22.51 | 22.33 | +0.18 | 14.89 | +7.62 | 22.10 | +0.41 | 14.66 | +7.85 |
| E | 19.36 | 19.89 | −0.53 | 12.45 | +6.91 | 22.10 | −2.74 | 14.66 | +4.70 |
| G | 23.40 | 22.81 | +0.59 | 15.36 | +8.04 | 22.10 | +1.30 | 14.66 | +8.74 |
| H | 21.45 | 21.81 | −0.36 | 14.36 | +7.09 | 22.10 | −0.65 | 14.66 | +6.79 |
| $\sum d^2$ | | 18.24 | | 1127.75 | | 46.59 | | 1156.08 | |

In Model II, the estimate, $\hat{\rho}_i$, for any particular row is obtained by subtracting $\hat{\mu}$ from the average of all the observations occurring in that row. In Model III, the estimate, $\hat{\gamma}_j$, for any particular column is obtained by subtracting $\hat{\mu}$ from the average of all the observations occurring in that column. Finally, in Model $A$, the estimates $\hat{\rho}_i$ and $\hat{\gamma}_j$ are identically the same as those obtained in Models II and III, respectively. Thus all parameter estimates are obtained by calculating the *marginal* averages (i.e., the row and column averages) and the grand average of the tabulated data.

The aim of the analysis is to test Models II, III, and IV, using the more general Model $A$ as a reference model. This is legitimate because, as has been explained in Sections 8.3 and 8.4, the more general model yields a valid estimate of error, even when a more restrictive hypothesis applies.

The sums of squares of residuals for Models $A$, II, III, and IV, together with their associated degrees of freedom, are listed in Table 9.14. From them, the reductions in the sum of squares appropriate for testing various hypotheses are derived, as shown in Table 9.15. This table is interpreted in the following way. The reduction in the sum of squares for rows (or participants) is due to the introduction of the $\rho_i$ terms in the model (see Table 9.12); the corresponding mean square is therefore the proper numerator for an $F$ test of the hypothesis that the $\rho_i$ are all zero. Similarly, the mean square for columns (or materials) is the numerator for an $F$ test of the hypothesis that all $\gamma_j$ are zero. In either case, the denominator of $F$ is the mean square for the reference model, $A$. The latter is termed *interaction*, or more correctly: *row by column interaction*. The term interaction has already been introduced in Section 9.2. Its precise meaning in the present context will be explained shortly. First, however, we must observe a very important feature of the analysis: namely, that the sums of squares for rows, columns, and interaction, when added together, yield exactly* the sum of squares marked "total." Thus, we have:

$$SS_{IV} = (SS_{III} - SS_A) + (SS_{II} - SS_A) + SS_A$$

which can be written

$$SS_{IV} - SS_{III} = SS_{II} - SS_A \qquad (9.28)$$

Now, examination of Table 9.12 shows that $SS_{IV} - SS_{III}$ corresponds to the introduction of the $\gamma_j$ into the model. Similarly, $SS_{II} - SS_A$ also corresponds to the introduction of the $\gamma_j$ into the model, but in the latter case the $\rho_i$ are already present, whereas in the former, the $\rho_i$ are not included in the model. What Eq. 9.28 means is therefore simply that the

---

* The slight discrepancy is due to rounding errors in the computations.

**TABLE 9.14**

| Model | Residuals | |
|:---:|:---:|:---:|
|  | DF | SS |
| IV | 27 | 1156.08 |
| III | 24 | 46.59 |
| II | 21 | 1127.75 |
| $A$ | 18 | 18.24 |

**TABLE 9.15**

| Source of variation | Origin | DF[a] | SS[b] | MS |
|:---|:---:|:---:|:---:|:---:|
| Total | IV | 27 | 1156.08 | 42.82 |
| Rows (participants) | III $- A$ | 6 | 28.35 | 4.72 |
| Columns (materials) | II $- A$ | 3 | 1109.51 | 369.84 |
| Interaction | $A$ | 18 | 18.24 | 1.01 |

*F tests*:

  for hypothesis $\rho_i = 0$ (all $i$):     $F = \dfrac{4.72}{1.01} = 4.67$

  for hypothesis $\gamma_j = 0$ (all $j$):     $F = \dfrac{369.84}{1.01} = 366$

---

  [a] For "Rows" and "Columns," a more correct term would be "reduction in the degrees of freedom."

  [b] For "Rows" and "Columns," a more correct term would be "reduction in the sum of squares."

presence of "column-terms" $\gamma_j$ has no effect whatsoever on the "row-terms" $\rho_i$. If Eq. 9.28 is rewritten in the equivalent form:

$$SS_{IV} - SS_{II} = SS_{III} - SS_A \qquad (9.29)$$

it shows, in an analogous manner, that the presence of row-terms has no effect on the column-terms. We have already mentioned that this condition occurs with certain types of data and have referred to it as *orthogonality*. In the present case, the orthogonal character of the row and column effects is also apparent from the fact, previously mentioned, that the *estimates* for $\rho_i$ are the same, whether or not the $\gamma_j$ are included in the model, and vice-versa.

The $F$ values calculated in Table 9.15 are both significant: that for "rows" on the 0.5 per cent level and that of "columns" at a much higher level of significance. The latter is hardly surprising since the hypothesis that the four materials are identical is absurd for the present data. The significance of the "row effects" is more interesting; it points to the existence of systematic differences between the results obtained by different participants. That this is indeed so can be seen, more directly perhaps, from an examination of the residuals in Table 9.13. Consider those for Model III. According to this model, no row effects would exist. But the residuals for this model do not show the randomness expected in a good fit: for example, the residuals for participant $C$ are all positive; so are the residuals for participant $G$. On the other hand, the residuals for participants $E$ and $H$ are all negative. Thus, $C$ and $G$ tend to get high results, while $E$ and $H$ tend to get low results. This pattern has disappeared in the additive model $A$, which allows for systematic differences between participants. The non-randomness of the residuals in Models II and IV needs hardly to be mentioned; both these models are, of course, utterly unacceptable.

An interesting question now arises. It has been established that Models II, III, and IV are all significantly worse than the additive model $A$. But is the latter the best possible? In order to be a good fit, this model must yield a set of residuals that is devoid of systematic patterns of any kind. Let us rewrite the residuals of Model $A$ in the original two-way array, as shown in Table 9.16. It may be noted that the marginal sums

TABLE 9.16   Residuals of Model $A$

|   | $Z$ | $Y$ | $X$ | $W$ |
|---|------|------|------|------|
| $A$ | −0.14 | +1.03 | +0.85 | −1.74 |
| $B$ | −0.86 | −0.87 | 0 | +1.73 |
| $C$ | −0.07 | +0.11 | −0.17 | +0.13 |
| $D$ | −0.35 | −0.82 | +0.99 | +0.18 |
| $E$ | +1.91 | −0.10 | −1.28 | −0.53 |
| $G$ | −0.45 | +0.52 | −0.66 | +0.59 |
| $H$ | −0.04 | +0.14 | +0.26 | −0.36 |

of the residuals are all zero, both row-wise and column-wise (except for small rounding errors). This is a characteristic of the additive model and would be true for any set of two-way data. On the other hand, we note that the residuals for material $Z$ are all negative, except for participant $E$, for which its magnitude is quite large. But the three other residuals for participant $E$ are all negative. This would indicate that the $\rho$ value for participant $E$ (of which the estimate is equal to $-2.21$) is insufficient to account for the high result obtained by this participant for material $Z$. Thus, the systematic effect of this participant would not be the same for all materials. There would therefore exist an *interaction* between the systematic effects of participants and the differences between materials. Model $A$ fails to account for such an interaction, since it allows only for a *constant* systematic effect for each participant, $\rho_i$, which is presumably the same for all materials.   It is for this reason that the sum of squares of the residuals yielded by Model $A$ are denoted as *row by column interaction* or, in the present case, as *participant-material-interaction*.

We might be tempted to dismiss the observed interaction effects as the mere interplay of chance effects.   Indeed this possibility always exists, and a mere examination of the residuals, such as we have just made, is rather too subjective to allow us to evaluate this possibility.   Is there a statistical technique that provides a more objective criterion?

The theory of the general linear hypothesis suggests that a test of the additive model, $A$, could be made if we succeeded in embedding this model within one of greater generality, i.e., one that allowed for a larger number of parameters.   Such a model is discussed in the following section.

## 9.4  A LINEAR MODEL

It is useful to give Model $A$ a geometric interpretation.   Let the "true values" of the thickness of wearing surface, for the four materials, $Z$, $Y$, $X$, and $W$ be represented by these same letters and let us plot these values* on the abscissa axis of Fig. 9.1.   On the ordinate we plot the values obtained for these four materials by participant $A$, i.e., 5.50, 13.40, 19.50, and 20.40.   We expect the points formed by the four pairs $(Z, 5.50)$, $(Y, 13.40)$, $(X, 19.50)$, and $(W, 20.40)$ to lie on a reasonably smooth curve. Any departures from a smooth curve must be due to experimental error.

We could, of course, make a similar plot for each of the other six participants.   If we plot the curves for all participants on the same graph, the values of the abscissa axis (the true values) are obviously the same for all participants, so that all points for material $Z$ will lie on the same vertical line, and similarly for the other materials.   What does Model $A$ imply with regard to these plots?

For participant $A$, the model reads:

$$y_{Aj} = \mu + \rho_A + \gamma_j + \varepsilon_{Aj} \tag{9.30}$$

For participant $B$, we have similarly:

$$y_{Bj} = \mu + \rho_B + \gamma_j + \varepsilon_{Bj} \tag{9.31}$$

Subtracting Eq. 9.31 from Eq. 9.30, we obtain:

$$y_{Aj} - y_{Bj} = (\rho_A - \rho_B) + (\varepsilon_{Aj} - \varepsilon_{Bj}) \tag{9.32}$$

* These true values are generally unknown.   The plot in question is a conceptual one, introduced only for the purpose of demonstrating the properties of the additive and the linear model.

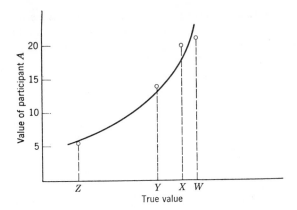

**Fig. 9.1**   The experimental values from any participant are functionally related to the true values; the scatter is due to experimental error.

Now $y_{Aj} - y_{Bj}$ represents the vertical distance between the points for $A$ and $B$, for material $j$. According to Eq. 9.32 this distance is equal to $\rho_A - \rho_B$, plus an error of measurement ($\varepsilon_{Aj} - \varepsilon_{Bj}$). But $\rho_A - \rho_B$ does not depend on the material $j$. Thus, *apart from experimental error*, the vertical distance between the points for $A$ and $B$ is the same for all materials. In other words, the smooth lines corresponding to participants $A$ and $B$

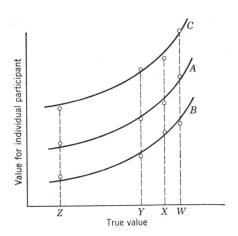

**Fig. 9.2**   The values obtained by different participants are represented by different curves.

are parallel to each other, except for the effects of experimental error. A similar situation applies to any two participants. Consequently, Model $A$ implies that the smooth lines corresponding to the 7 participants constitute a set of 7 parallel lines, except for the effects of experimental errors. Figure 9.2 illustrates this situation. It is seen that Model $A$, the additive model, is essentially an assumption of parallelism. Any departure from strict parallelism must, in this model, be due solely to experimental error. We now recall that the residuals of this model were denoted as *interaction*. Since the residuals reflect experimental errors, the interaction effects in an additive model may be interpreted as the *random* fluctuations that cause departure from strict parallelism.

There is no impelling reason for assuming that the lines for the various participants must necessarily be parallel. There exist many situations in which the systematic differences between two participants tend to increase with the magnitude of the measured property. In such cases, the lines for various participants would tend to "fan out" toward the right side of the graph. The problem is to construct a model representing such a state of affairs. Such a model is indeed available (3, 4, 5); it is represented by the following equation:

$$\text{Model } L: \qquad y_{ij} = \mu + \rho_i + \beta_i \gamma_j + \varepsilon_{ij} \qquad (9.33)$$

The symbol $L$ stands for *linear*. We will see below why this relation may be denoted as a *linear model*. Equation 9.33 may be written in the alternative form:

$$y_{ij} = \mu_i + \beta_i \gamma_j + \varepsilon_{ij} \qquad (9.34)$$

by making use of relation 9.22, by which:

$$\mu_i = \mu + \rho_i$$

The symbols $\mu$, $\rho_i$, and $\gamma_j$ retain the same meaning as in the additive model. The new symbol, $\beta$, having the subscript $i$, is associated with the *rows* of the table. Thus, in the example of Table 9.11 each participant is now characterized by *two* parameters, $\rho_i$ and $\beta_i$; each material is still characterized by a single parameter, $\gamma_j$.

Consider participants $A$ and $B$, for which

$$y_{Aj} = \mu + \rho_A + \beta_A \gamma_j + \varepsilon_{Aj} \qquad (9.35)$$

$$y_{Bj} = \mu + \rho_B + \beta_B \gamma_j + \varepsilon_{Bj} \qquad (9.36)$$

Subtracting Eq. 9.36 from Eq. 9.35, we obtain:

$$y_{Aj} - y_{Bj} = (\rho_A - \rho_B) + (\beta_A - \beta_B)\gamma_j + (\varepsilon_{Aj} - \varepsilon_{Bj}) \qquad (9.37)$$

Thus, in our present model, the vertical distance between the results for participants $A$ and $B$ is no longer the same for all materials. Quite apart from the effect of experimental error, $\varepsilon_{Aj} - \varepsilon_{Bj}$, the distance depends *systematically* on $\gamma_j$, because of the term $(\beta_A - \beta_B)\gamma_j$. Hence, the lines for the various participants are no longer parallel. The interaction, when defined as before in terms of residuals from the *additive model*, is now no longer due to random error alone; it contains a systematic component, due to the term $\beta_i\gamma_j$.

To see this more clearly, let us rewrite Eq. 9.33 in the following equivalent form:

$$y_{ij} = \mu + \rho_i + \gamma_j + [(\beta_i - 1)\gamma_j + \varepsilon_{ij}] \qquad (9.38)$$

Comparing this relation with Eq. 9.27, we see that the error term in the latter has been replaced by the sum of two terms: random error and the systematic, "non-additive" component $(\beta_i - 1)\gamma_j$. Thus, if for a given set of data, the correct model is 9.38, but ignoring this fact, we fit to them the additive model 9.27, we will obtain a set of residuals that will not show a purely random pattern; each residual will actually reflect, in addition to random error, the effect of the term $(\beta_i - 1)\gamma_j$.

A correct least squares fit of Eq. 9.33, or its equivalent forms, Eqs. 9.34 or 9.38, is not as simply obtained as in the case of the models previously discussed. This is due to the presence of the product term $\beta_i\gamma_j$; both factors in this term are parameters to be estimated from the data. Thus the model is not linear *in the parameters*. This difficulty can be circumvented by estimating $\rho_i$ and $\gamma_j$ exactly as in the additive model, and then estimating $\beta_i$ as though $\gamma_j$ were a known constant. Statistically, one obtains in this fashion a "conditional" fit, but it has been shown (5) that the analysis of variance corresponding to this fit remains valid. From the practical viewpoint the conditional fit is generally quite satisfactory; the larger the number of rows, the more it approaches the rigorous least squares fit.

We have introduced the linear model expressed by Eq. 9.33 to obtain a better fit in situations in which the residuals of the additive model fail to show a random pattern. In order to determine whether for any particular set of data, the linear model is a significantly better fit than the additive model, an analysis of variance is made, and the theory of the general linear hypothesis applied. Use is made of the fact that the additive model can be embedded in the linear model. Indeed the additive model is simply a special case of the linear model, with all $\beta_i$ equal to unity.

The mechanics of the test, illustrated for the data of Table 9.11, are as follows.

First, all $\hat{\rho}_i$ and $\hat{\gamma}_j$ are determined as for the additive model. Since this has already been done, no new calculations are involved so far. Note that in particular the $\hat{\gamma}_j$ are the averages of the columns of Table 9.11 minus the grand average $\hat{\mu}$. Their values are as follows.

| Material | $Z$ | $Y$ | $X$ | $W$ |
|----------|-----|-----|-----|-----|
| $\hat{\gamma}_j$ | $-9.061$ | $-2.330$ | $+3.945$ | $+7.444$ |

Next, we make the important observation that according to Eq. 9.34, for any given $i$ (i.e., for any given row), $y_{ij}$ is a *linear function* of $\gamma_j$ with an intercept equal to $\mu_i$ and a slope equal to $\beta_i$. It is for this reason that we have called this model a *linear* one.

An estimate for $\mu_i$ is obtained at once; $\hat{\mu}_i$ is simply the average of all observations in row $i$. To obtain an estimate of $\beta_i$, we may use the general theory presented in Section 7.5. A simpler method is discussed in Chapters 11 and 12 when the following formula (Eq. 11.4) is introduced:

$$\beta_i = \frac{\sum_j y_{ij}\hat{\gamma}_j}{\sum_j \hat{\gamma}_j{}^2} \tag{9.39}$$

For example, for the first row (participant $A$) we have the table

| $\hat{\gamma}_j$ | $-9.061$ | $-2.330$ | $+3.945$ | $+7.444$ | Average* 0 |
|------------------|----------|----------|----------|----------|------------|
| $y_{Aj}$ | 5.50 | 13.40 | 19.50 | 20.40 | 14.700 |

Thus we obtain for participant $A$ ($i = 1$):

$$\hat{\mu}_A = 14.700$$

$$\hat{\beta}_A = \frac{(5.50)(-9.061) + \cdots + (20.40)(7.444)}{(-9.061)^2 + \cdots + (7.444)^2} = 0.932$$

The remaining rows are similarly treated. Note that the denominator of $\hat{\beta}_i$ is the same for all rows and has to be computed only once. In fact, this denominator is also equal (except for rounding errors) to the sum of squares for columns, divided by the number of participants (see Table 9.15); in the present case:

$$\sum_j \hat{\gamma}_j{}^2 = \frac{1109.51}{7} = 158.50$$

Table 9.17 gives the numerical values obtained for all parameters: $\rho_i$, $\beta_i$, and $\gamma_j$.

---

* Theoretically, the average of the $\hat{\gamma}_j$ is zero. A small rounding error is present in this case.

**TABLE 9.17**

Estimates of parameters, linear model

| Participant | $\hat{\rho}_i$ | $\hat{\beta}_i$ | Material | $\hat{\gamma}_j$ |
|:---:|:---:|:---:|:---:|:---:|
| A | + 0.044 | 0.932 | Z | − 9.061 |
| B | + 0.297 | 1.144 | Y | − 2.330 |
| C | + 1.219 | 1.004 | X | + 3.945 |
| D | + 0.232 | 1.065 | W | + 7.444 |
| E | − 2.206 | 0.835 | | |
| G | + 0.707 | 1.030 | Average | 0 |
| H | − 0.293 | 0.990 | | |
| | | | $\hat{\mu} = 14.656$ | |
| Average | 0 | 1.000 | | |

**TABLE 9.18**

Residuals, linear model

| Participant | Z | Y | X | W |
|:---:|:---:|:---:|:---:|:---:|
| A | − 0.75 | 0.87 | 1.12 | − 1.24 |
| B | 0.44 | − 0.54 | − 0.56 | 0.66 |
| C | − 0.03 | 0.11 | − 0.19 | 0.10 |
| D | 0.24 | − 0.67 | 0.73 | − 0.30 |
| E | 0.42 | − 0.48 | − 0.63 | 0.69 |
| G | − 0.18 | 0.59 | − 0.78 | 0.37 |
| H | − 0.13 | 0.11 | 0.30 | − 0.28 |

Table 9.18 lists the residuals for the linear model. Finally, the analysis of variance in Table 9.19 allows us to test the significance of the improvement, if any, of the fit by the linear model over that of the additive model.

**TABLE 9.19**

| Model | Residuals DF | SS | MS |
|:---|:---:|:---:|:---:|
| Additive | 18 | 18.24 | 1.01 |
| Linear | 12 | 9.15 | 0.76 |
| Difference | 6 | 9.09 | 1.52 |

$$F = \frac{1.52}{0.76} = 2.00$$

The residual degrees of freedom for the linear model are readily obtained. We first observe that the average of the $\hat{\beta}_i$ is exactly unity (this can be proved mathematically). Thus, only 6 of the 7 $\hat{\beta}_i$ values are independent parameters. The total number of parameters estimated in the linear model is thus: 1 for $\mu$, 6 for $\rho_i$, 6 for the $\beta_i$, and 3 for the $\gamma_j$. This leaves $28 - 1 - 6 - 6 - 3 = 12$ degrees of freedom for the residuals, as indicated in Table 9.19. The $F$ value is not significant, even at the 10 per cent level. We conclude that there is not sufficient evidence for considering the linear model as a better fit than the additive model for the data in question. Nevertheless, both the $\hat{\rho}_i$ and the $\hat{\beta}_i$ values for participant $E$ are conspicuously low. Actually, this participant had had no previous experience with this measuring process. The linear model calculations have helped us to single out this participant with greater definiteness.

The sum of squares labeled "Difference" in Table 9.19 is, of course, due to the introduction of the $\beta_i$ in the model. The corresponding degrees of freedom can therefore be found directly as the number of $\beta$ values less one, in this case 6. The corresponding sum of squares can also be determined directly by means of the formula:

$$\text{SS}_\beta = \sum_i (\hat{\beta}_i - 1)^2 \sum_j \hat{\gamma}_j^2 \qquad (9.40)$$

We will return to the linear model when we deal with interlaboratory studies of test methods in a more systematic way. We will also encounter it in the fitting of families of curves. We will see that, unlike the present example, there exist many situations in which the linear model provides a much better fit, and a more natural interpretation of the data, than a simple additive model.

Finally, it is important to note that by simply transposing the rows of a table of data for its columns and vice-versa, a fit of the data by the linear model can be tried in terms of the *rows* of the original table (the *columns* of the transposed table). Thus a set of data that cannot be fitted adequately by the linear model as given in this chapter may still give a satisfactory fit by this model after transposition of rows and columns.

## 9.5  NOTES ON ANALYSIS AND COMPUTATIONS

It is useful to incorporate the analysis pertaining to the linear model, as shown in Table 9.19, with the analysis of variance shown in Table 9.15. The complete table is shown in Table 9.20.

This table shows more clearly the breakdown of the interaction sum of

**TABLE 9.20**

| Source of variation | DF | SS | MS |
|---|---|---|---|
| Total | 27 | 1156.08 | 42.82 |
| Rows (participants) | 6 | 28.35 | 4.72 |
| Columns (materials) | 3 | 1109.51 | 369.84 |
| Interaction (additive model) | 18 | 18.24 | 1.01 |
| Slopes | 6 | 9.09 | 1.52 |
| Residuals (linear model) | 12 | 9.15 | 0.76 |

squares into its two components, in accordance with the linear model: (a) the "slope" effect, corresponding to the term

$$(\beta_i - 1)\gamma_j$$

and (b) the "residuals" term, due to random error.

From a set of data such as Table 9.11 an analysis of variance table can be directly calculated, using formulas that obviate the calculation of residuals for several models. Briefly, the computations, for the general case, are as follows.

Denote by $y_{ij}$ the observation in cell $i, j$. Let the number of rows be represented by $p$ and the number of columns by $q$.

Denote by $R_i$ the *sum* of all observations in row $i$, and by $C_j$, the *sum* of all observations in column $j$. Let $G$ represent the *grand sum* (i.e., the sum of all $p \times q$ observations). Then the analysis of variance is given in Table 9.21, in which the sums of squares are calculated in accordance with Eqs. 9.41 through 9.46, and each mean square is the quotient of the corresponding sum of squares by the associated degrees of freedom.

**TABLE 9.21**

| Source of Variation | DF | SS | MS[a] |
|---|---|---|---|
| Total | $pq - 1$ | $SS_T$ | $MS_T$ |
| Rows | $p - 1$ | $SS_R$ | $MS_R$ |
| Columns | $q - 1$ | $SS_C$ | $MS_C$ |
| Interaction | $(p - 1)(q - 1)$ | $SS_{RC}$ | $MS_{RC}$ |
| Slopes | $p - 1$ | $SS_\beta$ | $MS_\beta$ |
| Residuals | $(p - 1)(q - 2)$ | $SS_\varepsilon$ | $MS_\varepsilon$ |

[a] $MS = SS/DF$ for each source of variation

Formulas for sums of squares:

$$SS_T = \sum_i \sum_j y_{ij}{}^2 - \frac{G^2}{pq} \tag{9.41}$$

$$SS_R = \sum_i \frac{R_i{}^2}{q} - \frac{G^2}{pq} \tag{9.42}$$

$$SS_C = \sum_j \frac{C_j{}^2}{p} - \frac{G^2}{pq} \tag{9.43}$$

$$SS_{RC} = SS_T - SS_R - SS_C \tag{9.44}$$

$$SS_\beta = \frac{1}{p} \left[ \sum_i (\hat{\beta}_i - 1)^2 \right] \left[ \sum_j \frac{(C_j{}^2)}{p} - \frac{G^2}{pq} \right] \tag{9.45}$$

$$SS_\varepsilon = SS_{RC} - SS_\beta \tag{9.46}$$

Each $\hat{\beta}_i$ is computed by the formula of Eq. 9.39, by observing that

$$\hat{\gamma}_j = \frac{C_j}{p} - \frac{G}{pq} \tag{9.47}$$

Equations 9.45 and 9.40 can be shown to be equivalent, by virtue of Eq. 9.47.

While the direct computation of the analysis of variance table constitutes a considerable short cut in computational labor, it is not advisable to draw final conclusions from the analysis without a detailed examination of all parameter estimates and of all residuals. One reason for this lies in the great sensitivity of the analysis of variance to even a single "outlying" observation. Furthermore, an analysis of variance table tends to concentrate on over-all effects; it fails to reveal details of structure in the data. In many cases, only a careful examination of individual parameter estimates and of residuals can guard against serious misinterpretations of the data. With the present wide-spread availability of high speed computing facilities, a complete tabulation of parameter estimates and of residuals for various models offers no serious problems. Such tabulations should be made whenever possible.

## 9.6 SUMMARY

Since laboratory experimentation is generally carried out in a more or less organized manner, the resulting data can in most cases be arranged in systematic patterns. For example, in equation of state studies, pressures may have been measured for given values of volume and temperature, and may accordingly be arranged in rectangular arrays, where volumes are represented by rows and temperatures by columns. Such data arrangements are denoted in this book as structured data.

A simple case of structured data is that in which the data are classified in terms of a single criterion. This case is illustrated here by a study of the electrochemical equivalent of silver, the criterion of classification being the type of treatment to which the sample of silver was subjected in order to remove dissolved oxygen. Since each treatment was carried out on two or more specimens, we may denote the data structure as a "between–within" classification (between and within treatments).

Factorial designs are schemes in which the effects of several criteria, or "factors," are studied simultaneously. The categories of each factor that are actually used in the experiment are called its levels. Thus in the previous example, each individual treatment constitutes a level of the factor "treatments." In the equation of state example, each temperature value used in the experiment is a level of the factor "temperature," and each volume value used in the experiment is a level of the factor "volume." In the analysis of factorial experiment, consideration is first given to the main effect of each factor, i.e., its average effect upon the measured value, the average being taken over all levels of all other factors. Also examined are the interactions between factors, i.e., the manner in which the effect of one factor depends on the levels of the other factors.

Of common interest to the physical scientist are structured data involving two criteria of classification. Such data are best represented by rectangular arrays. In analyzing any set of data one must first formulate a plausible model. For a two-way classification, the model stipulates the general manner in which the levels of the two factors affect the measured value. The simplest model is the additive one, in which the passage from one level to another in one of the factors has an effect that is independent of the level of the second factor. A purely additive model is completely free of interaction effects. However, most sets of two-way classification data show row-column interaction effects. Therefore it is advisable to first consider a more general model that allows for the presence of interactions. In this chapter such a general model is introduced and referred to as the linear model. This model contains as special cases both the additive model and the general concurrent model.* The basic feature of the linear model is that the entries in each row (or column) are, apart from random error, linearly related to the corresponding entries in any other row (or column). This leads to a geometrical representation by a family of straight lines, each line representing a different row (or column) of the table. The additive model is then represented by a family of parallel straight lines. (A concurrent model corresponds to a family of straight lines all of which pass through a common point.) A general analysis of variance is presented that allows one to decide whether, for any particular

---

* The concurrent model will be discussed in detail in Chapters 11 and 13.

set of data, the general linear model (when applicable) can be reduced to an additive model. Applications of these important concepts will be discussed in Chapters 11 and 13.

## REFERENCES

1. Craig, D. N., J. I. Hoffman, C. A. Law, and W. J. Homer, "Determination of the Value of the Faraday with a Silver-Perchloric Acid Coulometer," *J. Research Natl. Bur. Standards*, **64A**, 381–402 (Sept.–Oct. 1960).
2. Horowitz, E., J. Mandel, R. J. Capott, and T. H. Boone, "Evaluation of Micrometer and Microscopical Methods for Measuring Thickness of Floor Coverings," *Materials Research and Standards*, **1**, 99–102 (Feb. 1961).
3. Mandel, J., and F. L. McCrackin, "Analysis of Families of Curves," *J. Research Natl. Bur. Standards*, **67A**, 259–267 (May–June 1963).
4. Mandel, J., "The Measuring Process," *Technometrics*, **1**, 251–267 (Aug. 1959).
5. Mandel, J., "Non-Additivity in Two-Way Analysis of Variance," *J. Amer. Statistical Assn.*, **56**, 878–888 (Dec. 1961).

# chapter 10

# SOME PRINCIPLES OF

# SAMPLING

## 10.1  INTRODUCTION

In this chapter we first deal very briefly with the problem of judging the acceptability of a lot of merchandise on the basis of an examination of a sample taken from the lot.  We then proceed to some general sampling problems.  By "examination" of a sample we mean an evaluation of quality either by expert judgment or by application of physical or chemical tests.  We assume that the lot consists of discrete units such as automobile tires, floor tiles, surgeon's gloves (or pairs of gloves), medicinal pills, vacuum tubes, bricks, etc.  In most cases it is either totally impossible or economically prohibitive to test or examine each individual unit in the lot. Many tests are such that the specimen tested is destroyed during the testing process, but even if the test is non-destructive, sampling is obviously desirable.

The judgment based on a test may either be qualitative or quantitative. In the first case, the unit tested will be termed either "*defective*" or "*non-defective*," but no numerical measure will be associated with it. Thus a pill may be specified to contain a particular ingredient in an amount of not less than 0.55 grams and not more than 0.60 grams.  Any pill that fails to meet this requirement is "defective."  If the sample consists of 200 pills, a certain number of these may be found to be defective.  This number, expressed as a fraction of the total sample size (in this case, 200) is called the "fraction defective" in the sample.  Thus, if 5 of

224

the 200 pills are defective, the fraction defective in the sample is 5/200 = 0.025 or 2.5 per cent. Evidently, the lot (as distinguished from the sample) also has a certain fraction defective but this value is generally unknown.

The second case concerns situations in which the judgment based on the test is quantitative, i.e., a measured value, reported as such. For example, each automobile tire in a sample of tires may have been subjected to a road test of tread wear. The information obtained from testing a sample of 16 tires then consists of 16 values of rates of wear, expressed for example as grams of rubber lost during 1000 miles of road travel.

To differentiate between the qualitative and the quantitative case, the first is often referred to as *sampling by attributes*, and the second as *sampling by variables*.

## 10.2  SAMPLING BY ATTRIBUTES

The primary object of the theory of sampling by attributes is to provide a basis for rational action in regard to acceptance and rejection of a lot, on the basis of an examination of samples taken from the lot. The theory is a relatively simple application of the calculus of probability, but a full account is beyond the scope of this book. We will first state some basic facts of the theory, in the form of three principles, and then discuss the main concepts involved in the use of sampling plans.

*Principle I.* The information contained in a sample does not depend appreciably on the size of the lot from which the sample is taken (provided that the lot is not smaller than, say, 10 times the sample).

This principle is illustrated in Table 10.1.

**TABLE 10.1**  Effect of Lot Size

| Defectives in sample | Sample of 10 units, from a lot that is 10% defective | |
| | Lot of 10,000 units | Lot of 100 units |
| | Probability of occurrence of such a sample, in % | |
| --- | --- | --- |
| 0 | 34.9 | 33.0 |
| 1 | 38.7 | 40.8 |
| 2 | 19.4 | 20.1 |
| 3 | 5.7 | 5.2 |
| 4 | 1.12 | 0.76 |
| 5 or more | 0.16 | 0.07 |

There is a striking similarity between the probabilities in both cases, even though in the first case the sample represents only one tenth of one per cent of the lot, whereas in the second case the sample represents ten per cent of the lot, i.e., 100 times more percentage-wise. In other words, the behavior of a sample of ten, taken from a lot containing 10 per cent of defective units, is almost identically the same, whether the lot is 1000 times as large as the sample or only 10 times as large. Therefore the information that can be extracted from the sample with respect to its parent-lot, is also the same in both cases.

*Principle II.* The information contained in a sample increases when the size of the sample increases, regardless of the size of the lot.

An illustration of this almost self-evident principle is given in Table 10.2.

**TABLE 10.2**   Effect of Sample Size

| | Lot of 100,000 units, 10% defective |
|---|---|
| Sample size | Probability that sample % defective is between 5 and 15%, in % |
| 10 units | 39 |
| 20 units | 74 |
| 100 units | 92 |

We see that a sample of 100 units is much more likely to be a faithful image of the lot than one of 20 units, and that the latter is far superior to one of 10 units.

*Principle III.* The information contained in a sample does not depend on the *proportion* that the sample represents of the lot; *in any case*, larger samples contain more information than smaller ones.

A logical consequence of principles I and II, this principle is illustrated in Table 10.3.

**TABLE 10.3**   Inadequacy of Samples Proportional to Lot

| | Lots: 10% defective | |
|---|---|---|
| | Lot:    200 units Sample:  20 units | Lot:    100 units Sample:  10 units |
| Sample per cent defective | Probability of occurrence of such a sample, in % | |
| 0 | 10.8 | 33.0 |
| Between 0 and 20 | 76.9 | 40.8 |
| 20 or more | 12.2 | 26.2 |

It is apparent that the occurrence of samples very different from the parent-lot is much more frequent for the smaller sample than for the larger one, even though the proportion of sample to lot is the same in both cases.

## 10.3  SAMPLING PLANS

A sampling plan is a set of rules stating the size of the sample to be taken and the action to be taken with regard to acceptance of the lot after examination of the sample.  The purpose of a sampling plan is to insure that both the consumer and the producer are fairly treated.  The consumer is fairly treated if it does not happen "too often" that a "bad" lot is accepted; the producer is fairly treated if it does not happen "too often" that a "good" lot is rejected.  For purposes of action it is necessary to agree on the meaning of "too often."  For example, the two parties concerned may agree that "not too often" means: "not more than 5 times in a hundred."  Furthermore, it is necessary to define "good" and "bad" lots.  Once these terms are rigorously defined, it is possible to construct a sampling plan that will offer the desired assurance of fair treatment for both parties.  The mathematical concept that forms the basis of the construction and evaluation of sampling plans is the *operating characteristic curve*, often denoted as the *OC curve*.

On the abscissa of a rectangular graph, plot the *lot fraction defective*. Now consider a well-defined sampling plan and plot, on the ordinate of the graph, the *probability of acceptance* of a lot (whose fraction defective is the abscissa value) under this plan.  The curve thus obtained, by varying the lot fraction defective, is the *OC* curve (see Fig. 10.1).

Two points of the *OC* curve are of particular interest.  They are defined by their ordinates: the probability of acceptance corresponding to an "acceptable" lot and that corresponding to a "rejectable" lot.  The abscissa of the point corresponding to the first probability is called the *acceptable quality level* denoted *AQL*, and the abscissa of the point corresponding to the second probability is called the *rejectable quality level*, denoted *RQL*.  The *RQL* is also known as the *lot tolerance per cent defective* or *LTPD*.  Suppose that the probability of acceptance corresponding to the *AQL* is $1 - \alpha$.  Then the *AQL* is the fraction defective of a lot such that, under the sampling plan, this lot will be *rejected* with probability $\alpha$.  This *probability of rejection* of a lot of *AQL*-quality is called the *producer's risk*.  On the other hand, let $\beta$ represent the probability of acceptance corresponding to the *RQL*.  Then the *RQL* is the fraction defective of a lot such that, under the sampling plan, this lot will be *accepted* with probability $\beta$.  This *probability of acceptance* of a lot of

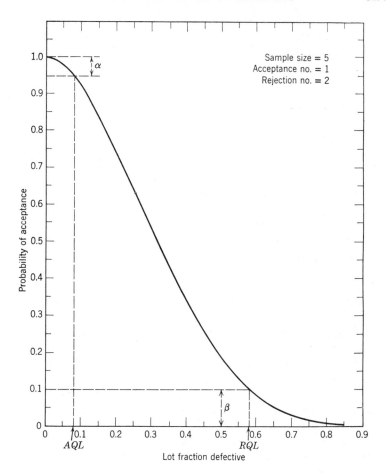

**Fig. 10.1** The operating characteristic (*OC*) curve of a sampling plan.

*RQL*-quality is called the *consumer's risk*. Values that have been used in this connection are 5 per cent for the producer's risk $\alpha$ and 10 per cent for the consumer's risk $\beta$.

Table 10.4 presents a set of sampling plans. The *OC* curve of plan *A* of this table is shown in Fig. 10.1. It is seen that a sampling plan specifies: (*a*) the size of the sample; (*b*) an "acceptance number"; and (*c*) a "rejection number."

These two terms are defined in Table 10.4. If the number of defectives in the sample is equal to the *acceptance number*, or less, the lot is accepted;

**TABLE 10.4** Sampling Plans

| Plan designation | Sample size | Acceptance number | Rejection number |
|:---:|:---:|:---:|:---:|
| A | 5 | 1 | 2 |
| B | 7 | 1 | 2 |
| C | 15 | 2 | 3 |
| D | 22 | 2 | 3 |
| E | 45 | 4 | 5 |

if the number of defectives in the sample is equal to the *rejection number*, or larger, the lot is rejected.

It follows from these definitions that if a lot is to be either definitely accepted or definitely rejected on the basis of the sample, the rejection number is necessarily equal to the acceptance number plus one. On the other hand, there exist plans in which this condition is not fulfilled. In that case, the plan must specify the action to be taken whenever the number of defectives in the sample is greater than the acceptance number but smaller than the rejection number. This action is to take a second sample of specified size and determine its number of defectives. If the plan terminates after the first sample, it is called a *single* sampling plan; if it terminates after the second sample, it is called a *double* sampling plan; if it specifies more steps it is called a *multiple* sampling plan. A *sequential* sampling plan is one in which the maximum number of samples to be taken from the lot before reaching a final decision in regard to its acceptance is not predetermined, but depends on the results given by the successive samples.

A detailed discussion of sampling plans is beyond the scope of this book. We wish, however, to stress some facts that are not always fully appreciated by users of sampling plans.

We first mention that in a table such as Table 10.4, the various sampling plans are often associated with lot sizes. Thus, the use of the plans in this table might be specified as follows:

| For lot size | Use plan |
|:---:|:---:|
| 100 to 180 | A |
| 181 to 500 | B |
| 501 to 800 | C |
| 801 to 3200 | D |
| 3201 to 8000 | E |

This association appears to contradict principle I, which states essentially the irrelevance of lot sizes with regard to the information supplied by samples.   Actually, the reason for the association is one of economics; in the case of a larger lot, there is economic justification for examining a larger sample and thereby obtaining *more* information about the lot. The plans in Table 10.4 become increasingly better as one proceeds downwards.   A "better" plan is one for which the $OC$ curve shows a steeper downward slope beyond the region characterizing very "good" lots.

Our second observation is that if samples are taken from successive lots of a manufactured product, *and if the manufacturing process is in statistical control*, then the application of a sampling plan will *not* result in a segregation of good lots from bad lots.   Indeed, on the average, the rejected lots will have exactly the same quality as the accepted lots.   Before resolving this apparent paradox, let us define *statistical control* as a state in which the successive lots differ from each other in a purely *random* manner, such that their fractions defective form a statistical population, and show no noticeable trends, cycles, or any other *systematic* variations. It is clear that if a process were definitely known to be in a state of statistical control, no sampling would be required once its fraction defective is known. Therefore, the use of sampling plans on lots taken from a process that is in statistical control results in nothing but a random segregation into lots that *will be called* acceptable or rejectable, but that are actually of the same average quality.   Nevertheless, sampling plans are useful and important. Their real function is to judge lots that do not originate from a single, controlled production process.   If, on the other hand, it is desired to achieve a state of statistical control in a manufacturing process, use should be made of the powerful control chart technique (see Chapter 5).

### 10.4   SAMPLING BY VARIABLES

Suppose that the buyer of a shipment of a zinc alloy is interested in an alloy containing at least 42 per cent zinc.   If the shipment meets this requirement, he is eager to accept it.   On the other hand, should the zinc content of the alloy be less, he would then be less interested in the purchase.   The smaller the zinc content of the shipment, the greater is his reluctance to buy it.   If the zinc content is as low as 41 per cent, he wishes to have only a small probability of accepting it.   (There exists always a chance that such a poor lot, because of the uncertainty of the zinc determination, will appear to be of better quality.)

To be specific, let us suppose that for a true zinc content of 42.0 per cent we require the probability of acceptance of the lot to be high, say 95 per cent.   If the true zinc content is 41.0 per cent, we require the probability of acceptance to be only 1 per cent.   These two conditions define a sampling plan and determine the entire $OC$ curve.

Let $N$ be the number of determinations to be run, and let us suppose for the moment, that the standard deviation $\sigma$ of the Zn-analysis is exactly known. The sampling plan consists in giving a value for $N$ and in stipulating that the lot be accepted whenever the average $\bar{x}$ of the $N$ determinations is at least equal to $k$ per cent Zn, where a numerical value for $k$ is also stipulated. Our problem is to determine the values of $N$ and $k$.

*a.* Suppose that the true zinc content of the shipment is 42.0 per cent. Then the sample average, $\bar{x}$, is normally distributed with mean $\mu = 42.0$ and standard deviation $\sigma/\sqrt{N}$. The probability of acceptance is the probability that

$$\bar{x} > k \tag{10.1}$$

which is the probability that the standard normal deviate,

$$t = \frac{\bar{x} - 42.0}{\sigma/\sqrt{N}}$$

is greater than

$$\frac{k - 42.0}{\sigma/\sqrt{N}}$$

Since this probability is to be 0.95, we infer from the normal table (one-tail area), that

$$\frac{k - 42.0}{\sigma/\sqrt{N}} = -1.64 \tag{10.2}$$

*b.* Consider now the alternative situation: that the true zinc content of the shipment is 41.0 per cent. In that case, we require that the probability of acceptance be 0.01; i.e., the probability is 0.01 that

$$\bar{x} > k$$

This time, the standard normal deviate is: $t = (\bar{x} - 41.0)/(\sigma/\sqrt{N})$. Thus, the probability is to be 0.01 that $(\bar{x} - 41.0)/(\sigma/\sqrt{N})$ be greater than

$$\frac{k - 41.0}{\sigma/\sqrt{N}}$$

From the normal table (one-tail area) we conclude that

$$\frac{k - 41.0}{\sigma/\sqrt{N}} = +2.33 \tag{10.3}$$

Equations 10.2 and 10.3 provide a system of two equations in two unknowns, $k$ and $N$. Dividing the first by the second, we obtain

$$\frac{k - 42.0}{k - 41.0} = -\frac{1.64}{2.33} = -0.704$$

from which we obtain:

$$k = 41.58 \qquad (10.4)$$

Substituting this value in Eq. 10.2 and solving for $N$, we get

$$N = \left[\frac{-1.64\sigma}{41.58 - 42.0}\right]^2 = 15.2\sigma^2 \qquad (10.5)$$

Assuming, for example, that $\sigma$ is equal to 0.45, we have

$$N = 3.08$$

Thus, the sampling plan would consist in specifying that 3 determinations be made, and that the shipment be accepted whenever the average of the 3 values found is not less than 41.58.

It is interesting to note that a sampling plan by variables such as the one just discussed has an $OC$ curve analogous to that of an "attributes" plan. It is obtained by plotting, for each value of the zinc content of the shipment, the probability of its acceptance.

If $\sigma$ is not known, the solution of the problem is no longer tractable by the elementary procedures used above. An interesting treatment is given by Chand (2). It is, however, worth observing that the numerical value for $k$ in the preceding treatment was obtained without making use of $\sigma$ (this quantity cancelled out in the division of Eq. 10.2 by Eq. 10.3). Equation 10.5 then shows $N$ to be proportional to $\sigma^2$. Thus, one might use $k$ as given by an equation obtained as was Eq. 10.4 and then decide on a value of $N$, using a conservatively guessed value for $\sigma^2$. Such a procedure is, of course, only approximate but would not be misleading unless the guessed value of $\sigma^2$ were quite wrong.

## 10.5  OBJECTIVES AND METHODS OF SAMPLING

The magnitude of a sample, as well as the procedure of obtaining it, depend of course on the objectives of the sampling process, and vary in fact a great deal with these objectives. Many technical discussions are concerned with sampling problems arising in a very special problem: the estimation of the *average* of a population. Another specific problem is the estimation of the *variance* of a population. In the scientific and technical field, these objectives are certainly not the only ones and perhaps not even the most important. The situations arising in scientific research are often extremely complex; a statistical formulation, in order to be appropriate, would generally involve a large number of populations. For example, in the case of the data described in Section 9.3, the number of populations varied with the postulated model and the main problem was precisely to determine which model was most appropriate.

In questions of process control, or of sampling made for the purpose of determining conformance with specifications, we are more genuinely concerned with sampling problems of the usual type: the estimation of the average or of the variance of a given population.    To be realistic, a solution to these problems should not ignore the cost of sampling and of testing.    The technical aspects of this matter fall outside the scope of this book.    We will, however, briefly discuss two specific problems: the size of the samples required for determining averages and standard deviations.

## 10.6  SAMPLE SIZES FOR ESTIMATING AVERAGES

Suppose we wish to estimate the average $\mu$ of a population whose standard deviation $\sigma$ is known.

A random sample of size $N$ will yield the estimate $\bar{x}$, the standard error of which is

$$\sigma_{\bar{x}} = \frac{\sigma}{\sqrt{N}}$$

If we wish $\sigma_{\bar{x}}$ to be less than some given number, say $E$, we must have

$$\frac{\sigma}{\sqrt{N}} < E$$

or

$$N > \frac{\sigma^2}{E^2} \qquad (10.6)$$

This relation shows that the desired precision requires that the size of the sample, $N$, be equal to the smallest integer that is larger than $\sigma^2/E^2$.

Consider now the same type of problem, assuming this time that $\sigma$ is not exactly known, but that an estimate of $\sigma$ is available.    From Section 6.8 we know that the length of a confidence interval for $\mu$ is given by $L = 2t\hat{\sigma}/\sqrt{N}$, from which it follows that:

$$N = \frac{4t^2\hat{\sigma}^2}{L^2} \qquad (10.7)$$

In this formula, $\hat{\sigma}$ is an estimate of $\sigma$, based on information available at the time the experiment is planned.    If the estimate $\hat{\sigma}$ is based on $m$ degrees of freedom, the value of $t$ to be used in this formula is the critical value read from the table of Student's $t$ for $m$ degrees of freedom, at the level of significance corresponding to the required confidence coefficient. Thus, given a required value of $L$ and a confidence coefficient, $N$ can be computed.    At the conclusion of the experiment, the best estimate of $L$ may well be different from that required, because the experiment itself has provided $N - 1$ additional degrees of freedom for the estimation of $\hat{\sigma}$.

As an illustration, suppose that a chemical determination of zinc is to be made on a given alloy, and that it is desired to bracket the amount of zinc to within $\pm 0.5$ per cent zinc, with 95 per cent confidence. Previous experience provides the estimate $\hat{\sigma} = 0.45$, based on 21 degrees of freedom. How many determinations are required to obtain the desired precision?

From Student's table, we obtain for a level of significance of 0.05 ($= 1 - 0.95$), and 21 DF: $t = 2.080$. Applying Eq. 10.7, for $L = 2 \times 0.5 = 1.0$, we obtain:

$$N = \frac{4(2.080)^2 \cdot (0.45)^2}{(1.0)^2} = 3.5$$

We conclude that 4 determinations are required. Suppose that the values obtained in this experiment were:

$$42.37, \quad 42.18, \quad 42.71, \quad 42.41$$

Their average is 42.42 and the estimated standard deviation is 0.22, based on 3 degrees of freedom. Combining this estimate with the one previously available, we obtain the pooled estimate:

$$\hat{\sigma} = \left[\frac{21(0.45)^2 + 3(0.22)^2}{21 + 3}\right]^{1/2} = 0.43$$

Student's $t$ can now be taken with $21 + 3 = 24$ degrees of freedom. For DF $= 24$, and a 0.05 level of significance, the table gives the value $t = 2.064$. Thus, the length of the confidence interval actually obtained is

$$L = 2(2.064)(0.43)/\sqrt{4} = 0.89$$

## 10.7   SAMPLE SIZES FOR ESTIMATING VARIANCES

Suppose that a sample is to be taken from a normal population for the purpose of estimating its variance (or standard deviation). This problem may arise in analytical chemistry, when a new analytical method has been developed and the investigator wishes to determine its reproducibility under the degree of control of environmental conditions that can be achieved in his laboratory. If he runs $N$ replicate determinations, he will obtain an estimate $\hat{\sigma}$ based on $N - 1$ degrees of freedom. His problem, roughly stated, is to choose $N$ in such a way that $\hat{\sigma}$ will not differ from the true standard deviation $\sigma$ by more than a given factor. The word "factor" is used deliberately because the pertinent statistical theory involves the *ratio* of $\hat{\sigma}$ to $\sigma$.

Consider a series of samples, all of size $N$, taken from a single normal population of variance, $\sigma^2$. Let

$$\hat{\sigma}_1^2, \ \hat{\sigma}_2^2, \ \hat{\sigma}_3^2, \ \ldots$$

represent the estimates of $\sigma^2$ yielded by the successive samples. These estimates themselves constitute a statistical population. It can be shown that the average of this population is $\sigma^2$, the true value of the variance. However, the population of the $\hat{\sigma}^2$ is not normal. It is, in fact, a skew distribution, closely related to the well-known statistical distribution known as the *chi-square distribution*, denoted $\chi^2$.

Chi-square is actually a family of distributions differing from each other in terms of a parameter known as its "degrees of freedom." (In this respect $\chi^2$ is analogous to the family of Student's *t* distributions.) To each value of $n$, the degrees of freedom, there corresponds a specific $\chi^2$ distribution.

Returning now to the distribution of the $\hat{\sigma}^2$, we state that the quantity $[n\hat{\sigma}^2/\sigma^2]$, where $n$ is the number of degrees of freedom available for $\hat{\sigma}^2$ (in our case, $n = N - 1$), has the $\chi^2$ distribution with $n$ degrees of freedom. Thus, we write

$$\frac{n\hat{\sigma}^2}{\sigma^2} = \chi_n^2 \tag{10.8}$$

The critical values of $\chi_n^2$ have been tabulated for various values of $n$ (or DF) and for a number of levels of significance. A brief tabulation is given in Table V of the Appendix. For example, for $n = 12$, $\chi^2$ will exceed the tabulated value 21.0261 five per cent of the time; it will exceed the value 5.2260 ninety-five per cent of the time. Consequently, $\chi_{n=12}^2$ will lie between 5.2260 and 21.0261 ninety per cent of the time, as is evident from the following diagram:

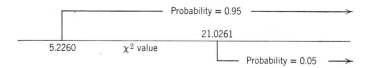

Now, using Eq. 10.8 we infer from what has just been stated that for $n = 12$, the *ratio* $\hat{\sigma}^2/\sigma^2$, being equal to $\chi_n^2/n$, or $\chi_{12}^2/12$, will lie, with 90 per cent probability, between the values $(5.2260)/12$ and $(21.0261)/12$ that is, between 0.4355 and 1.7552. Thus, the ratio $\hat{\sigma}/\sigma$ will lie, with 90 per cent probability, between the square roots of these quantities, i.e., between 0.66 and 1.32. Having obtained the estimate $\hat{\sigma}$ from the sample of size 12, one can then obtain a confidence interval for $\sigma$ in the following way:

From $0.66 < \hat{\sigma}/\sigma < 1.32$ it follows that $\hat{\sigma}/1.32 < \sigma < \hat{\sigma}/0.66$ which, after substitution of $\hat{\sigma}$ by its numerical value, gives the desired confidence interval.

The problem we have just solved is, in a way, the reverse of that of determining a sample size. Using algebraic notation, we have shown that the ratio $[\hat{\sigma}/\sigma]^2$ (where $\hat{\sigma}$ is estimated with $n$ degrees of freedom) is contained, with a probability* equal to $\beta$, between the two quantities:

$$\frac{\chi_1^2}{n} \quad \text{and} \quad \frac{\chi_2^2}{n}$$

where $\chi_1^2$ and $\chi_2^2$ are critical values of $\chi^2$ with $n$ degrees of freedom, chosen at levels of significance such that their difference is equal to $\beta$.

Now let us examine the reverse problem: determination of a sample size. Suppose, for example, that we wish to estimate $\hat{\sigma}$ to within 20 per cent of the true value, $\sigma$, and that we wish to achieve this approximation with 90 per cent confidence. Let $N$ be the unknown number of determinations (the sample size) from which $\hat{\sigma}$ is derived. Then, the number of degrees of freedom, $n$, is $N - 1$; $n = N - 1$.

We require that

$$1 - 0.20 < \frac{\hat{\sigma}}{\sigma} < 1 + 0.20$$

or

$$0.80 < \frac{\hat{\sigma}}{\sigma} < 1.20$$

This requirement can be written

$$0.64 < \frac{\hat{\sigma}^2}{\sigma^2} < 1.44$$

Since $\beta = 0.90$, we divide the complementary quantity $\alpha = 1 - \beta = 1 - 0.90 = 0.10$ into two equal parts. This gives the levels of significance 0.05 and 0.95. We now try to find, in Table V, under the column heading 0.95, a value such that its quotient by the corresponding $n$ *is not less* than 0.64. By trial and error we find that the smallest value of $n$ for which this occurs is approximately $n_1 = 40$. Similarly, we look under the column heading 0.05 for a value whose quotient by the corresponding $n$ *does not exceed* 1.44. This time the smallest $n$ to satisfy this condition is $n_2 = 35$. The desired value of $n$ is the larger of these two ($n_1$ and $n_2$), i.e., $n = 40$; and hence: $N = n + 1 = 41$.

The method we have just developed is the more rigorous way of obtaining confidence limits for $\sigma$, as well as deriving the sample size required for achieving a desired precision for $\hat{\sigma}$. There exists a simpler, though approximate, method. It is based on the fact that the distribution of a

---

* The limits define a probability interval, not a confidence interval, since the bracketed value $[\hat{\sigma}/\sigma]^2$ is a random variable, and the limits are fixed constants.

sample estimate, $\hat{\sigma}$, based on $n$ degrees of freedom, is approximately normal with a mean approximately equal to $\sigma$ and a variance approximately equal to

$$V(\hat{\sigma}) = \frac{\sigma^2}{2n} \qquad (10.9)$$

where $\sigma$ is the true standard deviation of the original population. It follows from this theorem that the ratio $\hat{\sigma}/\sigma$ is also approximately normally distributed with mean unity and variance equal to

$$V\left(\frac{\hat{\sigma}}{\sigma}\right) = \frac{V(\hat{\sigma})}{\sigma^2} = \frac{1}{2n} \qquad (10.10)$$

Returning now to the problem above, we wish to find a sample size $N$ such that a 95 per cent probability interval for $\hat{\sigma}/\sigma$ be given by the limits 0.8 and 1.2. Using the theory of the normal distribution, we have for the length of the 95 per cent probability interval:

$$L = (1.2 - 0.8) = 0.4 = 2(1.96)\,\frac{1}{\sqrt{2n}}$$

from which we derive:

$$n = \frac{1}{2}\left[\frac{2(1.96)}{0.4}\right]^2 = 48$$

Hence, the required sample size is $N = 48 + 1 = 49$. The agreement with the value derived from the more rigorous theory is quite satisfactory, since the estimation of a sample size should be considered as a guideline fixing an order of magnitude rather than an exact value.

The large sample size required to achieve the modest accuracy of $\pm 20$ per cent is striking, and provides a good illustration of the poor reliability of estimates of variability based on small samples. Now, the scientist concerned with measurement problems is often confronted with the estimation of variability. It is a deplorable misconception, based on ignorance or disregard of elementary statistical facts, that small samples are sufficient to obtain satisfactory estimates of variability. We will discuss the problem in greater detail in the following section and also in our treatment of interlaboratory tests of measuring techniques (Chapter 13).

## 10.8 THE POWER OF STATISTICAL TESTS

Although the topics we have so far dealt with in this chapter may appear to be somewhat disconnected, they are actually related to each other in their common concern with the relationship between the size of a sample and the amount of information it provides. In every case the *type* of information that is desired must be stated precisely; statistical theory then provides the sample size required to obtain this information. In acceptance sampling, the desired information concerns a *decision* (to accept or to reject), which is itself related to the desire of obtaining a certain quality of product. In the problems of Sections 10.6 and 10.7, the desired information related to the *location* (average) or *spread* (standard deviation) of a statistical population. The major unknown, in all these problems, is the sample size.

It is possible to rephrase these problems in the language of significance testing. This is simple in the case of estimating an average or a standard deviation. For example, in the former case, a certain value for the population mean may be taken as the "null-hypothesis." The problem may then be stated in the following form: "How large a sample is required to *detect*, with given probability, a given departure from the null-hypothesis?" A similar statement can be made for a standard deviation, or, in fact, for any estimated parameter of a statistical population. The verb "detect" denotes, in this connection, the *rejection* of the null-hypothesis at the adopted level of significance, and consequently the *acceptance* of the existence of a departure from this null-hypothesis. When the probability of detection is plotted versus the magnitude of the departure from the null-hypothesis, the resulting curve is known as the *power function* of the statistical test.

The relation between the power function and the $OC$ curve, as defined in Section 10.3, is easy to perceive; in acceptance sampling we are also concerned with departures from a stated standard quality, and the $OC$ curve may be considered as giving the probability of detecting such departures.

The power of a statistical test increases, of course, with the sample size. Therefore, the general problem of calculating sample sizes may be treated from the power viewpoint. In most cases such a treatment requires new statistical functions, known as *non-central* distribution functions, because they involve probabilities calculated on the assumption that one or more parameters do not have the "central" values stipulated by the null-hypothesis. Power functions for many statistical tests have been tabulated and presented as families of curves (3, 5, 6). The treatment we have given in Sections 10.6 and 10.7 circumvents the use of power functions by formulating the problem in terms of the *length of confidence intervals*, rather than as the probability of detecting departures from the null-hypothesis. From the viewpoint of the physical scientist, the length of a confidence interval is often a more directly useful concept than a probability of detection, since the very formulation of a null-hypothesis is often a rather artificial approach to a physical problem. For this reason no illustration is given here of the calculation of power. The interested reader is referred to references 1 and 4.

Whichever approach is used, power function or length of confidence interval, the important point is that the planning of an experiment (or of a sampling procedure) involves as an indispensible element, the posing of the question of *adequacy*: is the sample of sufficient size to achieve its aim? In this connection it should be recalled that the level of significance offers no guarantee of sufficiency in this sense. The reader is urged at this point to reread Section 8.3 and to note that the concept of power (or the related concept of the length of confidence intervals) constitutes precisely the logical element that was missing when only the level of significance was considered. Briefly stated, the level of significance relates to what happens when the null-hypothesis is true; the power relates to what happens when it is not true. If the method of confidence intervals is used, the dichotomy—true or false null-hypothesis—is no longer made; it is replaced by an *entire set of values* all of which could qualify as acceptable estimates for the unknown parameter. The word "acceptable," as used here, means "not contradicted by the results of the experiment." Here the power concept enters through the observation that samples of insufficient size generally lead to excessively long confidence intervals. Under these conditions, even values that are quite different from the correct value will be considered acceptable. By increasing the sample size, we decrease the length of the confidence interval and thereby reduce the set of acceptable estimates to any desired degree of precision.

## 10.9  SUMMARY

The acceptance or rejection of a lot of merchandise submitted for purchase is generally based on an examination of a portion (sample) of the lot either by visual inspection or by physical or chemical testing. The theory of sampling is concerned with appropriate methods for the selection of samples from lots and for making inferences about the quality of a lot on the basis of an examination of a sample taken from it. The size of a lot or of a sample is the number of units of product contained in it. When the quality of a lot is defined by the relative number of defective units it contains, and the sample is also judged by its number of defectives, the sampling is said to be by attributes.

Statistical theory shows that the adequacy of a random sample depends primarily on its size and only to a very small extent on the fraction that the sample represents of the lot.

A sampling plan is a set of rules stating the size of the sample to be taken and the action to be taken with regard to acceptance of the lot after examination of the sample. Sampling plans are characterized by their $OC$ (operating characteristic) curves. These are the curves that give, for each quality level of the lot, the probability that the lot will be accepted under the given plan. In practice, $OC$ curves are often described in terms of two of their points, defined in terms of an acceptable quality level and a rejectable quality level. With these two reference qualities are associated two probability values defined as the producer's risk and the consumer's risk.

In many cases, the quality of manufactured products can be evaluated quantitatively, rather than by a mere count of conforming and of defective units. In those cases, the sampling is said to be by variables. Thus, when the lot is evaluated on the basis of the chemical composition of a sample, a sampling plan by variables is generally indicated. Methods are available for the derivation of sampling plans by variables that will have preselected characteristics.

In laboratory work, sampling problems often arise in connection with the number of replicate measurements that are needed to achieve a desired degree of precision. The latter may refer either to a measured property or to the estimation of the reproducibility of a measuring process. Thus, the two problems to be solved consist in the determination of sample sizes for the estimation, with a given degree of precision, of averages, and of variances. Formulas are presented for the calculation of sample sizes in both problems.

The problem of sample size is closely connected with that of the power of statistical tests. The concept of power is analogous to that of the $OC$ curve of a sampling plan. Power is defined as the probability of detecting a false hypothesis. The more erroneous the hypothesis, the greater will be the probability of discovering its falsehood, and the smaller will be the sample size required for this discovery. Thus sample size and power are related concepts. In this book we deal with power only in terms of the length of confidence intervals: other things being equal, a shorter confidence interval reflects larger power.

## REFERENCES

1. Anderson, R. L., and T. A. Bancroft, *Statistical Theory in Research*, McGraw-Hill, New York, 1952, pp. 114ff.
2. Chand, U., "On the Derivation and Accuracy of Certain Formulas for Sample Sizes and Operating Characteristics of Nonsequential Sampling Procedures," *J. Research Natl. Bur. Standards*, **47**, No. 6, 491–501 (Dec. 1951).
3. Dixon, W. J., and F. J. Massey, *Introduction to Statistical Analysis*, McGraw-Hill, New York, 2nd ed., 1957.

4. Eisenhart, C., M. W. Hastay, and W. A. Wallis, eds., *Selected Techniques of Statistical Analysis*, McGraw-Hill, New York, 1947.

5. Natrella, M. G., *Experimental Statistics National Bureau of Standards Handbook*, **91**, August, 1963. (For sale by the Superintendent of Documents, U.S. Government Printing Office, Washington 25, D.C.)

6. *Symbols, Definitions, and Tables, Industrial Statistics and Quality Control*, Rochester Institute of Technology, Rochester, New York, 1958.

## chapter 11

# THE FITTING OF CURVES AND SURFACES

### 11.1 INTRODUCTION

For a meaningful discussion of the subject of this chapter, a preliminary clarification of terms and objectives is indispensable. In the physical sciences, curve fitting generally denotes the determination of a mathematical expression of the type

$$F(X, Y) = 0 \tag{11.1}$$

where $X$ and $Y$ are two variables for which the values are obtained from theory or by measurement, or as a result of deliberate choice.

For example, the isotherms of equations of state for ionized gas mixtures may be constructed from theoretical knowledge. At a given temperature both the compression and the specific volume are separately calculated, and the problem consists in finding an expression of the type of Eq. 11.1 to relate compression to specific volume. Examples of this type are, however, infrequent. A more common situation arises when the values for one of the variables, for example $X$, are fixed, and the corresponding values of $Y$ measured. Here two variants are possible; the values of $X$ may either be preselected, or they may themselves be obtained by measurement, at each "setting" of the equipment. Suppose that the birefringence of a certain type of optical glass is to be measured at different compression loads. One way of doing this is to apply loads of preselected magnitudes, and measure the birefringence at each load. An alternative procedure is to compress the specimens by increasing amounts and measure both the load and the birefringence at each setting. Whichever experimental procedure is used, the final aim is to obtain a relation between load and birefringence, as represented by Eq. 11.1, where $X$ and $Y$ now represent these two physical quantities. For mathematical convenience, it is customary to characterize an equation of this type in two steps: first, as to *functional type* (linear, quadratic, exponential, sum of two exponentials, sinusoidal, etc.); and secondly, by the values of the *parameters* occurring in the expression. For example, suppose the equation in question is

$$Y - \alpha - \beta X = 0$$

The functional type may be specified by simply stating that the relation is linear. The second step then consists in determining the values of the parameters $\alpha$ and $\beta$. Generally, then, the equation may be written

$$F(X, Y \mid \alpha, \beta, \ldots) = 0 \tag{11.2}$$

where the vertical line separates the variables from the parameters.

We now turn to the basic question: what does a scientific worker expect to accomplish by fitting a curve? Or, stated differently: what is the real purpose of representing a set of related quantities $X$ and $Y$ by a mathematical function?

The answer varies with the general objectives of the investigation.

*a.* In many cases, the functional type of the relation is known from theory, and the purpose of the fit is to derive values for the parameters. There even exist situations in which not only the functional type but also the values of the parameters are known from theory. In such cases the purpose of the fit is to confirm the theory from which the model, including the value for each parameter, is derived.

The cases cited so far involve relatively straightforward methods of analysis. Using the method of least squares, the parameters and their standard errors are estimated. If desired, confidence intervals can also be constructed for the parameters. However, it must be emphasized that the fit depends on the assumptions made in regard to the nature of the experimental errors in the determination of $X$ and $Y$. In particular, it is necessary to have information, or to make plausible assumptions, about the standard deviations of these errors, their relation to each other, and their dependence on the magnitude of the measured values.

*b.* A more frequent situation is that in which no particular functional type can be postulated *a priori*. When this is the case, the fitting procedure generally serves one of two purposes: it is made either to gain further insight into the experimental process under investigation, or to provide an empirical representation of the data.  In the former case, the data are expected to provide clues that may form the starting point of a new theory. For example, the data may suggest a certain type of discontinuity which might then be interpreted as a "transition point" of some particular type. Or a departure from linearity may suggest the presence of a formerly unsuspected factor.  A skillful representation of the data by an appropriately selected function may, in such cases, facilitate their interpretation. Intuition and scientific insight play a major role in such fitting attempts, and it is difficult to formulate general rules.  We will see, however, that the widening of the investigation from the determination of a single curve to that of a family of curves will generally permit us to follow a more systematic approach.

When the fitting process is carried out merely to provide a convenient summary of the data, the situation is somewhat different.  In such cases the objective is to replace a large body of data, which it may be difficult to always have at hand when needed, by a compact formula, easily accessible at all times, and readily communicable.  When the fitting is purely empirical, the closeness of fit is dictated primarily by our needs.  Whether or not we have fitted the data to within their experimental error is relatively unimportant as long as we have approximated them to a degree consistent with their practical use.  This is an important point, often overlooked in discussions on curve fitting.

## 11.2  RESTRICTIVE CONDITIONS IN EMPIRICAL FITS

There exist many situations in which a curve is to be fitted in the absence of any precise theoretical knowledge regarding the nature of the relationship.  But even in such cases, the investigator generally has enough information to eliminate a vast class of functions, thus narrowing down the field in which ingenuity, or trial and error, remain the only guides.  For example, it might be evident from the nature of the problem that the unknown function $y = f(x)$ is constantly increasing (or constantly decreasing) as $x$ increases.  In some cases, it might be safely assumed that as $x$ becomes very large, $y$ must approach an "asymptotic" value, $y_0$.  In other cases, the nature of the problem may make it evident that values of $x$ that are situated at the same distance, either to the left or to the right, from some value $x_0$ will result in the same value of $y$, thus yielding a symmetrical curve.

We may denote information of this type as "restrictive conditions" because of the restrictions they impose on the equations that might reasonably be considered for the fitting process.  The so-called "boundary conditions" of physical laws can often be used as restrictive conditions.

In general, such information, however incomplete, should not be ignored in empirical fitting procedures. On the other hand, an empirical fit is not necessarily useless because it violates known boundary conditions. The essential point is to be aware of the limits of applicability of the empirical formula.

## 11.3  FITTING BY POLYNOMIALS

A widely used method for the empirical fit of curves is the use of poly-nomials. By choosing a polynomial of sufficiently high degree, it is generally possible to achieve a good empirical fit for any reasonably smooth set of data. The fitting process is straightforward: the method of least squares is used with appropriate weight factors and the uncertainty of the parameter estimates is obtained either in terms of standard errors or confidence intervals.

The reader is referred to Chapter 7 for the general theory and to text-books on statistics* for computational details (1, 2, 5). We will, however, deal briefly with two simple cases of polynomial fitting: the straight line with zero intercept and the general straight line (with an intercept not necessarily equal to zero). Straight line fitting will be dealt with in much greater detail in Chapter 12. At this point we merely mention the formulas applying to the simplest models.

1. *Straight line with zero intercept:*

The model is

$$y_i = \beta x_i + \varepsilon_i \tag{11.3}$$

where $x_i$ is a set of values for the controlled variable, $y_i$ the corresponding measurements; and $\varepsilon_i$ is the experimental error associated with $y_i$. Assum-ing that the errors are mutually independent, have zero mean and a common standard deviation, the slope $\beta$ is estimated by

$$\hat{\beta} = \frac{\sum_i x_i y_i}{\sum_i x_i^2} \tag{11.4}$$

2. *General straight line:*

The model for the more general case is

$$y_i = \alpha + \beta x_i + \varepsilon_i \tag{11.5}$$

The same assumptions hold as for model 1.

---

* In textbooks on statistics, matters relating to curve fitting are often found under the title "regression," a term that has only historical significance.

The parameters $\alpha$ and $\beta$ are best estimated in reverse order, as follows:

$$\beta = \frac{\sum_i x_i y_i - \frac{1}{N}\left(\sum_i x_i\right)\left(\sum_i y_i\right)}{\sum_i x_i^2 - \frac{1}{N}\left(\sum_i x_i\right)^2} \tag{11.6}$$

and

$$\hat{\alpha} = \frac{1}{N}\left(\sum_i y_i - \beta \sum_i x_i\right) \tag{11.7}$$

where $N$ is the number of $(x_i, y_i)$ pairs (i.e., the number of points).

If we denote by $\bar{x}$ and $\bar{y}$ the average values of the $x_i$ and $y_i$, respectively, we can write Eq. 11.7 as follows:

$$\hat{\alpha} = \frac{\sum y}{N} - \beta\frac{\sum x}{N} = \bar{y} - \beta\bar{x} \tag{11.8}$$

We now obtain for the estimated value of $y_i$:

$$\hat{y}_i = \hat{\alpha} + \beta x_i$$
$$= \bar{y} - \beta\bar{x} + \beta x_i$$

or

$$\hat{y}_i - \bar{y} = \beta(x_i - \bar{x}) \tag{11.9}$$

This equation shows that when $x_i = \bar{x}$, we also have $y_i = \bar{y}$. Thus, the fitted line passes through the point $(\bar{x}, \bar{y})$. This point is called the *centroid* of the line. Equation 11.9, which represents the fitted line in terms of the centroid and the slope, is a useful form, preferable in many applications to the representation involving the intercept.

For examples of straight line fitting, the reader is referred to Chapter 8, Table 8.9, where Model I′ represents a straight line fit with zero intercept, and Model II a general straight line fit.

Polynomial fits are subject to a severe limitation: They may violate the restrictive conditions that are known to apply in a particular case. For example, in any polynomial, $y$ approaches either plus infinity or minus infinity as $x$ tends to infinity; and polynomials have maxima, minima, and inflection points that may be incompatible with the physical reality underlying the data. In particular, if a function is known to be monotonic, or to have finite asymptotes, polynomials are generally not well suited for the fitting procedure.

A function $y = f(x)$ is said to be *monotonic* in a certain range if, as $x$ increases within this range, the change of $y$ is always of the same sign, or zero, but not positive in one place of the range and negative in another. Of course, any function having a maximum or a minimum inside a certain range cannot be monotonic.

Strictly speaking, a monotonic function could remain constant over a certain part of the range, but we will consider monotonic functions that either increase steadily or decrease steadily, though the rate of change may be very small in certain parts of the range.

We will consider first a rather simple monotonic function, show when and how it can be used, and then generalize it to cover a wider range of applications.

## 11.4 FITTING THE FUNCTION $y = Ax^B$

In considering the function

$$y = Ax^B \tag{11.10}$$

we will assume that $x$ is always positive. Let us also assume, for the moment, that $A$ is positive. In that case the function is represented by the three curves I, II, or III shown in Fig. 11.1, according as $B < 0, 0 < B < 1$, or $B > 1$. (The cases $B = 0$ and $B = 1$ are degenerate: for $B = 0$, we have a horizontal straight line; for $B = 1$, we have a straight line going through the origin.)

For Type I ($B < 0$), both the $x$ axis and the $y$ axis are asymptotes. In fact this type is a generalization of the equilateral hyperbola $xy = A$, as can be seen by writing Eq. 11.10 in the form

$$yx^{-B} = A$$

where $-B$ is now positive.

Type II ($0 < B < 1$) is concave towards the $x$ axis, and its tangent at the origin coincides with the $y$ axis.

Type III ($B > 1$) is convex towards the $x$ axis, which is its tangent at the origin.

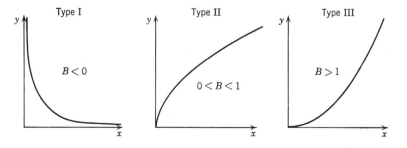

**Fig. 11.1** The function $y = Ax^B$.

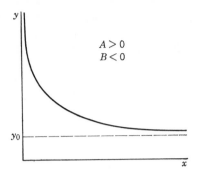

**Fig. 11.2**   The function $y - y_0 = Ax^B$.

We now observe that the function $y = -Ax^B$ is simply the mirror image of $y = Ax^B$ with respect to the $x$ axis. Thus, the behavior of Eq. 11.10 when $A$ is negative is derived at once from the cases already discussed.

Given a set of data, it is generally possible to ascertain whether Eq. 11.10 will provide a satisfactory fit. A plot of $y$ versus $x$ should be made wherever possible prior to an empirical fitting process. Taking logarithms of both sides of the equation we obtain*:

$$\log y = \log A + B \log x \tag{11.11}$$

Thus, if Eq. 11.10 applies, a plot of $\log y$ versus $\log x$ should yield a straight line of slope $B$ and intercept $\log A$.

Figure 11.2 suggests a simple way of generalizing the model represented by Eq. 11.10. We have seen that for Type I, both axes are asymptotes. It is readily conceivable that the physical situation underlying the data requires that, as $x$ increases, $y$ approaches asymptotically some value different from zero, say a positive value $y_0$. For example, if $y$ represents the temperature of a cooling body, and $x$ represents time, the value that $y$ will reach ultimately is the temperature surrounding the body. This temperature is not necessarily zero in the scale selected for temperature measurements. Mathematically this case is a trivial generalization of Eq. 11.10; all that is required is a parallel displacement of the $x$ axis by a distance equal to $y_0$ (see Fig. 11.2). The equation becomes:

$$y - y_0 = Ax^B \tag{11.12}$$

---

* If $A < 0$, then $y < 0$. In this case, rewrite the equation in the form $(-y) = (-A)x^B$ before taking logarithms.

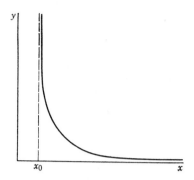

**Fig. 11.3**   The function $y = A(x - x_0)^B$.

Similarly, if the vertical asymptote for Type I is to be located at a value $x = x_0$, rather than at the $y$ axis (see Fig. 11.3), the equation is

$$y = A(x - x_0)^B \tag{11.13}$$

If both asymptotes are distinct from the co-ordinate axes, the equation is:

$$y - y_0 = A(x - x_0)^B \tag{11.14}$$

This model is illustrated in Fig. 11.4, for $A > 0$ and $B < 0$.

Any one of Eqs. 11.12, 11.13, or 11.14 could, of course, also occur in connection with types II and III.   Now, while these generalizations are

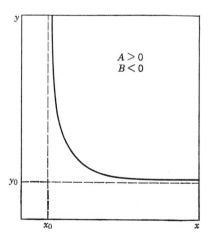

**Fig. 11.4**   The function $y - y_0 = A(x - x_0)^B$.

trivial from the purely mathematical viewpoint, they are far from trivial in the actual fitting process. The reason for this is that since $x_0$ and $y_0$ are in general not exactly known, the simple device of rectifying the curve by taking logarithms cannot be directly used.

Let us first observe that the fitting of Eq. 11.12 is essentially the same problem as that of fitting Eq. 11.13. This is at once recognized by rewriting Eq. 11.12 in the form:

$$x = \left(\frac{1}{A}\right)^{1/B} (y - y_0)^{1/B}$$

Denoting $(1/A)^{1/B}$ by $A'$ and $(1/B)$ by $B'$, this equation becomes

$$x = A'(y - y_0)^{B'}$$

which is of the type of Eq. 11.13 except for an interchange of $x$ and $y$.

We will now discuss the fitting of Eq. 11.13.

## 11.5 FITTING THE FUNCTION $y = A(x - x_0)^B$

The unknown parameters are $A$, $B$, and $x_0$. The fit is made by first obtaining an estimate for $x_0$, using the following procedure.

1. Plot the data on rectangular co-ordinates.

2. Draw a smooth curve through the plotted points, using French curves or similar devices.

3. Select two values $x_1$ and $x_2$, near the opposite extremes of the data range, and read the corresponding values $y_1$ and $y_2$ from the smooth curve.

4. Calculate the quantity $y_3 = \sqrt{y_1 y_2}$ and locate it on the $y$ axis; read the corresponding value $x_3$ from the smooth curve. The three points $(x_1, y_1)$, $(x_2, y_2)$, and $(x_3, y_3)$ are on the curve to be represented by the equation $y = A(x - x_0)^B$. Therefore:

$$y_1 = A(x_1 - x_0)^B \tag{11.15}$$

$$y_2 = A(x_2 - x_0)^B \tag{11.16}$$

$$y_3 = A(x_3 - x_0)^B \tag{11.17}$$

Multiplying Eqs. 11.15 and 11.16 and subtracting the square of Eq. 11.17 we obtain:

$$y_1 y_2 - y_3{}^2 = A^2[(x_1 - x_0)(x_2 - x_0)]^B - A^2[(x_3 - x_0)^2]^B \tag{11.18}$$

Because of the selection of $y_3$ as equal to $\sqrt{y_1 y_2}$, the left side of Eq. 11.18 is zero. Therefore the right side is also zero. Dividing by $A^2$ and simplifying, we then have:

$$(x_1 - x_0)(x_2 - x_0) - (x_3 - x_0)^2 = 0$$

from which we obtain the estimate:

$$x_0 = \frac{x_1 x_2 - x_3{}^2}{x_1 + x_2 - 2x_3} \tag{11.19}$$

For each of the original values of $x$ we can now compute a corresponding quantity $u$:

$$u = x - x_0 \tag{11.20}$$

Equation 11.13 then becomes:

$$y = A u^B \tag{11.21}$$

which is the model dealt with in Section 11.4.

## 11.6   THE FUNCTION $y - y_0 = A(x - x_0)^B$

We note that differentiation of this model gives:

$$\frac{dy}{dx} = AB(x - x_0)^{B-1} \tag{11.22}$$

the second member of which is of the same type as in Eq. 11.13. This suggests the following procedure:

1. Plot the data on rectangular co-ordinates.
2. Draw a smooth curve through the plotted points, using French curves or similar devices.
3. Select a set of $x$ values, covering the entire range of interest.

With some practice, the decision regarding the number and the spacing of the points selected in each case will offer no difficulties.

4. Using the smooth curve, read the corresponding set of $y$ values.
5. Denoting the selected points by $(x_1, y_1), (x_2, y_2), \ldots, (x_N, y_N)$, construct Table 11.1.
6. Plot $\Delta y / \Delta x$ versus $\bar{x}$ (see Table 11.1).   Let

$$z = \Delta y / \Delta x \tag{11.23}$$

## TABLE 11.1

| $\bar{x}$ | $\Delta x$ | $\Delta y$ | $\Delta y / \Delta x$ |
|---|---|---|---|
| $\frac{1}{2}(x_1 + x_2)$ | $x_2 - x_1$ | $y_2 - y_1$ | $(y_2 - y_1)/(x_2 - x_1)$ |
| $\frac{1}{2}(x_2 + x_3)$ | $x_3 - x_2$ | $y_3 - y_2$ | $(y_3 - y_2)/(x_3 - x_2)$ |
| $\frac{1}{2}(x_3 + x_4)$ | $x_4 - x_3$ | $y_4 - y_3$ | $(y_4 - y_3)/(x_4 - x_3)$ |
| $\vdots$ | $\vdots$ | $\vdots$ | $\vdots$ |
| $\frac{1}{2}(x_{N-1} + x_N)$ | $x_N - x_{N-1}$ | $y_N - y_{N-1}$ | $(y_N - y_{N-1})/(x_N - x_{N-1})$ |

Since $\Delta y/\Delta x$ can be considered as an approximation to $dy/dx$, we have approximately, according to Eq. 11.22:

$$z = \Delta y/\Delta x \approx (AB)(\bar{x} - x_0)^{B-1} \qquad (11.24)$$

which is again of the type of Eq. 11.13. Thus, $B - 1$ and $AB$, as well as $x_0$, may be estimated.

7. Compute, for each value of $x$ in the original set of data, the quantity $A(x - x_0)^B$ and subtract it from the corresponding $y$ value. Each such difference is an estimate of $y_0$. Therefore, a better estimate for $y_0$ is obtained by taking the average of all these differences. We will illustrate the fitting procedure in Section 11.8. First, however, we make a number of important comments.

a. Proper attention must be given to algebraic signs (plus or minus) in the successive steps of the procedure. The sign to be used is always obvious from the shape of the curve. The procedure makes use of graphical as well as numerical methods. Therefore, frequent reference should be made to the plots in carrying out the fitting process.

b. Unlike the method of least squares, the procedure we have just discussed is not entirely objective. It involves visual judgment in the drawing of graphs as well as other non-objective features.

Nevertheless, the values obtained for the four parameters, $x_0$, $y_0$, $A$, and $B$, by the proposed procedure may give a satisfactory fit without further adjustments. If a better fit is desired, or if a weighting process is indicated, the estimates obtained may be used as a set of "initial values" and corrections calculated by least squares, using the methods described in Chapter 7.

c. While the method outlined above requires some graphing and computations, it is not particularly time consuming and may well be justified in careful research work. On the other hand, the method is not practical for the routine fitting of numerous sets of data.

## 11.7 THE EXPONENTIAL

Consider the equation $y - y_0 = A(x - x_0)^B$ and suppose that $A > 0$ and that $B < 0$. Then the curve belongs to Type I.

Suppose now that for a particular set of data, $x_0$ is located very far to the left, in the negative region of the $x$ axis, while the entire range of $x$ values for the data is in the positive region and much closer to the origin than $x_0$. We have:

$$x - x_0 = (-x_0)\left(1 - \frac{x}{x_0}\right)$$

Both quantities $(-x_0)$ and $[1 - (x/x_0)]$ are positive.   We can therefore take logarithms:

$$\ln (x - x_0) = \ln (-x_0) + \ln \left(1 - \frac{x}{x_0}\right)$$

Now, according to our assumptions $x/x_0$ is small, in absolute value, in comparison with unity.   Therefore we have

$$\ln \left(1 - \frac{x}{x_0}\right) \approx - \frac{x}{x_0}$$

Hence:

$$\ln (x - x_0) \approx \ln (-x_0) - \frac{x}{x_0}$$

from which we derive, by taking antilogarithms:

$$x - x_0 \approx (-x_0)e^{-x/x_0} \tag{11.25}$$

Introducing this approximation in the model we obtain:

$$y - y_0 \approx A(-x_0)^B e^{-(B/x_0)x}$$

The quantities $A(-x_0)^B$ and $B/x_0$ are both positive constants.   Let us denote them by $C$ and $k$, respectively.   Then we have:

$$y - y_0 \approx Ce^{-kx} \tag{11.26}$$

Thus our general model (Eq. 11.14) includes the negative exponential function 11.26 as a special case: namely, when $B$ is negative and $x_0$ is very large and negative.

Once it has been ascertained that Eq. 11.26 is appropriate, the actual fitting process is carried out as follows:

1. Plot the data on rectangular co-ordinates.

2. Draw a smooth curve through the plotted points, using French curves or similar devices.

3. Select two $x$ values, $x_1$ near the left end of the range and $x_2$ near the right end.

4. Calculate a third $x$ value, $x_3$, such that

$$x_3 = \tfrac{1}{2}(x_1 + x_2) \tag{11.27}$$

5. Obtain the values $y_1$, $y_2$, and $y_3$ corresponding to $x_1$, $x_2$, and $x_3$, from the smooth curve.

6. It is now readily seen, using the same reasoning as in Section 11.5, that

$$y_0 = \frac{y_1 y_2 - y_3{}^2}{y_1 + y_2 - 2y_3} \tag{11.28}$$

7. Once a value has been obtained for $y_0$, the quantity $y - y_0$ can be numerically calculated for each $y$. Taking logarithms of Eq. 11.26, we obtain:

$$\ln (y - y_0) = \ln C - kx \tag{11.29}$$

This equation represents a straight line in terms of $\ln (y - y_0)$ versus $x$, with a slope equal to $(-k)$ and an intercept equal to $\ln C$. Thus, the fit is completed by estimating the slope and the intercept in accordance with Section 11.3 and deriving $k$ and $C$ from these estimates.

We have discussed at some length the fitting of Eqs. 11.10, 11.12, 11.13, and 11.14, because of their relative simplicity and general usefulness. It is hardly necessary to mention that a vast number of other algebraic relations exist that can be used with success for empirical fitting processes. There exist few reliable guidelines to help one in selecting a good formula. Experience and intuition are indispensable. So far, no comprehensive statistical theory has been developed in connection with the empirical fitting of experimental data.

## 11.8   A NUMERICAL EXAMPLE

Columns 1 and 2 of Table 11.2 present measurements at 34.50 °C and 1 atmosphere of the refractive index of benzene at various wavelengths (4).

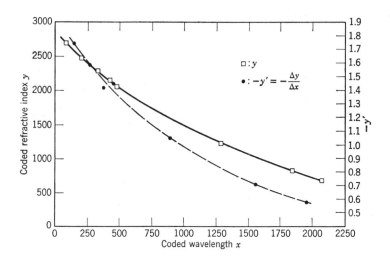

**Fig. 11.5**   Working curves for fitting the data of Table 11.2.

For ease of calculation both the refractive index values and the wavelengths are coded as indicated (columns 3 and 4); i.e.,

$$x = \lambda - 4600$$

$$y = 10^5(n - 1.48)$$

The coded data are plotted in Fig. 11.5 (solid line). The data show no visible scatter; therefore, the slopes can be estimated from the differences of consecutive values, as shown in Table 11.3. A plot of $(-y')$ versus $\bar{x}$ is shown as a dotted line in Fig. 11.5.

Selecting two values near the extremes: $-y_1' = 1.6$, $-y_2' = 0.6$. We read from the curve $x_1 = 240$, $x_2 = 1900$. A value $(-y_3')$ is now calculated as

$$(-y_3') = \sqrt{(-y_1')(-y_2')} = \sqrt{0.96} = 0.98$$

From the curve we read the corresponding value $x_3 = 980$.

**TABLE 11.2** Refractive Index of Benzene

| (1)<br>Wavelength,<br>$\lambda$ | (2)<br>Refractive index,<br>$n$ | (3)<br><br>$x$ | (4)<br><br>$y$ | (5)<br>$n$,<br>calculated |
|---|---|---|---|---|
| 6678 | 1.48684 | 2078 | 684 | 1.48690 |
| 6438 | 1.48822 | 1838 | 822 | 1.48829 |
| 5876 | 1.49221 | 1276 | 1221 | 1.49226 |
| 5086 | 1.50050 | 486 | 2050 | 1.50050 |
| 5016 | 1.50151 | 416 | 2151 | 1.50147 |
| 4922 | 1.50284 | 322 | 2284 | 1.50282 |
| 4800 | 1.50477 | 200 | 2477 | 1.50473 |
| 4678 | 1.50690 | 78 | 2690 | 1.50684 |

$\lambda$ in Angstroms; $x = \lambda - 4600$; $y = 10^5 (n - 1.48)$.

We then have

$$x_0 = \frac{x_1 x_2 - x_3^2}{x_1 + x_2 - 2x_3} = -2802$$

Table 11.4 gives the values of $\log(\bar{x} - x_0)$ and $\log(-y')$ as calculated from Table 11.3, using $x_0 = -2802$. Figure 11.6 is a plot of $\log(-y')$ versus $\log(\bar{x} - x_0)$ and shows that the points fall close to a straight line. This gives us some reassurance that the model is reasonable for these data. The equation of this straight line is (cf. Eq. 11.24):

$$\log(-y') = \log(-AB) + (B - 1)\log(\bar{x} - x_0)$$

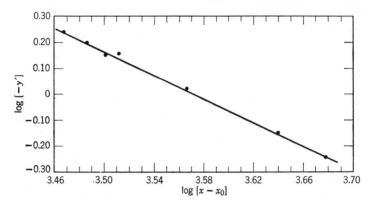

**Fig. 11.6** A further working curve for fitting the data of Table 11.2.

The slope of the line is $(B - 1)$ and its intercept is $\log(-AB)$. A visual fit was made and yielded the line shown in Fig. 11.6. In order to determine the slope and the intercept of the line we select two points on it:

|  | $\log(\bar{x} - x_0)$ | $\log(-y')$ |
|---|---|---|
| First point | 3.460 | 0.255 |
| Second point | 3.640 | -0.155 |

**TABLE 11.3**

| $x$ | $y$ | $\bar{x}$ | $\Delta x$ | $\Delta y$ | $-\dfrac{\Delta y}{\Delta x} = -y'$ |
|---|---|---|---|---|---|
| 2078 | 684 |  |  |  |  |
|  |  | 1958 | -240 | 138 | 0.575 |
| 1838 | 822 |  |  |  |  |
|  |  | 1557 | -562 | 399 | 0.710 |
| 1276 | 1221 |  |  |  |  |
|  |  | 881 | -790 | 829 | 1.049 |
| 486 | 2050 |  |  |  |  |
|  |  | 451 | -70 | 101 | 1.443 |
| 416 | 2151 |  |  |  |  |
|  |  | 369 | -94 | 133 | 1.415 |
| 322 | 2284 |  |  |  |  |
|  |  | 261 | -122 | 193 | 1.582 |
| 200 | 2477 |  |  |  |  |
|  |  | 139 | -122 | 213 | 1.746 |
| 78 | 2690 |  |  |  |  |

**TABLE 11.4**

| $\log(\bar{x} - x_0)$ | $\log(-y')$ |
|:---:|:---:|
| 3.678 | −0.249 |
| 3.640 | −0.149 |
| 3.566 | 0.021 |
| 3.512 | 0.159 |
| 3.501 | 0.151 |
| 3.486 | 0.199 |
| 3.468 | 0.241 |

The slope of the fitted line is

$$B - 1 = \frac{-0.155 - 0.255}{3.640 - 3.460} = -2.28$$

For the intercept we obtain:

$$\log(-AB) = 0.255 - (-2.28)(3.460) = 8.144$$

Hence

$$AB = -1.393 \times 10^8$$

From these values we derive: $B = -1.28$, $A = 1.088 \times 10^8$. We now have the equation

$$y = y_0 + 1.088 \times 10^8 (x + 2802)^{-1.28}$$

from which we obtain

$$y_0 = y - 1.088 \times 10^8 (x + 2802)^{-1.28}$$

Substituting, in this formula, $x$ and $y$ by the data given in Table 11.2 we obtain for $y_0$ the values:

$$-1383, \ -1384, \ -1382, \ -1377, \ -1373, \ -1375, \ -1373, \ -1371$$

the average of which is $-1377$.

Thus we finally have:

$$y = -1377 + 1.088 \times 10^8 (x + 2802)^{-1.28}$$

The only remaining task is to replace $x$ and $y$ by their expressions in terms of $\lambda$ and $n$. After some simplification this gives the empirical formula:

$$n = 1.46623 + \frac{1088}{[\lambda - 1798]^{1.28}}$$

The last column in Table 11.2 shows the agreement between values calculated by this formula and the measured values of $n$. The fit is rather good, but could probably be improved by a least squares adjustment of the parameter estimates by the methods discussed in Section 7.7. It is important to note that no claim can be made for the validity of this formula outside the range of wavelength values in Table 11.2.

## 11.9  THE USE OF EMPIRICAL FORMULAS

It has been noted already that empirical formulas may occasionally aid in the theoretical understanding of physical phenomena. Two examples may be cited: (1) the discovery by Balmer, in 1885, that wavelengths of the hydrogen spectrum could be represented by the formula

$$\lambda = 3645 \frac{n^2}{n^2 - 4}$$

where $n$ is an integer; and (2) Moseley's law (1913) according to which the square root of the frequency of certain spectral lines is a linear function of the atomic number. Both relationships are of great theoretical importance. The practice of examining data for the purpose of discovering relationships needs no apology; what is empirical today may well become part of a coherent theory tomorrow.

The most natural use of empirical formulas is for quick reference, particularly in engineering. On the other hand they are of great value, even in scientific research, for interpolation procedures. With regard to extrapolation, i.e., use of the formula beyond the range of values for which it was established, extreme caution is indicated. It should be emphasized in this connection that statistical measures of uncertainty, such as standard errors, confidence regions, confidence bands, etc., apply only to the region covered by the experimental data; the use of such measures for the purpose of providing statistical sanctioning for extrapolation processes is nothing but delusion.

While extrapolation is always hazardous, there exist situations in which the danger of obtaining wrong results is greatly reduced. Suppose for example that *it is known* that the curve representing a certain relationship must be of the general shape shown in Fig. 11.1, Type III. Specifically, we assume that (a) the curve is known to pass through the origin; (b) data are not available near the origin; (c) an experimental set of data is available for a region $a < x < b$, where $a$ is a certain distance to the right of the origin. Then an empirical fit through the origin to the experimental data could be used without great risk for extrapolation in the region $o < x < a$, because of our knowledge regarding the general shape of the curve. Except for situations of this or a similar type, extrapolation procedures using empirical curves should be accepted with a great deal of mental reserve.

## 11.10  FAMILIES OF CURVES

In Section 11.8 we have studied the relationship between the refractive index of benzene and wavelength, at a temperature of 34.50 °C and a pressure of 1 atmosphere.  If the temperature or the pressure, or both, were changed, a different curve would ensue for refractive index versus wavelength.  Suppose we keep the temperature constant at 34.50 °C but vary the pressure.  Then there corresponds to each pressure a definite curve, and the totality of these curves, for all values of pressure, constitute a *family of curves*.  Mathematically, this family of curves is represented by the relation

$$n = f(\lambda; P) \tag{11.30}$$

where $P$, representing pressure, is the "parameter" of the family.

Table 11.5 gives measured values of the refractive index of benzene, coded as indicated, for various wavelengths and for five values of pressure (4).  The first row, for $P = 1$ atm, is identical with the data given in Table 11.2.

**TABLE 11.5**  Refractive Index[a] of Benzene at 34.50 °C.   Effect of Wavelength and of Pressure

| Pressure, atm | Wavelength, Angstroms | | | | | | | |
|---|---|---|---|---|---|---|---|---|
| | 6678 $A$ | 6438 $B$ | 5876 $C$ | 5086 $D$ | 5016 $E$ | 4922 $F$ | 4800 $G$ | 4678 $H$ |
| $1 = $ I | 684 | 822 | 1221 | 2050 | 2151 | 2284 | 2477 | 2690 |
| $246.2 = $ II | 1879 | 2025 | 2438 | 3286 | 3391 | 3532 | 3724 | 3946 |
| $484.8 = $ III | 2882 | 3029 | 3445 | 4316 | 4421 | 4557 | 4762 | 4986 |
| $757.2 = $ IV | 3844 | 3991 | 4418 | 5305 | 5415 | 5555 | 5761 | 5992 |
| $1107.7 = $ V | 4903 | 5052 | 5489 | 6401 | 6514 | 6657 | 6867 | 7102 |

[a] Coded as follows: entry in table $= y = 10^5 (n - 1.48)$, where $n$ represents refractive index.

The problem we deal with in the following sections is the empirical fitting of a family of curves, such as is exemplified by the data in Table 11.5. Before proceeding with the discussion of this problem we note two important points.

In the first place, we observe that Eq. 11.30 represents a function of two variables, $\lambda$ and $P$.  Geometrically we can represent such a function in three-dimensional space, as is shown in Fig. 11.7, by locating on the

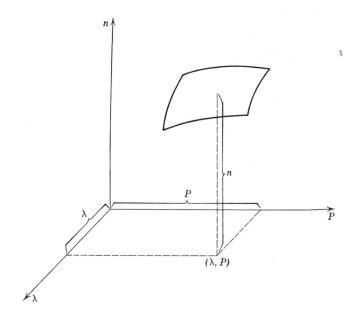

**Fig. 11.7**  A surface in three-dimensional space.

vertical axis, the value $n$ corresponding to each pair $\lambda$, $P$ (identified by a point in the $\lambda$, $P$ plane), and plotting the corresponding point $\lambda$, $P$, $n$. Thus we obtain a *surface S* as the geometrical representation of Eq. 11.30. We see that the fitting of families of curves is identical with the problem of fitting surfaces.

Our second remark concerns the novel feature that the variation of the parameter $P$ introduces in the fitting process.  After all, one could fit independently each row in Table 11.5, just as we did in Section 11.8 for the set of values corresponding to $P = 1$.  In this way the fitting process for the entire family of curves would be a mere sequence of repetitions of the curve fitting procedure for individual $P$ values.  It is easily seen that this way of attacking the problem fails to solve the basic question.  For instead of obtaining an empirical expression for Eq. 11.30, which is a function of two variables, we obtain a set of functions of one variable each (namely $\lambda$), without a definite link between the various functions in this set.  The real problem in fitting a one-parameter family of curves is one of finding a *single* expression for a function of two variables, rather than that of finding a set of functions of one variable each.  We will present one approach to this problem in the following section.

## 11.11   FITTING A FAMILY OF CURVES: INTERNAL STRUCTURE

We will consider Table 11.5 in the following manner. The table is a "two-way classification," in the sense that $y$, the coded value of $n$, is classified in accordance with two criteria, $P$ and $\lambda$. The values of $P$ we consider as "labels" for the rows, and the values of $\lambda$ as "labels" for the columns. Thus the values of $P$ and $\lambda$ are "marginal labels." The values of $y$ constitute the "body" of the table.

We begin by ignoring the numerical labels, for example, by replacing them by Roman numerals to represent the rows and by capital letters to represent the columns. Thus, the rows are now identified by I through V and the columns by $A$ through $H$, as indicated in Table 11.5. The reason for this step is to concentrate our attention on the *internal structure* of the "body" of the table, before establishing its relationship with the "labels." That such an internal structure actually exists, quite apart from the values of the "labels," is apparent from an examination of the body of the table and will become more evident as we proceed.

One way of studying the internal structure of the body of the table is to try to fit it by appropriate "models," such as was done in Chapter 9. In particular we should consider the additive model and the linear model* (Eqs. 9.27 and 9.33). Since the analysis of variance for the linear model includes that of the additive model as a special case, we go directly to this more comprehensive analysis. The results are shown in Table 11.6. Two features of the analysis are of particular interest:

1. A comparison of the mean squares for "slopes" and "residuals" ($F = 1922$) shows unmistakably that the linear model is a significantly better representation of the data (their "body") than the additive one.

2. The estimate of the variance of residual error for the linear model is 3.05. This corresponds to a standard deviation of 1.75 in the coded

**TABLE 11.6**   Refractive Index [a] of Benzene at 34.50 °C, Analysis of Variance

| Source of variation | Degrees of freedom | Sum of squares | Mean square |
|---|---|---|---|
| Rows (pressure) | 4 | 91,027,765 | 22,756,941 |
| Columns (wavelength) | 7 | 23,143,363 | 3,306,195 |
| Interaction (add. model) | 28 | 23,546.2 | 840 |
| Slopes | 4 | 23,473.0 | 5,868 |
| Residual (lin. model) | 24 | 73.2 | 3.05 |

[a] Data of Table 11.5.

* The linear model can be applied in two ways: in terms of the column-averages or in terms of the row-averages. For the present purpose it is advisable to try both ways and to select the alternative that yields the best approximation to straight line fits.

scale, i.e., $1.75 \times 10^{-5}$ in the original scale. Now the data are given with five decimal places and the error of measurement is known to be of the order of a few units in the fifth place. Thus the fit by the linear model is, on the whole, within experimental error. To ascertain whether the fit is uniformly good over the entire body of the data, it is necessary to study the individual residuals, given in Table 11.7. This table shows that the linear model is indeed an excellent fit for the data over their entire range.

**TABLE 11.7** Refractive Index[a] of Benzene at 34.50 °C. Residuals from Linear Model Fit

|     | A | B | C | D | E | F | G | H |
|-----|------|-------|-------|-------|-------|-------|-------|-------|
| I   | 0.40 | −0.22 | −0.13 | −0.03 | −0.86 | −0.70 | 1.17  | −0.35 |
| II  | −2.71| 0.66  | 3.24  | −1.57 | −1.33 | 0.37  | −1.65 | −0.36 |
| III | −0.50| 1.22  | −0.84 | 1.47  | −0.25 | −0.27 | 1.23  | 0.41  |
| IV  | 0.94 | −0.43 | −0.38 | −0.54 | 0.48  | −0.95 | −0.26 | 1.14  |
| V   | 1.86 | −1.24 | −1.90 | 0.67  | 1.96  | −0.23 | −0.49 | −0.85 |

[a] Data of Table 11.5.

The equation representing the fitted model is:

$$y_{ij} = \mu_i + \beta_i(x_j - \bar{x}) + \varepsilon_{ij} \tag{11.31}$$

where $y_{ij}$ is the table-entry for row $i$ and column $j$ and $\varepsilon_{ij}$ the corresponding error; $\mu_i$ and $\beta_i$ are parameters associated with the rows and $x_j$ is a parameter associated with the columns. We know (see Section 9.4) that $\hat{\mu}_i$ is the row average for row $i$, while $x_j$ is the column average for column $j$, and $\bar{x}$ is the average of the $x_j$. The parameter estimates are given in Table 11.8. We could terminate at this point the analysis of the internal

**TABLE 11.8** Parameter Estimates for Refractive Index Data

| Row, i | $\hat{\mu}_i$ | $\hat{\beta}_i$ | Column, j | $x_j$ |
|--------|----------|---------|-----------|--------|
| I      | 1797.375 | 0.95342 | A         | 2838.4 |
| II     | 3027.625 | 0.98092 | B         | 2983.8 |
| III    | 4049.750 | 0.99919 | C         | 3402.2 |
| IV     | 5035.125 | 1.02043 | D         | 4271.6 |
| V      | 6123.125 | 1.04604 | E         | 4378.4 |
|        |          |         | F         | 4517.0 |
| Average| 4006.600 | 1.00000 | G         | 4718.2 |
|        |          |         | H         | 4943.2 |
|        |          |         | $\bar{x} = 4006.6$ | |

structure of the body of the data. The number of parameters required for the fit is 4 for the $\hat{\mu}_i$, 4 for the $\hat{\beta}_i$, 7 for the $(x_j - \bar{x})$ and 1 for $\bar{x}$, i.e., a total of 16. This leaves 24 degrees of freedom for the estimation of residual error. However, in the present instance it is possible to gain considerably more insight into the data by pursuing the analysis a little further, as will be shown in the following section.

## 11.12   THE CONCURRENT MODEL

An examination of Table 11.8 reveals a striking feature: the values of $\hat{\beta}_i$ and $\hat{\mu}_i$ are highly correlated. This is perhaps not surprising since both parameters are associated with the same variable, pressure, although it is by no means a mathematical necessity.

Let us examine the relationship between $\hat{\beta}_i$ and $\hat{\mu}_i$ by plotting one versus the other, as is shown in Fig. 11.8. We see that the points fall very close to a straight line. Let us assume for the moment that the relationship between $\hat{\mu}_i$ and $\hat{\beta}_i$ is exactly linear, and let us examine the mathematical consequences of this assumption.

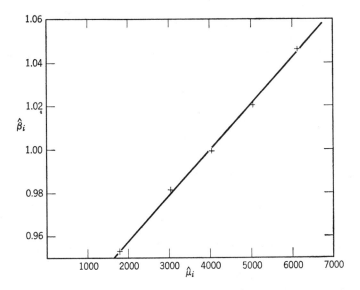

**Fig. 11.8**  Relation between $\hat{\mu}$ and $\hat{\beta}$ in the linear model fit of the data of Table 11.5. The straight line implies a concurrent model.

If we determine the straight line relation between $\hat{\beta}_i$ and $\hat{\mu}_i$ by least squares we obtain the equation (cf. Eq. 11.9):

$$\hat{\beta}_i - 1 = \alpha(\hat{\mu}_i - \bar{x}) + \text{error} \qquad (11.32)$$

where $\alpha$ is the slope of the line.

The reason for writing the constants (1) and ($\bar{x}$) is that a least squares fit for a straight line passes through the centroid of the points, i.e., through the point $\bar{\beta}$ and $\bar{\mu}$; and we know that $\bar{\beta} = 1$ and $\bar{\mu} = \bar{x}$. If we consider the error term in Eq. 11.32 as a negligible quantity we have

$$\hat{\beta}_i = 1 + \alpha(\hat{\mu}_i - \bar{x}) \qquad (11.33)$$

Introducing this value of $\hat{\beta}_i$ into the model given by Eq. 11.31, and replacing the parameters and $y$ values by their estimates, we obtain:

$$\hat{y}_{ij} = \hat{\mu}_i + [1 + \alpha(\hat{\mu}_i - \bar{x})][x_j - \bar{x}] \qquad (11.34)$$

This equation represents a family of straight lines ($y_{ij}$ versus $x_j$) with parameter $\hat{\mu}_i$. Furthermore, the equation is linear in the parameter $\hat{\mu}_i$ as can be seen by rewriting it in the form:

$$\hat{y}_{ij} = (1 - \alpha\bar{x})(x_j - \bar{x}) + (1 + \alpha x_j - \alpha\bar{x})\hat{\mu}_i \qquad (11.35)$$

If we make the term containing $\hat{\mu}_i$ equal to zero, $\hat{y}_{ij}$ becomes equal to $(1 - \alpha\bar{x})(x_j - \bar{x})$ *and this quantity is independent of i.* Consequently all the lines of the family pass through the point for which this happens, i.e., for $x_j$ such that

$$1 + \alpha x_j - \alpha\bar{x} = 0$$

or

$$x_j = \bar{x} - \frac{1}{\alpha}$$

When $x_j$ has this value, we obtain from Eq. 11.35:

$$\hat{y}_{ij} = (1 - \alpha\bar{x})\left(\bar{x} - \frac{1}{\alpha} - \bar{x}\right) = \bar{x} - \frac{1}{\alpha}$$

Thus, the point whose co-ordinates are both equal to $\bar{x} - 1/\alpha$ is a common point for all lines of the family. Such a point is denoted as a *point of concurrence.* Let $y_0$ denote the common value for both co-ordinates of the point of concurrence; then:

$$y_0 = \bar{x} - \frac{1}{\alpha} \qquad (11.36)$$

Subtracting Eq. 11.36 from Eq. 11.34 we obtain:

$$\hat{y}_{ij} - y_0 = \left(\hat{\mu}_i - \bar{x} + \frac{1}{\alpha}\right) + (x_j - \bar{x})[1 + \alpha(\hat{\mu}_i - \bar{x})]$$

$$= \left(\hat{\mu}_i - \bar{x} + \frac{1}{\alpha}\right)[1 + \alpha(x_j - \bar{x})]$$

$$= \alpha\left(\hat{\mu}_i - \bar{x} + \frac{1}{\alpha}\right)\left(x_j - \bar{x} + \frac{1}{\alpha}\right)$$

Hence

$$\hat{y}_{ij} - y_0 = \alpha(\hat{\mu}_i - y_0)(x_j - y_0) \tag{11.37}$$

From this equation, together with the preceding discussion, we derive an important statement: if $\hat{\beta}_i$ and $\hat{\mu}_i$ are linearly related to each other, then the bundle of straight lines represented by the linear model are *concurrent* (at a point whose co-ordinates are both equal to $y_0$); and the equation representing the bundle is essentially of the *multiplicative type*. By this we mean that each entry $y_{ij}$ in the body of the table is essentially (except for a fixed additive constant) the *product* of two quantities, one depending only on the row and the other depending only on the column.

Equation 11.37 represents a *concurrent model*. In the particular case in which $y_0 = 0$, it represents the *multiplicative model*:

$$y_{ij} = \alpha\mu_i x_j \tag{11.38}$$

We can test the adequacy of the concurrent model for our data by calculating $\hat{y}_{ij}$ from Eq. 11.37 for each $i$ and $j$ and comparing the results with the values of Table 11.5. The only additional quantity required for this calculation is $\alpha$, the slope of the line represented in Fig. 11.8. From this figure a good estimate for $\alpha$ could be obtained even by a visual fit. Let us, however, use the least squares estimate from the linear regression of $\hat{\beta}_i$ on $\hat{\mu}_i$. This estimate is obtained from the formula:

$$\hat{\alpha} = \frac{\sum_i (\hat{\beta}_i - 1)(\hat{\mu}_i - \bar{x})}{\sum_i (\hat{\mu}_i - \bar{x})^2} \tag{11.39}$$

which for the present data, gives: $\hat{\alpha} = 2.11019 \times 10^{-5}$. Using this value, Eq. 11.36 gives the value of $y_0$:

$$\hat{y}_0 = 4006.6 - \frac{1}{2.11019 \times 10^{-5}} = -43,382.5$$

Introducing these values for $\hat{\alpha}$ and $\hat{y}_0$ in Eq. 11.37 we obtain the expression for the fit:

$$\hat{y}_{ij} = -43,382.5 + 2.11019 \times 10^{-5}(\hat{\mu}_i + 43,382.5)(x_j + 43,382.5) \tag{11.40}$$

The values calculated by this relation are shown in Table 11.9. A comparison of this table with Table 11.5 shows that the fit by the concurrent model is truly excellent. From a practical viewpoint this fit is more useful than that obtained by applying the linear model without taking account of the relationship between $\hat{\mu}_i$ and $\hat{\beta}_i$. Indeed, we have seen that the number of independent parameters occurring in Eq. 11.31 is 16. On the other hand, the concurrent model (Eq. 11.37) involves only 13 independent parameters: 4 for the $\hat{\mu}_i$, 7 for the $(x_j - \bar{x})$, 1 for $\bar{x}$, and 1 for $\hat{\alpha}$. The advantage would increase with increasing number of rows, since there are as many $\beta_i$ as there are rows; and the existence of a relation such as Eq. 11.33 replaces them all by only one new parameter, $\alpha$ (in addition to their average, unity). We will show in Section 11.13 how the number of parameters can be further reduced.

**TABLE 11.9**   Refractive Index of Benzene at 34.50 °C; Calculated Values by Concurrent Model

|     | A | B | C | D | E | F | G | H |
|-----|------|------|------|------|------|------|------|------|
| I   | 684  | 822  | 1221 | 2050 | 2152 | 2284 | 2476 | 2690 |
| II  | 1884 | 2026 | 2436 | 3287 | 3392 | 3527 | 3724 | 3945 |
| III | 2880 | 3026 | 3445 | 4315 | 4421 | 4561 | 4762 | 4987 |
| IV  | 3842 | 3990 | 4418 | 5306 | 5415 | 5557 | 5762 | 5992 |
| V   | 4902 | 5055 | 5492 | 6400 | 6512 | 6656 | 6866 | 7102 |

Since the concurrent model involves 13 parameters, the number of degrees of freedom for the residuals equals $40 - 13$, or 27. In general, if a table of $p$ rows and $q$ columns can be fitted by the model expressed by Eq. 11.37, the number of degrees of freedom for residual error is

$$(DF)_\varepsilon = pq - p - q \qquad (11.41)$$

To prove this formula, subtract from the total number of observations $pq$, the number of parameters: 1 for $\bar{x}$, $(p - 1)$ for the $\hat{\mu}_i$, $q - 1$ for the $(x_j - \bar{x})$, and 1 for $\hat{\alpha}$; i.e., $pq - 1 - (p - 1) - (q - 1) - 1$ or $pq - p - q$.

It should be observed that the analysis so far has made no use whatsoever of the numerical values of either the pressure or the wavelength. To complete the fitting process we must express the parameters occurring in the fitted model—in this case the concurrent model—as functions of pressure and wavelength. This constitutes no new problem since all the parameters are functions of one variable only; the $\hat{\mu}_i$ depend on pressure only, and the $x_j$ are functions of the wavelength only. We will complete the fitting process in the following section.

## 11.13 FITTING THE PARAMETERS

Table 11.10 lists the parameters occurring in the internal structure fit (by the concurrent model), together with the corresponding values of pressure and wavelength.

Let us first deal with the $x_j$. These values show a behavior similar to that of the refractive index values of Table 11.2, and can therefore be fitted in a similar way.

**TABLE 11.10**

| Pressure, $P_i$ | $\hat{\mu}_i$ | Wavelength, $\lambda_j$ | $x_j$ |
|---|---|---|---|
| 1 | 1797.4 | 6678 | 2838.4 |
| 246.2 | 3027.6 | 6438 | 2983.8 |
| 484.8 | 4049.8 | 5876 | 3402.2 |
| 757.2 | 5035.1 | 5086 | 4271.6 |
| 1107.7 | 6123.1 | 5016 | 4378.4 |
| | | 4922 | 4517.0 |
| | | 4800 | 4718.2 |
| | | 4678 | 4943.2 |

Using calculations analogous to those of Section 11.8 we obtain in a straightforward way:

$$x_j = -755 + 2.818 \times 10^6(\lambda_j - 2145)^{-0.792} \tag{11.42}$$

Similarly, we obtain for $\hat{\mu}_i$,

$$\hat{\mu}_i = 33,148 - 2.6913 \times 10^5(P_i + 1650)^{-0.290} \tag{11.43}$$

Substituting these expressions in Eq. 11.40, we obtain:

$$y_{ij} = -43,382 + 2.1102 \times 10^{-5}(76,530 - 2.6913 \times 10^5 G_i) \\ \times (42,627 + 2.818 \times 10^6 H_j) \tag{11.44}$$

where the symbols $G_i$ and $H_j$ represent the quantities:

$$G_i = (P_i + 1650)^{-0.290} \tag{11.45}$$

$$H_j = (\lambda_j - 2145)^{-0.792} \tag{11.46}$$

Table 11.11 lists the values of $y_{ij}$ calculated in accordance with Eq. 11.44. If we remember that the values in Table 11.11 are coded as indicated in the footnote of Table 11.5, we see that most of the calculated values agree with the observed values to within 1 or 2 units in the fourth decimal place. Thus Eq. 11.44 may be used for interpolation in the ranges of pressure and wavelength covered by the data for all purposes for which this accuracy is sufficient.

**TABLE 11.11**  Refractive Index of Benzene.  Calculated Values* using an Empirical Fit

| $P_i$ \ $\lambda_j$ | 6678 | 6438 | 5876 | 5086 | 5016 | 4922 | 4800 | 4678 |
|---|---|---|---|---|---|---|---|---|
| 1 | 677 | 816 | 1238 | 2066 | 2157 | 2289 | 2469 | 2667 |
| 246.2 | 1884 | 2027 | 2461 | 3311 | 3405 | 3541 | 3726 | 3929 |
| 484.8 | 2882 | 3028 | 3472 | 4341 | 4437 | 4575 | 4764 | 4972 |
| 757.2 | 3827 | 3976 | 4382 | 5315 | 5413 | 5554 | 5747 | 5960 |
| 1107.7 | 4903 | 5055 | 5518 | 6425 | 6525 | 6669 | 6867 | 7084 |

\* $y_{ij}$ values, calculated by Eq. 11.44.

Before leaving this example, an important observation must be made. The fit given by Eqs. 11.44, 11.45, and 11.46 is entirely empirical; no theoretical meaning can be attached to it.  It is well known, however, that the relationship between refractive index, density (which is a function of pressure), and wavelength of pure compounds has been studied in terms of various theoretical models.  One of these is embodied in the Lorentz–Lorenz formula, which for our purpose may be written in the following form:

$$\frac{n^2 - 1}{n^2 + 2}\frac{1}{D} = f(\lambda)$$

where $D$ stands for density, and $f(\lambda)$ is a specified function of the wavelength $\lambda$.

If we express the density as a function of pressure, say $D = \varphi(P)$, we can write the Lorentz–Lorenz formula in the form

$$\frac{n^2 - 1}{n^2 + 2} = \varphi(P)f(\lambda)$$

Thus, for a given substance, the quantity $(n^2 - 1)/(n^2 + 2)$ is expressible as a product of two quantities, one a function of pressure only, and the other a function of wavelength only.  It is therefore interesting to repeat the previous analysis in terms of the transformed variable $(n^2 - 1)/(n^2 + 2)$. Validity of the Lorentz–Lorenz formula then implies that such an analysis must lead to a purely multiplicative model, i.e., a concurrent model in which the point of concurrence is the origin.

It is not necessary to discuss this particular example further, since the applicability of the Lorentz–Lorenz formula, or of any other theoretical model, is a matter that requires the informed judgment of physicists, and cannot be settled by the statistical analysis of one set of data.  We point out, however, that the method of analysis presented in this chapter can be one of the tools by which a physical theory may be tested.

## 11.14　FITTING OF NON-LINEAR MODELS

The linear model, when applied in terms of the column averages, predicates that the elements in each row of the two-way table are linearly related to the corresponding column averages. Evidently this condition is not always fulfilled. Neither can we assume that the elements of each column are necessarily linearly related to the corresponding row averages. If neither of these conditions applies, the linear model will not provide a satisfactory representation of the data. In those cases, one may resort to a plausible generalization of the linear model, by assuming for example that the elements of each row are related by a quadratic expression to the corresponding column averages. Thus, instead of Eq. 11.31 one would write:

$$y_{ij} = A_i + B_i(x_j - \bar{x}) + C_i(x_j - \bar{x})^2 + \varepsilon_{ij} \tag{11.47}$$

where the subscript $i$ indicates that the parameters of the quadratic expression may vary from row to row. For further details and generalizations the reader may consult reference 3, in which a numerical illustration is given for a fit by a quadratic model.

It is also possible to introduce weighting factors into the fitting process, in case the variance of $\varepsilon_{ij}$ cannot be assumed to be the same for all cells of the two-way table. This matter is also dealt with in reference 3.

## 11.15　SUMMARY

The problem of fitting mathematical expressions to data that can be represented by curves or surfaces arises in different types of situations. In one such situation the expression to be fitted is known and the fitting process is carried out for the purpose of obtaining satisfactory values for the parameters involved in the expression. The method of least squares is well suited for this problem. Another case is one in which no expression can be formulated prior to the analysis of the data, and the objective of the fitting process is to obtain such an expression from an examination of the data. An empirical expression of this type is useful mainly as a compact representation of the data. Occasionally, however, it also has theoretical value, namely when it reveals new features of the data that may be explained by theoretical considerations.

In empirical fitting one should be guided by restrictive conditions, i.e., by whatever is known about the general nature of the relationships between the quantities, such as the existence or non-existence of maxima, minima, asymptotes, or changes of curvature. Curve fitting by polynomials is advantageous in routine fitting problems, since it can be done objectively, requiring no judgment other than that of specifying the degree of the polynomial. In many situations, however, polynomials do not satisfy the restrictive conditions, and use must be made of other mathematical functions. The simplest form of polynomial is the straight line; the fitting of straight lines under various circumstances is discussed in more detail in Chapter 12.

Monotonic functions can often be fitted by the general expression:

$$y = y_0 + A(x - x_0)^B$$

which involves 4 parameters: $x_0$, $y_0$, $A$, and $B$. A general method is presented for fitting such an expression to a set of $(x, y)$ data. It is also shown that the exponential function

$$y = y_0 + Ce^{-kx}$$

may be considered as a special case of the equation above, and a method for fitting the exponential is presented.

A measured quantity that depends on two quantitative factors can be represented either as a surface in three-dimensional space or as a family of curves. When the data are in the form of a two-way classification table, in which the rows represent the levels of one factor and the columns the levels of the other factor, the fitting process can be accomplished in two steps. First, the internal structure of the data in the body of the table is determined, by trying to fit an appropriate model to these data. Useful for this purpose are the additive model, the more general linear model (applied either in terms of the rows or in terms of the columns), and the concurrent model. A general method is presented for selecting the most appropriate of these models. Should it be impossible to fit the data satisfactorily by any of these methods, then one may make use of non-linear models, the simplest of which is the quadratic model. This analysis of the internal structure of the data leads to a representation of the data in the body of the table in terms of parameters, the number of which is generally considerably smaller than the number of measured values. Each set of parameters is a function either of the rows or of the columns, but not of both. The second step then consists in representing each set of parameters as a function of the levels of the rows or of the columns, whichever applies. By this method a surface fitting process is reduced to one of fitting a number of functions of one variable each, yet the method provides a single algebraic expression for the function of two variables. It is therefore superior to the often used procedure of separately fitting each curve in a family of curves.

### REFERENCES

1. Anderson, R. L., and T. A. Bancroft, *Statistical Theory in Research*, McGraw-Hill, New York, 1952.

2. Bennett, C. A., and N. L. Franklin, *Statistical Analysis in Chemistry and the Chemical Industry*, Wiley, New York, 1954.
3. Mandel, J., and F. L. McCrackin, "Analysis of Families of Curves," *J. Research Natl. Bur. Standards*, **67A**, 259–267 (May–June 1963).
4. Waxler, R. M., Private communication.
5. Williams, E. J., *Regression Analysis*, Wiley, New York, 1959.

*chapter 12*

# THE FITTING OF STRAIGHT
# LINES

## 12.1 INTRODUCTION

The apparently simple problem of determining the most suitable straight line passing through a given set of points, and the associated question of evaluating the precision of the fitted line, give rise to a host of statistical problems of varying complexity. To deal with all known cases would far exceed the boundaries of this book and to enumerate them all would merely lead to confusion. We will limit ourselves to four specific cases, covering many important applications.

These four situations will be denoted as (1) the classical case; (2) the case where both variables are subject to error; (3) the Berkson case; and (4) the case with cumulative errors.

All problems of straight line fitting involve the basic linear equation

$$Y = \alpha + \beta X \tag{12.1}$$

The statistical aspect of the problem arises through the presence of "error terms," i.e., of fluctuating quantities associated with $X$ or $Y$, or both. In this chapter we will denote by $\varepsilon$ an error associated with $X$, and by $\delta$ an error associated with $Y$. The statistical fitting of a straight line is generally referred to as *linear regression*, where the word "regression" has only historical meaning. If $x$ represents the actual measurement corresponding to $X$, we have

$$x = X + \varepsilon \tag{12.2}$$

Similarly, representing by $y$ the measurement corresponding to $Y$, we write

$$y = Y + \delta \tag{12.3}$$

Thus we are dealing with equations of the type:

$$y - \delta = \alpha + \beta(x - \varepsilon) \tag{12.4}$$

The basic problem may now be formulated in the following manner: given a set of $N$ pairs of associated $x$ and $y$ measurements, $x_i$ and $y_i$, we wish to infer from these values, the best possible estimates of $\alpha$ and $\beta$. We also want to know how reliable these estimates are as representations of the true values $\alpha$ and $\beta$. We will see that further questions of interest are likely to arise in practical situations.

## 12.2   THE CLASSICAL CASE, AND A NUMERICAL EXAMPLE

The first two columns of Table 12.1 list the results of a spectrophotometric method for the determination of propylene in ethylene–propylene copolymers (3). For the purpose of the present exposition we will assume that the $y$ values* are linearly related to the $x$ values, and that the latter

**TABLE 12.1**   Analysis of Propylene in Ethylene–Propylene Copolymers

| Data | | Fitted | Residual |
|---|---|---|---|
| $X = (\%\ \text{Propylene})$ | $y = (R)\dagger$ | $\hat{y}$ | $d$ |
| 10 | 2.569 | 2.539 | $+0.030$ |
| 20 | 2.319 | 2.320 | $-0.001$ |
| 31 | 2.058 | 2.078 | $-0.020$ |
| 40 | 1.911 | 1.880 | $+0.031$ |
| 50 | 1.598 | 1.660 | $-0.062$ |
| 100 | 0.584 | 0.562 | $+0.022$ |

$$S_X = 251 \qquad\qquad S_y = 11.039$$
$$u = 30{,}365 \qquad\quad \hat{\beta} = -0.02197$$
$$w = 14.697281 \qquad \hat{\alpha} = 2.759$$
$$p = -667.141$$

$$\hat{\sigma}^2 = \frac{1}{4}\left[\frac{14.697281 - 14.657570}{6}\right] = 0.001655$$

$$\dagger R = \log\left[100\,\frac{A_{915\text{cm}^{-1}}}{A_{892\text{cm}^{-1}}}\right]$$

* Each $y$ value is the common logarithm of 100 times the ratio of the absorption peaks corresponding to 915 cm$^{-1}$ and 892 cm$^{-1}$.

represent the correct per cent propylene contents of the six copolymers used in this study. This implies that all $\varepsilon$ values are zero, since no error is involved in the $x$'s. Thus, the linear relation which constitutes the statistical model for our data is

$$y_i = \alpha + \beta X_i + \delta_i \tag{12.5}$$

where $X_i$, $y_i$ is the $i$th pair of associated $X$ and $y$ values and $\delta_i$ is the error associated with the measurement $y_i$.

We now make the further assumptions that the $\delta_i$ corresponding to the various $y_i$ all belong to the same statistical population; that they are statistically independent; and that the mean of this population is zero. The variance of the $\delta$, which we denote by $\sigma^2$, is a fixed quantity but its value will generally not be known. These assumptions, which constitute the *classical case*, are entirely reasonable in the example we are discussing. The independence of the $\delta_i$ follows from the fact that each $y$ measurement resulted from an independent experiment. The assumption that they all belong to the same population is supported by experimental evidence (not shown here), and the assumption that the mean of the population of $\delta_i$ is zero is really not an assumption at all, but rather a matter of definition. Indeed if this mean were any quantity different from zero, say 0.065, we could write

$$\delta_i = 0.065 + \delta_i'$$

and

$$y_i = \alpha + \beta X_i + 0.065 + \delta_i'$$
$$= (\alpha + 0.065) + \beta X_i + \delta_i'$$

Thus, $\delta_i'$ could then properly be called the error of $y_i$, and the constant 0.065 would be absorbed in the intercept of the straight line. Since the mean of $\delta_i$ is 0.065, the mean of $\delta_i'$ is zero, and we are back to our initial assumptions.

Under the assumptions made, we can apply the method of least squares with the simplifying condition that the "weights" corresponding to the various $y_i$ are all equal. Since the method of least squares requires only relative weights, we may then set all weights equal to unity. It is readily verified that under these simple assumptions, the method of least squares can be expressed by the following computational procedure.

1. From the $N$ pairs of $(X, y)$ values, compute the following quantities:

$$S_x = \sum_i X_i \qquad S_y = \sum_i y_i$$
$$U = \sum_i X_i^2 \qquad W = \sum_i y_i^2 \qquad P = \sum_i X_i y_i \tag{12.6}$$

2. From these, derive

$$u = N \cdot U - (S_x)^2$$

$$w = N \cdot W - (S_y)^2$$

$$p = N \cdot P - (S_x)(S_y) \qquad (12.7)$$

3. Then the estimate of $\beta$ is

$$\hat{\beta} = p/u \qquad (12.8)$$

The estimate of $\alpha$ is

$$\hat{\alpha} = \frac{1}{N}(S_y - \hat{\beta}S_x) \qquad (12.9)$$

and the estimate of $\sigma^2$ is

$$\hat{V}(\delta) = \frac{1}{N-2}\left[\frac{w - p^2/u}{N}\right] \qquad (12.10)$$

Applying these formulas to the data in Table 12.1 we obtain

$$\hat{\beta} = -0.02197; \quad \hat{\alpha} = 2.759; \quad \hat{V}(\delta) = 0.001655$$

To obtain a better understanding of the quantity $\hat{V}(\delta)$, let us proceed in accordance with the methods explained in Chapters 7 and 8.  Using the estimates $\hat{\alpha}$ and $\hat{\beta}$, we compute, for each $X_i$, the value

$$\hat{y}_i = \hat{\alpha} + \hat{\beta}X_i \qquad (12.11)$$

This value represents, of course, the ordinate of the point lying on the fitted line and corresponding to the abscissa $X_i$.  It differs from $y_i$ (given by Eq. 12.5) by the residual $d_i$:

$$d_i = y_i - \hat{y}_i \qquad (12.12)$$

Table 12.1 also lists the $\hat{y}_i$ and $d_i$ for each value of $x_i$.  From the results of Section 7.6 (Eq. 7.30) we know that the sum of squares of the residuals, divided by an appropriate number of degrees of freedom, yields an estimate for the variance of $\delta$.  The degrees of freedom are obtained by subtracting from the total number of observed $y$ values, the number of parameters for which estimates were calculated.  In the present case, this number of parameters is two ($\alpha$ and $\beta$).  Hence, the degrees of freedom for the estimation of the error $\delta$ is $N - 2$.  We therefore have:

$$\hat{V}(\delta) = \frac{1}{N-2}\sum_i d_i^2 \qquad (12.13)$$

Comparing this equation with Eq. 12.10 we see that the quantity

$(w - p^2/u)/N$ must be equal to $\sum_i d_i^2$. This is indeed the case, as may be verified either algebraically, or arithmetically for any set of data. In practice it is always advisable to compute the individual residuals, as is done in Table 12.1 and to derive from them the estimate of $V(\delta)$ using Eq. 12.13. The reasons for this are threefold. In the first place, if the given set of data contained a gross error, the value of the residual corresponding to the faulty value will tend to be abnormally large, thus revealing that something is amiss. Secondly, in the event that the true model underlying the data is a curve rather than a straight line (as shown in Fig. 12.1) it is seen that the residuals will tend to follow a definite pattern. In the case illustrated by the figure, the residuals are first large and positive, then become smaller, then negative, then positive again. For data of high precision, such a pattern may be difficult to detect by visual inspection of a graph, but it will become apparent from an examination of the numerical values and the succession of signs (e.g., positive, negative, positive) of the residuals.

Finally, the third reason for computing individual residuals is that in Eq. 12.10, the quantities $w$ and $p^2/u$ are generally large, while their difference is often exceedingly small in comparison. Thus, slight round-

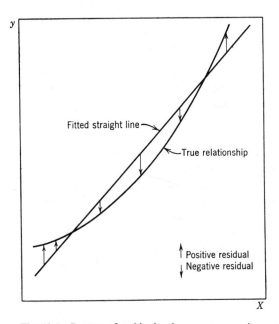

**Fig. 12.1**  Pattern of residuals when curvature exists.

ing errors in $w$ or $p^2/u$ may seriously affect their difference. Formula 12.13 is therefore a safer method for computing $\hat{V}(\delta)$, unless one carries enough places in the computation of $w$ and $p^2/u$ to make certain that any rounding errors are small with respect to their difference. This point can be verified for the data of Table 12.1, for which $w = 14.697281$ and $p^2/u = 14.657570$, while their difference is equal to 0.039711.

The estimate of the error-variance, $\hat{V}(\delta)$, is of direct interest in many applications. Its interpretation generally touches on interesting aspects of the basic problem underlying the data. Thus, in the case of the copolymer data, if the possibility of curvature has been dismissed, there still remain two distinct factors that can contribute to $V(\delta)$. The first of these factors is "experimental error," in the sense of random fluctuations in the measured value that result from fluctuations in environmental conditions. The second factor consists in the possibility that the points corresponding to different samples of copolymer do not really fall along a straight line even in the absence of experimental errors. Such a situation can readily arise as the result of sample-differences that are not directly related to their propylene content, but have nevertheless an effect on the measured value. In the language of analytical chemistry, this factor would be denoted as "interfering substances." Thus, the copolymer which contains 40 per cent propylene, yielded a point situated above the fitted line; this may be due either to random experimental error, or to the presence in this sample of a substance that tends to increase the absorption at 915 cm$^{-1}$ or to decrease the absorption at 892 cm$^{-1}$. Of course there exists the third possibility that the "known" value, 40 per cent propylene, itself is in error, but the consideration of such a possibility violates our basic assumption that the $x$ values are not subject to error.

How can we distinguish between the two factors that may contribute to $V(\delta)$? If $V(\delta)$ is due solely to random fluctuations in the measuring process, we should obtain errors of the same magnitude by applying the process repeatedly to the same sample. Thus, a comparison of $\hat{V}(\delta)$ with the variance of "replication error" is useful for assessing the contribution of the latter to the observed scatter about the line. If, in such a comparison, we find that $V(\delta)$ is so much larger than the estimated variance of replication error that the latter cannot be its only cause, we may attribute the excess to the presence of variability factors that are inherent in these samples.

Let us now turn to another important use of the quantity $\hat{V}(\delta)$. As we have pointed out, $V(\delta)$ is a measure of the scatter of the experimental points about the straight line that represents them, and so of course is its estimate $\hat{V}(\delta)$. It is clear that the success of the fitting procedure increases as $V(\delta)$ decreases. Thus, the precision of the parameter estimates $\hat{\alpha}$ and

$\hat{\beta}$ must depend on $V(\delta)$. The formulas establishing these relationships are:

$$V(\hat{\beta}) = \frac{N}{u} V(\delta) \qquad (12.14)$$

$$V(\hat{\alpha}) = \frac{U}{u} V(\delta) \qquad (12.15)$$

These are relations between true (population) variances. By replacing $V(\delta)$ by its estimate, the same relations give the estimates $\hat{V}(\hat{\beta})$ and $\hat{V}(\hat{\alpha})$. Thus:

$$\hat{V}(\hat{\beta}) = \frac{N}{u} \hat{V}(\delta) \qquad (12.16)$$

$$\hat{V}(\hat{\alpha}) = \frac{U}{u} \hat{V}(\delta) \qquad (12.17)$$

Very often the quantities $\beta$ and $\alpha$ have intrinsic physical meaning. Equations 12.8 and 12.9 provide estimates for these parameters and Eqs. 12.16 and 12.17 give us information about the reliability of these estimates. In other cases, it is not so much the value of the slope or of the intercept of the line that matters most, but rather the line itself. The data we used as an illustration (Table 12.1) provide an example of this case. Here the line is to be used as a calibration curve, relating a spectroscopic measurement to the propylene-content of a copolymer sample. This is done by reading the abscissa corresponding to that point on the line for which the ordinate is the measured spectroscopic value. The error that may possibly affect such an abscissa value is composed of two parts. The first is due to experimental error in the measured value (ordinate); the other to uncertainty in the fitted calibration line. It is clear that this second source of error is related to the uncertainty of the estimates $\hat{\alpha}$ and $\hat{\beta}$ and it can in fact be derived from the quantities $V(\hat{\alpha})$ and $V(\hat{\beta})$. Figure 12.2 clarifies the situation. The solid line represents the "true" relationship, the dashed line is the least squares fit; $y$ is a measurement made on a sample for which $X$ is to be estimated. Since $y$ itself is subject to experimental error, we must also consider the true value $E(y)$ corresponding to this measurement. Using the true value $E(y)$, and the true relationship, we would obtain the true value $X$. All that is available to us, however, is $y$ (instead of $E(y)$) and the least squares fit (instead of the true line). As a consequence, we obtain the estimate $\hat{X}$, instead of the true value $X$.

Algebraically, $\hat{X}$ is given by the relation:

$$\hat{X} = \frac{(y - \hat{\alpha})}{\hat{\beta}} \qquad (12.18)$$

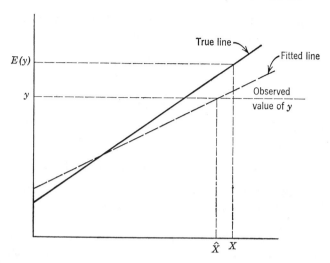

**Fig. 12.2**    Relationship between true and fitted lines, and between true and estimated *X*.

which results from the fact that the point $(\hat{X}, y)$ lies on the line whose equation is

$$y = \hat{\alpha} + \hat{\beta}X$$

The right side of Eq. 12.18 involves three random variables: $y$, $\hat{\alpha}$, and $\hat{\beta}$. Furthermore, this equation is not linear in $\hat{\beta}$. One could of course derive an approximate standard deviation for $\hat{X}$ by the law of propagation of error. But $\hat{\alpha}$ and $\hat{\beta}$ are not statistically independent, and their correlation must be taken into account. Even then, the result would only be an approximation. A better approach to the problem of determining the uncertainty of $\hat{X}$ is to derive a confidence interval for the true value $X$. This can be done in a rigorous manner, using a theorem due to Fieller, to be found in reference 6.

We here mention the result without derivation. To obtain a confidence interval for $X$, first solve the following quadratic equation in $v$:

$$\left[\hat{\beta}^2 - \frac{NK^2}{u}\right]v^2 - [2\hat{\beta}(y - \bar{y})]v + \left[(y - \bar{y})^2 - \frac{N+1}{N}K^2\right] = 0 \tag{12.19}$$

where $\bar{y} = S_y/N$ and the meaning of $K$ is discussed below. This gives two values for $v$; denote them as $v_1$ and $v_2$. Next, obtain the two values:

$$X_1 = v_1 + \bar{X} \tag{12.20}$$

$$X_2 = v_2 + \bar{X} \tag{12.21}$$

It can be shown that these values are the limits of a confidence interval for $X$, provided that the following condition is fulfilled:

$$\hat{\beta}^2 u > NK^2 \tag{12.22}$$

The quantity $K$ is defined for our present purpose by the relation

$$K^2 = t_c^2 \hat{V}(\delta) \tag{12.23}$$

where $t_c$ is the critical value of Student's $t$ at the appropriate level of significance. It may be recalled here that

$$\text{level of significance} = 1 - (\text{confidence coefficient}) \tag{12.24}$$

Thus, if a 95 per cent confidence interval is required for $X$, use the $t$ value at the 5 per cent level of significance (since $0.05 = 1 - 0.95$).

Let us apply these formulas to the propylene data of Table 12.1. Suppose that an ethylene–propylene copolymer of unknown composition is analyzed and a value of $R = 0.792$ is found. We wish to estimate the propylene content of the sample and derive a 95 per cent confidence interval for it.

We have:

$$K^2 = t^2 \hat{V}(\delta) = (2.776)^2 (0.001655) = 0.01275$$

$$\hat{\beta}^2 - \frac{NK^2}{u} = (-0.02197)^2 - \frac{6}{30,365}(0.01275) = 4.8018 \times 10^{-4}$$

$$2\hat{\beta}(y - \bar{y}) = 2(-0.02197)(0.792 - 1.8398) = 0.04604$$

$$(y - \bar{y})^2 - \frac{N+1}{N} K^2 = (0.792 - 1.8398)^2 - \frac{7}{6}(0.01275) = 1.0830$$

Hence the quadratic equation* is (after multiplication of all terms by $10^4$):

$$4.8018v^2 - 460.4v + 10830 = 0$$

which yields:

$$v_1 = 41.39 \quad \text{and} \quad v_2 = 54.48$$

From these values we obtain:

$$X_1 = 83.2 \quad \text{and} \quad X_2 = 96.3$$

The point estimate for $X$ is given by

$$\hat{X} = \frac{y - \hat{\alpha}}{\hat{\beta}} = \frac{0.792 - 2.759}{-0.02197} = 89.5$$

Thus the propylene content of the unknown sample is estimated to be 89.5 per cent; a 95 per cent confidence interval for its true value is given by

$$83.2 < X < 96.3$$

* Condition (12.22) is fulfilled, since $\hat{\beta}^2 u = 14.6566$ and $NK^2 = 0.0765$.

We now make two important remarks. In the first place, Eq. 12.19 is based on the assumption that the measurement of $y$ for the unknown sample is made with the same precision as that with which the individual points in the calibration curve were obtained, i.e., with a variance equal to $V(\delta)$. If several replicate measurements are made for the unknown sample, and their results averaged, this is of course no longer true. Suppose that $y$ is the average of $n$ replicates. Then the only modification in the theory consists in replacing, in Eq. 12.19, the quantity $(N + 1)/N$ by $1/n + 1/N$. Thus, the quadratic equation to be solved in that case is:

$$\left[\beta^2 - \frac{NK^2}{u}\right]v^2 - [2\beta(y - \bar{y})]v + \left[(y - \bar{y})^2 - \left(\frac{1}{n} + \frac{1}{N}\right)K^2\right] = 0$$

$$(12.25)$$

It is interesting to observe what happens when $n$ becomes very large. The expression $1/n + 1/N$, for $n = \infty$, approaches the value $1/N$. The quadratic equation then yields a somewhat shorter confidence interval, but the gain may be small or even insignificant. The reason for this is of course that even though the value $y$, for the unknown sample, is now measured without error, the use of the calibration curve for the estimation of the corresponding $X$ is still subject to all the uncertainty associated with the estimated calibration line. We have here a quantitative expression for the qualitatively obvious fact that the precision of measurements involving the use of calibration lines is limited by any lack of precision in the calibration line itself. This brings us to the second point.

It is in the nature of calibration lines that they are used repeatedly, either in relation to a single problem involving several determinations or for a variety of different problems. Any errors in the calibration line itself are thus perpetuated in all the determinations for which it is used. Suppose that for each such determination a 95 per cent confidence interval is derived. It is readily seen that the sense of security generally given by confidence intervals may be illusory in such a situation. Indeed, if through the interplay of chance effects the estimated calibration line happened to be appreciably different from the true line, the proportion of cases in which the confidence intervals would fail to include the true values of the unknown samples would then be appreciably larger than the expected 5 per cent. The reason for this is that this expected probability pertains to *independent* experiments, not to a series of determinations using the same (erroneous) calibration line, and therefore vitiated by the *same* errors. How serious this is depends on the importance of the requirement that for *all* determinations based on the use of the same calibration line the confidence intervals bracket their true values. It is an interesting fact that it is possible to achieve this more demanding requirement in the framework of the preceding theory.   This is done as follows.

Suppose that a given calibration line, say the one derived from Table 12.1, is to be used for $k = 25$ determinations, after which the measuring process will be recalibrated.   We wish to derive confidence intervals for each of the 25 determinations in such a way that the probability will be 95 per cent that *all 25* intervals bracket the corresponding true values.

To accomplish this, find the confidence interval for each of the $k$ determinations (in our example $k = 25$), by means of Eq. 12.19 in which, however, $K$ is now defined by the following relation

$$K^2 = (k+2)F_c \hat{V}(\delta) \tag{12.26}$$

where $F_c$ is the critical value of the $F$ distribution with $k + 2$ and $N - 2$ degrees of freedom.   Using again the data of Table 12.1 as an illustration, and taking $k = 25$, we find, for a 95 per cent confidence coefficient:

$$F_c = F \text{ with 27 and 4 degrees of freedom} = 5.76$$

Hence

$$K^2 = 27(5.76)(0.001655) = 0.25739$$

The quadratic equation in $v$, for $R = 0.792$, now becomes

$$4.3182v^2 - 460.4v + 7976 = 0$$

yielding

$$v_1 = 21.8 \quad \text{and} \quad v_2 = 84.8$$

from which we obtain the limits

$$X_1 = 63.6 \quad \text{and} \quad X_2 = 126.7$$

This calculation can be made for each of the 25 determinations.   The probability will be 95 per cent that *all 25* confidence intervals bracket their respective true $X$ values.

The confidence interval now obtained is so large as to be virtually valueless. The reason for this lies in the stringency of the requirements. The example shows that it is unrealistic to expect with 95 percent probability that as many as 25 intervals will bracket the corresponding true values, because the intervals that will satisfy this requirement will be too long to be of practical value.

## 12.3   JOINT CONFIDENCE REGION FOR SLOPE AND INTERCEPT

The concept of a confidence interval has a useful generalization. Consider a set of points to which a straight line is to be fitted.   Even though the method of least squares, under the assumptions made in Section 12.2, yields a unique line, other straight lines could also qualify as reasonable fits.　The reason for this is that the least squares solution is but a statistical estimate of the true line.　One of the great advantages of making a least squares analysis is that it yields not only an estimate of the true line, but also a range of possibilities for its location.　Since the location of a straight line requires two parameters, such as the intercept and the slope, a region of possible locations of the true line is equivalent to a set of possible pairs of values for these parameters.

Suppose that the straight line fitted by least squares to a set of points has intercept $\hat{\alpha}$ and slope $\hat{\beta}$.　Consider a second line, whose slope is slightly *larger* than $\hat{\beta}$.　It is then readily seen that if this alternative line is to qualify as a reasonable fit to the same set of points, its intercept should in general be *smaller* than $\hat{\alpha}$.　Thus, the collection of lines that qualify as reasonable fits to the same set of points must satisfy the condition that a larger slope tends to be associated with a smaller intercept, and vice versa. In other words, there is a negative correlation between the estimates of the slope and the intercept.　We can represent each qualifying line as a pair of values for $\alpha$ and $\beta$, or graphically, as a point on a rectangular plot in which the abscissa represents the intercept $\alpha$, and the ordinate the slope $\beta$.　Such a plot is shown in Fig. 12.3.　Statistical theory shows that the collection of qualifying points (pairs of $\alpha$ and $\beta$ values), for a given

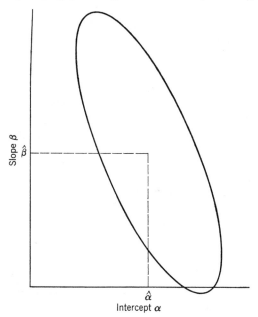

**Fig. 12.3**　Joint confidence ellipse for the slope and the intercept of a fitted straight line.

confidence coefficient, is contained within an ellipse slanted in such a way that its main axis runs in the north-west to south-east direction. This slanted position reflects of course the negative correlation between the estimates for slope and intercept.

The mathematical meaning of the confidence ellipse is quite analogous to that of a simple confidence interval. Adopting for example a confidence coefficient of 95 per cent, the probability that the area inside the ellipse will contain the point $(\alpha, \beta)$ (representing the parameters of the true line) is 95 per cent. The center of the ellipse is the point $(\hat{\alpha}, \hat{\beta})$, i.e., the point representing the least squares fit. To a larger confidence coefficient corresponds a larger ellipse. Thus the price for greater confidence of including the "true" point is a greater region of uncertainty.

The equation of the confidence ellipse is

$$U(\beta - \hat{\beta})^2 + 2S_x(\beta - \hat{\beta})(\alpha - \hat{\alpha}) + N(\alpha - \hat{\alpha})^2 = 2F\hat{V}(\delta) \quad (12.27)$$

where $F$ is the critical value of the $F$ distribution with 2 and $N - 2$ degrees of freedom, at a significance level equal to

$$1 - \text{(confidence coefficient)}$$

The joint confidence ellipse for the slope and intercept of a straight line has interesting and useful applications. We will describe an application in the theory of the viscosity of dilute solutions (9).

If $c$ represents the concentration of a dilute solution, $\eta_{sp}$ its specific viscosity, and $[\eta]$ its intrinsic viscosity, the following equation often applies:

$$\frac{\eta_{sp}}{c} = [\eta] + k'[\eta]^2 c \quad (12.28)$$

where $k'$ is a constant characterizing the solute–solvent system.

Suppose that the specific viscosities of a series of solutions of different concentrations are measured. Then the quantity defined by

$$y = \frac{\eta_{sp}}{c} \quad (12.29)$$

is, according to Eq. 12.28, a linear function of $c$. The slope of the straight line is $k'[\eta]^2$ and its intercept is $[\eta]$. Writing Eq. 12.28 as

$$y = \alpha + \beta c \quad (12.30)$$

we have

$$\alpha = [\eta] \quad (12.31)$$

$$\beta = k'[\eta]^2 \quad (12.32)$$

An estimate for $k'$ is therefore obtained by dividing the slope by the square of the intercept:

$$k' = \frac{k'[\eta]^2}{[\eta]^2} = \frac{\beta}{\alpha^2} \quad (12.33)$$

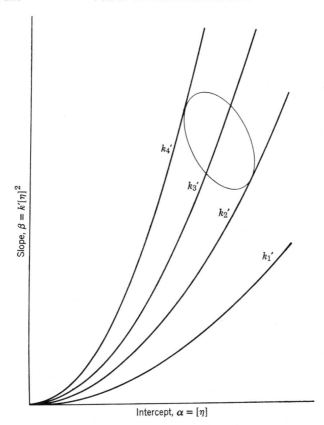

**Fig. 12.4** Construction of confidence interval for non-linear function ($k'$). The parabolas represent the relation $\beta = k'\alpha^2$ for various values of $k'$. The confidence interval for $k'$ consists of all values of $k'$ contained between $k'_2$ and $k'_4$.

This relation provides us not only with a method for estimating $k'$, but also with a procedure for obtaining a confidence interval for $k'$. We now explain how such a confidence interval is derived from the joint confidence ellipse for $\alpha$ and $\beta$.

From Eq. 12.33 we derive

$$\beta = k'\alpha^2 \qquad (12.34)$$

For any given value of $k'$, this equation represents a parabola in the $\alpha$, $\beta$ plane (Fig. 12.4).

Different values of $k'$ result in different parabolas some of which intersect the ellipse, while others do not. A value of $k'$ is acceptable, if it corresponds to a parabola which contains acceptable combinations of $\alpha$

and $\beta$, i.e., points inside the ellipse. Consequently, the totality of all acceptable $k'$ values is composed of those values for which the corresponding parabolas intersect the ellipse. In Fig. 12.4 this is the set of $k'$ values contained between $k_2'$ and $k_4'$. Thus, $k_3'$ is acceptable, while $k_1'$ is not. The interval of $k'$ values contained between $k_2'$ and $k_4'$ is the desired confidence interval for the parameter $k'$.

In the example we have just discussed, the function of interest is the quotient of the slope by the square of the intercept. A similar line of reasoning shows that confidence intervals can be derived, by means of the joint confidence ellipse, for any function of these two parameters. For example, if our interest had been in the quantity

$$Z = \alpha - \beta$$

we would have considered the family of straight lines

$$\beta = -Z + \alpha$$

As $Z$ takes on different values, this relation generates a family of parallel lines, whose common slope is equal to unity, and whose intercepts are equal to $-Z$. We would then have considered the set of $Z$ values for which the corresponding straight lines intersect the ellipse. This set would have provided the desired confidence interval for $Z$.

A particularly interesting aspect of this use of the joint confidence ellipse is that regardless how often it is used for the derivation of confidence intervals for functions of the slope and intercept, and regardless of the diversity of such functions, the probability remains at least 95 per cent (the confidence coefficient selected) that *all* confidence intervals thus derived are jointly valid, i.e., that all these intervals bracket the true values for their parameters.

## 12.4  CONFIDENCE BANDS FOR A CALIBRATION LINE

It is a noteworthy fact that many problems related to the setting of confidence limits for quantities derived from linear calibration curves, including some of the situations discussed in the preceding sections, can be solved by the same basic procedure. This procedure is best described in terms of a graphical presentation of the calibration line. In Fig. 12.5, the least squares line fitted to a set of data such as that of Table 12.1 is shown as a solid line. Also shown are the two branches of a hyperbola situated symmetrically with respect to the least squares line. The exact location of the hyperbola depends on the problem to be solved.

1. Estimation of $X$ corresponding to a given value of $y$.

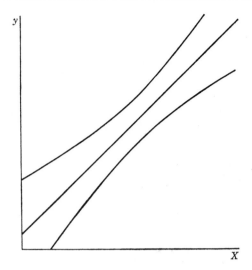

**Fig. 12.5**   Confidence band for a straight line.   The straight line in the figure is the least squares fit.   The region between the two branches of the hyperbola is a confidence band for the true line.

Let $y$ be the average of $n$ replicate determinations.   Then the equation of the hyperbola is:

$$y - \hat{\alpha} - \hat{\beta}X = \pm K \left[ \frac{1}{n} + \frac{1}{N} + \frac{N(X - \bar{X})^2}{u} \right]^{1/2} \tag{12.35}$$

where $K = \sqrt{t_c^2 \hat{V}(\delta)}$ and $t_c$ is the critical value of Student's $t$, with $N - 2$ degrees of freedom, at a level of significance equal to

$$1 - \text{(confidence coefficient)}$$

Having obtained the hyperbola, draw a horizontal line through the value $y$ laid off on the ordinate.   The abscissa of the intersection of this line with the least squares fitted line is the estimated value for $X$.   The confidence interval for $X$ is given by the abscissas corresponding to the intersection of the horizontal line with the two branches of the hyperbola.

A special case is that for which $n = \infty$.   In this case $y$, being the average of an infinite number of replicates, is really the expected value $E(y)$.   Equation 12.35 is applicable, omitting the term $1/n$ (which is zero for $n = \infty$).

2. Repeated use of the calibration line for the estimation of the $X$ values corresponding to a series of $k$ values of $y$.

Assume again that each $y$ is the average of $n$ replicates. The equation of the appropriate hyperbola is:

$$y - \hat{\alpha} - \hat{\beta}X = \pm K \left[\frac{1}{n} + \frac{1}{N} + \frac{N(X - \bar{X})^2}{u}\right]^{1/2} \qquad (12.36)$$

where $K = \sqrt{(k+2)F_c \hat{V}(\delta)}$.

$F_c$ is the critical value of the $F$ distribution with $k + 2$ and $N - 2$ degrees of freedom, at a level of significance equal to

$$1 - (\text{confidence coefficient})$$

The procedure for using the hyperbola is exactly the same as in case 1. The case $n = \infty$ is treated as before.

3. Determining the uncertainty of the calibration line.

Draw the following hyperbola:

$$y - \hat{\alpha} - \hat{\beta}X = \pm K \left[\frac{1}{N} + \frac{N(X - \bar{X})^2}{u}\right]^{1/2} \qquad (12.37)$$

where $K = \sqrt{2F_c \hat{V}(\delta)}$.

$F_c$ is the critical value of the $F$ distribution with 2 and $N - 2$ degrees of freedom, at a level of significance equal to

$$1 - (\text{confidence coefficient})$$

The uncertainty band for the true line is the area located between the two branches of the hyperbola. If a vertical line is drawn at any given value of $X$, its intersection with the fitted straight line yields an estimate of $E(y)$; and its points of intersection with the two branches of the hyperbola yield a confidence interval for $E(y)$.

## 12.5 BOTH VARIABLES SUBJECT TO ERROR

So far we have assumed that $X$ is known without error, while $y$ is a measured value subject to uncertainty. The illustration we have used (data of Table 12.1) involved measurements made on copolymer samples of varying composition, and we have assumed that the true composition of each sample is known. Actually there was reason to believe, in that study, that these "true" values are themselves subject to uncertainty. This became apparent when the series of measurements were repeated on the samples after vulcanizing them. The actual measurements, both for the raw and the vulcanized samples, are given in Table 12.2. The data for the raw samples are of course identical with those given in Table 12.1.

Figure 12.6 is a plot of the measurements on the vulcanized samples versus the corresponding measurements on the raw samples. Apparently

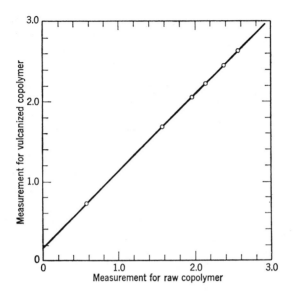

**Fig. 12.6**   Comparison of results of analysis on raw and vulcanized rubbers.

the relation is very closely linear, and we wish to fit a straight line to the data by the method of least squares.   However, we can no longer assume that one of the two variables (measurement on the raw or on the vulcanized samples) is known without error, since both are measured quantities.   It is actually known, from information not given here, that the standard deviation of experimental error for both measurements is the same and that its estimate is: $\hat{\sigma} = 0.0090$.   The problem is to fit a straight line to a set of $(x, y)$ points when the error standard deviation is the same for $x$ and $y$.

We will first solve a more general problem and then return to the data of Table 12.2 to illustrate the procedure.

Consider a set of $(X, Y)$ pairs satisfying the relation

$$Y_i = \alpha + \beta X_i$$

and the corresponding measurements

$$x_i = X_i + \varepsilon_i$$

$$y_i = Y_i + \delta_i$$

where $\varepsilon_i$ and $\delta_i$ are errors of measurement.

Let $\sigma_\varepsilon$ and $\sigma_\delta$ be the standard deviations of the errors $\varepsilon$ and $\delta$, respectively.

Our problem is to estimate $\alpha$ and $\beta$ and to obtain estimates for $\sigma_\varepsilon$ and $\sigma_\delta$. The solution is obtained by applying the generalized method of least squares (5) which we have described in Section 7.9. It will be recalled that this solution requires that "weights" corresponding to $x$ and $y$ be known. We also know that it is sufficient to have the relative values of the weights. Assume therefore that we know the ratio

$$\lambda = \frac{V(\varepsilon)}{V(\delta)} \tag{12.38}$$

Then the quantities $\lambda$ and 1 are proportional to $V(\varepsilon)$ and $V(\delta)$, and their reciprocals $1/\lambda$ and 1 are proportional to the weights of $\varepsilon$ and $\delta$, respectively. We can therefore take $1/\lambda$ and 1, or their multiples, 1 and $\lambda$, as the weights of $\varepsilon$ and $\delta$.

Let $\hat{X}_i$ and $\hat{Y}_i$ be estimates for $X_i$ and $Y_i$. Deming's method consists in minimizing the weighted sum of squares:

$$S = \sum_i [(x_i - \hat{X}_i)^2 + (y_i - \hat{Y}_i)^2 \lambda] \tag{12.39}$$

subject to the conditions:

$$\hat{Y}_i = \hat{\alpha} + \hat{\beta}\hat{X}_i \tag{12.40}$$

where $i$ takes on all values from 1 to $N$ (the number of $(x, y)$ pairs given). Introducing these values of $\hat{Y}_i$ in $S$, we have

$$S = \sum_i [(x_i - \hat{X}_i)^2 + (y_i - \hat{\alpha} - \hat{\beta}\hat{X}_i)^2 \lambda]$$

This quantity is to be minimized with respect to $\hat{\alpha}$, $\hat{\beta}$, and all $\hat{X}_i$. We will omit the details of calculation and state the results: let $\bar{x}$ and $\bar{y}$ be the averages of the $x_i$ and $y_i$ respectively, and let the quantities $u$, $w$, and $p$ be defined as follows (cf. Eqs. 12.7):

$$u = N \sum_i (x_i - \bar{x})^2 \tag{12.41}$$

$$w = N \sum_i (y_i - \bar{y})^2 \tag{12.42}$$

$$p = N \sum_i (x_i - \bar{x})(y_i - \bar{y}) \tag{12.43}$$

Then $\hat{\beta}$ is the solution of the following quadratic equation:

$$(\lambda p)\hat{\beta}^2 + (u - \lambda w)\hat{\beta} - p = 0 \tag{12.44}$$

That is*:

$$\hat{\beta} = \frac{\lambda w - u + \sqrt{(u - \lambda w)^2 + 4\lambda p^2}}{2\lambda p} \tag{12.45}$$

---

* It can be shown that only the solution with the positive sign before the radical is acceptable.

Furthermore:

$$\hat{\alpha} = \bar{y} - \hat{\beta}\bar{x} \tag{12.46}$$

The estimates $\hat{X}_i$ and $\hat{Y}_i$ are obtained as follows: let

$$d_i = y_i - \hat{\alpha} - \hat{\beta}x_i \tag{12.47}$$

Then

$$\hat{X}_i = x_i + \frac{\lambda\hat{\beta}}{1 + \lambda\hat{\beta}^2}\, d_i \tag{12.48}$$

$$\hat{Y}_i = y_i - \frac{1}{1 + \lambda\hat{\beta}^2}\, d_i \tag{12.49}$$

Finally, estimates for $V(\varepsilon)$ and $V(\delta)$ are given by the formulas:

$$\hat{V}(\varepsilon) = \frac{\lambda}{1 + \lambda\hat{\beta}^2} \frac{\sum\limits_i d_i^2}{N - 2} \tag{12.50}$$

$$\hat{V}(\delta) = \frac{1}{1 + \lambda\hat{\beta}^2} \frac{\sum\limits_i d_i^2}{N - 2} \tag{12.51}$$

The quantity $\sum d_i^2$ can be calculated by means of Eq. 12.47; it can also be obtained directly by means of the following formula:

$$\sum_i d_i^2 = \frac{w - 2\hat{\beta}p + \hat{\beta}^2 u}{N} \tag{12.52}$$

It can be shown that the quantity $\sum d_i^2$ is equal to the minimum value of $S$ (Eq. 12.39); i.e., the value of $S$ in which $\hat{X}_i$ and $\hat{Y}_i$ are calculated by means of Eqs. 12.48 and 12.49. The quantity $\sum\limits_i d_i^2/(N - 2)$ can be treated as a variance estimate with $N - 2$ degrees of freedom.

Applying the theory we have just described to our numerical example (Table 12.2), we have first, according to our assumption: $\lambda = 1$. From the data of Table 12.2 we derive:

$$\bar{x} = 1.8398 \qquad \bar{y} = 1.9283$$

$$u = 14.697281 \qquad w = 13.963976 \qquad p = 14.324972$$

Using these values, we obtain:

$$\hat{\beta} = 0.9747 \qquad \hat{\alpha} = 0.135$$

$$\sum d_i^2 = 0.000339$$

and consequently:

$$\hat{V}(\varepsilon) = \hat{V}(\delta) = \frac{0.8475 \times 10^{-4}}{1 + 0.95} = 0.434 \times 10^{-4}$$

Hence

$$\hat{\sigma}_\varepsilon = \hat{\sigma}_\delta = 0.0066$$

**TABLE 12.2** Comparison of Results for Raw and Vulcanized Ethylene–Propylene Copolymers

| $x$ (raw) | $y$ (vulcanized) | $d = y - \hat{\alpha} - \hat{\beta}x$ |
|---|---|---|
| 2.569 | 2.646 | 0.0070 |
| 2.319 | 2.395 | −0.0003 |
| 2.058 | 2.140 | −0.0010 |
| 1.911 | 2.000 | 0.0023 |
| 1.598 | 1.678 | −0.0154 |
| 0.584 | 0.711 | 0.0068 |
| $\hat{\beta} = 0.9747$ | $\hat{\alpha} = 0.135$ | $\hat{\sigma}_\varepsilon = \hat{\sigma}_\delta = 0.0066$ |

It is interesting to compare the value for $\hat{\sigma}_\varepsilon$ resulting from this analysis with two other estimates of error. The first of these is the estimate (mentioned earlier) which was obtained from a comparison of replicate measurements made on the same set of samples using three different instruments; its value is $\hat{\sigma}_\varepsilon = 0.0090$. The second is the estimate obtained in Section 12.2 from a fit of the data for the raw samples against the values assigned by the manufacturers: $\hat{\sigma}_\varepsilon = \sqrt{0.001655} = 0.0407$. Agreement between our present value and the internal estimate derived from replication is quite satisfactory, but the estimate obtained in Section 12.2 is considerably higher, implying a much poorer precision. Now, the measurements of the vulcanized samples were made several months after those of the raw samples. If we perform for the vulcanized samples an analysis similar to that of Section 12.2, we obtain an equally poor fit. Yet, despite the lag in time between the two sets of measurements, their agreement is consistent with the much smaller measure of experimental error obtained from replication. A natural conclusion is that the values given by the manufacturers are themselves subject to error.

## 12.6  CONTROLLED $X$-VARIABLE: THE BERKSON CASE

An important aspect of straight-line fitting was discovered by Berkson (1). To understand it, let us return momentarily to the situation treated in the preceding section.

This situation is represented by the three relations:

$$Y_i = \alpha + \beta X_i$$

$$x_i = X_i + \varepsilon_i$$

$$y_i = Y_i + \delta_i$$

which give rise to:

$$y_i = Y_i + \delta_i = \alpha + \beta X_i + \delta_i$$
$$= \alpha + \beta(x_i - \varepsilon_i) + \delta_i$$

or

$$y_i = (\alpha + \beta x_i) + (\delta_i - \beta \varepsilon_i) \tag{12.53}$$

Now, the variance of $\delta_i - \beta \varepsilon_i$ is equal to $V(\delta) + \beta^2 V(\varepsilon)$, if, as we assumed, $\varepsilon_i$ and $\delta_i$ are statistically independent. This variance is independent of $i$. It looks therefore as though the classical case were applicable, and one may wonder why the more complicated treatment of Section 12.5 was necessary.

Actually, the error term in Eq. 12.53 is not independent of the quantity $\alpha + \beta x_i$. Indeed, from $x_i = X_i + \varepsilon_i$ it follows that

$$x_i - \varepsilon_i = X_i$$

Now, $X_i$ is a fixed quantity (not a random variable); hence any fluctuation of $\varepsilon_i$ is accompanied by an equivalent fluctuation of $x_i$. Thus $\varepsilon_i$ and $x_i$ are correlated and so are the two quantities $(\alpha + \beta x_i)$ and $(\delta_i - \beta \varepsilon_i)$. This correlation invalidates the solution based on the classical case, in which the error term is assumed to be independent of the true values $X$ and $Y$, as well as of the error $\delta$ associated with the measurement of $Y$.

What Berkson pointed out is that many experiments are carried out in such a way that this bothersome correlation no longer exists. This happens when the variable $X$ is a "controlled" quantity, in the sense that for each measurement $y_i$, the corresponding value of $X$ is "set" at an assigned value $X_i$, or at least as close to $X_i$ as is experimentally feasible.

Suppose, for example, that we are studying the resistance of plastic pipe to stresses caused by hydrostatic pressure. Identical specimens of the pipe are subjected to a series of assigned hydrostatic pressures, $P_1, P_2, \ldots, P_N$, and for each specimen the time to failure, $T$, is experimentally determined. Let us assume that the logarithm of $T$ is a linear function of the logarithm of stress, $\log S$. Then:

$$\log T = \alpha + \beta \log S$$

Now, while we can "set" the pressure $P$ at assigned values, such as $10, 20, 30, \ldots$ pounds per square inch, the resulting stresses will not be maintained at fixed values throughout the life of the pipe specimens, because of variation in temperature, deformation of the pipe, etc. Nor will these fluctuations be the same for all specimens, since their lifetimes will differ. Therefore, the *nominal* stresses $S_1, S_2, \ldots$, are really only "aimed-at" values of stress, while the *actual* values of stress might be somewhat different. Of course, the lifetime is governed by the *actual* stress, not the nominal one, but the latter is the only one known to us. Representing $\log S$ by $X$ and $\log T$ by $Y$, we have

$$Y = \alpha + \beta X$$

Let $X_i$ be the *nominal* value of $X$, and $x_i$ its *actual* value. The difference

$$\varepsilon_i = x_i - X_i \tag{12.54}$$

is an error unknown to us. The "response" $Y_i$ is a linear function of the *actual* value of $X$, i.e., $x_i$:

$$Y_i = \alpha + \beta x_i \tag{12.55}$$

In addition, there may be an *error of measurement*, $\delta_i$, in the determination of $Y_i$:

$$\delta_i = y_i - Y_i \tag{12.56}$$

where $y_i$ is the *measured* value of $Y_i$.

Thus $Y_i = y_i - \delta_i$, and consequently, from Eq. 12.55:

$$y_i = \alpha + \beta x_i + \delta_i \tag{12.57}$$

Replacing $x_i$ by its value derived from Eq. 12.54, we obtain:

$$y_i = \alpha + \beta(X_i + \varepsilon_i) + \delta_i$$

or

$$y_i = (\alpha + \beta X_i) + (\delta_i + \beta \varepsilon_i) \tag{12.58}$$

This relation is formally analogous to Eq. 12.53 but its physical meaning is entirely different. Indeed, in Eq. 12.58, $X_i$ is the *nominal* value, well known to the experimenter, since it is one of the values at which he "set" the variable $X$. *The error term $\delta_i + \beta \varepsilon_i$ is now statistically independent of $X_i$*, since the deviation $\varepsilon_i$ of the actual from the nominal value is due to changes in temperature, etc., which occurred independently of the nominal value $X_i$. Therefore, the requirements of the classical case are fulfilled by Eq. 12.58, and the regression of $y$ on $X$ will yield the correct least squares estimates for $\alpha$ and $\beta$. The only difference with the situation treated in Section 12.2 is that, in accordance with Eq. 12.58, the scatter about the fitted straight line will now provide a measure for the variance of $\delta_i + \beta \varepsilon_i$, rather than for $\delta_i$ alone.

$$V(\delta_i + \beta \varepsilon_i) = V(\delta) + \beta^2 V(\varepsilon) \tag{12.59}$$

Denoting the residuals of the regression by $d_i$:

$$d_i = y_i - \hat{\alpha} - \hat{\beta} X_i \tag{12.60}$$

we have:

$$\hat{V}(\delta_i + \beta \varepsilon_i) = \frac{\sum d_i^2}{N - 2} \tag{12.61}$$

or

$$\hat{V}(\delta) + \beta^2 \hat{V}(\varepsilon) = \frac{\sum d_i^2}{N - 2} \tag{12.62}$$

If the error of measurement of $Y$, $V(\delta)$, can be estimated independently, Eq. 12.62 will provide an estimate for $V(\varepsilon)$. However, in most cases it will not be necessary to separate $V(\delta)$ and $V(\varepsilon)$.

One may interpret Eq. 12.58 in the following way. The error of the response $y_i$ is composed by two parts: the *error of measurement* $\delta_i$ and an error *induced* in $y$ as a result of the failure of the actual value of $X_i$ to be identical with its nominal value. This induced portion of the error is $(+\beta\varepsilon_i)$.

From a practical point of view, the Berkson model is of great importance. It assures us that in performing experiments at preassigned values of the independent variables—the common, though not the only way in laboratory experimentation—we may apply the method of least squares for straight line fitting as though the controlled variables were free of error, even though this may not be the case. It is only in the interpretation of the residuals that a difference exists between this case and the one in which the independent variables are really free of error. In the Berkson model, the residual is due in part only to an error of measurement in $y$ (the dependent variable); the other part is the error *induced* in $y$ as a result of discrepancies between the nominal and the actual values of the independent variables.

## 12.7 STRAIGHT LINE FITTING IN THE CASE OF CUMULATIVE ERRORS

There are situations in curve fitting in which, by the very nature of the experiment, the error associated with each point includes the errors associated with all previous points.

An instructive example is provided by the data obtained by Boyle in 1662 (2) in the celebrated experiment by which the law named after him was first established. A schematic of his instrument is shown in Fig. 12.7. A fixed mass of air is contained in the graduated closed end of the U-tube. By adding increasing amounts of mercury through the open end of the other leg of the tube, the level of the mercury in the closed leg is made to coincide successively with each of the evenly spaced graduation marks. These marks provide a measure for the volume of the entrapped air, while the pressure is given by the sum of the atmospheric pressure, measured separately, and the difference between the levels of mercury in the two legs of the U-tube. Boyle's original data are given in Table 12.3. Our task is to fit to them the law expressed by the equation:

$$pV = K \tag{12.63}$$

where $p$ represents the pressure, $V$ the volume of the gas, and $K$ is a constant. The values of $V$ are nominally predetermined by the graduation

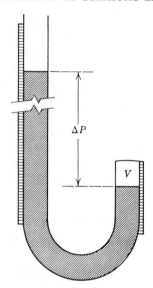

**Fig. 12.7**  Schematic of Boyle's apparatus.

marks in the closed leg of the tube.  This does not imply that they are necessarily without error, since the marks may not occur exactly where they should.  We have seen, however, in Section 12.6 that it is legitimate

**TABLE 12.3**  Boyle's Data

| Volume (graduation mark) | Pressure (inches mercury) | Volume (graduation mark) | Pressure (inches mercury) |
|---|---|---|---|
| 48 | $29\frac{2}{16}$ | 23 | $61\frac{5}{16}$ |
| 46 | $30\frac{9}{16}$ | 22 | $64\frac{1}{16}$ |
| 44 | $31\frac{15}{16}$ | 21 | $67\frac{1}{16}$ |
| 42 | $33\frac{8}{16}$ | 20 | $70\frac{11}{16}$ |
| 40 | $35\frac{5}{16}$ | 19 | $74\frac{2}{16}$ |
| 38 | 37 | 18 | $77\frac{14}{16}$ |
| 36 | $39\frac{5}{16}$ | 17 | $82\frac{12}{16}$ |
| 34 | $41\frac{10}{16}$ | 16 | $87\frac{14}{16}$ |
| 32 | $44\frac{3}{16}$ | 15 | $93\frac{1}{16}$ |
| 30 | $47\frac{1}{16}$ | 14 | $100\frac{7}{16}$ |
| 28 | $50\frac{5}{16}$ | 13 | $107\frac{13}{16}$ |
| 26 | $54\frac{5}{16}$ | 12 | $117\frac{9}{16}$ |
| 24 | $58\frac{13}{16}$ | | |

to perform a least-squares analysis in which all the error is assumed to occur in the $p$ measurements and none in the nominally predetermined $V$ values. To facilitate the analysis we write Eq. 12.63 in the form:

$$\frac{1000}{p} = \frac{1000}{K} V \tag{12.64}$$

According to this equation, a plot of $1000/p$ versus $V$ should yield a straight line, going through the origin, with a slope equal to $1000/K$. Writing

$$X = V \tag{12.65}$$

$$y = \frac{1000}{p} \tag{12.66}$$

$$\beta = \frac{1000}{K} \tag{12.67}$$

and expressing all measurements in the decimal number system, we obtain the data in the first two columns of Table 12.4, and we state that these data should conform to the model equation:

$$y = \beta X \tag{12.68}$$

**TABLE 12.4**   Analysis of Boyle's Data Assuming Berkson Model

| $X = V$ | $y = 1000/p$ | $d = y - \hat{\beta}X$ | $X = V$ | $y = 1000/p$ | $d = y - \hat{\beta}X$ |
|---|---|---|---|---|---|
| 48 | 34.335 | 0.244 | 23 | 16.310 | −0.025 |
| 46 | 32.720 | 0.049 | 22 | 15.610 | −0.015 |
| 44 | 31.311 | 0.061 | 21 | 14.911 | −0.004 |
| 42 | 29.851 | 0.021 | 20 | 14.147 | −0.058 |
| 40 | 28.319 | −0.090 | 19 | 13.491 | −0.003 |
| 38 | 27.027 | 0.038 | 18 | 12.841 | 0.057 |
| 36 | 25.437 | −0.131 | 17 | 12.085 | 0.011 |
| 34 | 24.024 | −0.124 | 16 | 11.380 | 0.016 |
| 32 | 22.631 | −0.096 | 15 | 10.745 | 0.092 |
| 30 | 21.248 | −0.059 | 14 | 9.956 | 0.013 |
| 28 | 19.876 | −0.011 | 13 | 9.275 | 0.042 |
| 26 | 18.412 | −0.054 | 12 | 8.506 | −0.017 |
| 24 | 17.003 | −0.043 | | | |

$$\hat{\beta} = \frac{\sum Xy}{\sum X^2} = 0.7102, \qquad \hat{\sigma}_{\hat{\beta}} = 0.00054, \qquad \hat{V}(\hat{\beta}) = \frac{\sum d^2/(N-2)}{\sum (X - \bar{X})^2} = 29 \times 10^{-8}$$

The theory of least squares provides the slope-estimate:

$$\hat{\beta} = \frac{\sum yX}{\sum X^2} \tag{12.69}$$

and the residuals:

$$d_i = y_i - \hat{\beta}X_i \tag{12.70}$$

The residuals are listed in the third column of Table 12.4, and the value of $\hat{\beta}$ is given at the bottom of the table. It is at once apparent that the residuals show a serious lack of randomness in the succession of their algebraic signs. How is this to be interpreted?

Usually we would infer from a set of residuals of this type that the $x$, $y$ data, instead of conforming to a straight line, lie actually on a curve. This would mean that Eq. 12.68 does not represent the correct model. But then it follows that Eq. 12.63 too is incorrect. However, from our extensive present knowledge of the compression properties of air we can readily infer that under the conditions of the experiment (room temperature, and a pressure ranging from one to four atmospheres) the ideal gas law applies well within the limits of precision of Boyle's data. It is therefore inconceivable that these data would show a systematic departure from Eq. 12.63 as a result of an inadequacy of the assumed model. It is, of course, true that Boyle made no attempt to control the temperature during the experiment. A gradual change of temperature might cause a disturbance such as that shown by the residuals of our fit. There exists however a much simpler explanation of the behavior of these residuals, an explanation that, far from requiring an additional hypothesis (such as a gradual change of temperature) is in fact suggested by the way in which the experiment was performed.

Consider the successive volume graduation marks in the short end of the U-tube. The marks were made at intervals of 1 inch by means of a ruler. If the tube had been perfectly cylindrical, and the graduation marks made without error, the mark labels would be rigorously proportional to the corresponding volumes inside the tube. But any imperfections in the bore of the tube or in the marking process will cause a disturbance in this strict proportionality. As a result, the volumes contained between successive marks will not all be equal. Let us represent their true values (in an arbitrary unit of volume) by:

$$1 + \varepsilon_1, \; 1 + \varepsilon_2, \; 1 + \varepsilon_3, \; \ldots$$

Then the true volumes corresponding to marks $1, 2, 3, \ldots$, are:

$$1 + \varepsilon_1, \; 2 + (\varepsilon_1 + \varepsilon_2), \; 3 + (\varepsilon_1 + \varepsilon_2 + \varepsilon_3), \; \ldots \tag{12.71}$$

It is seen that the errors for increasing volumes are *cumulative*: the second includes the first, the third the first and the second, etc. Series 12.71 represents the *true* values of the volumes, the *nominal* values of which are 1, 2, 3, etc. Assuming for the moment that pressure was measured with a negligible error, the values of pressure corresponding to these nominal volumes of measure 1, 2, 3, etc., reflect nevertheless the errors in these volumes, and so do their reciprocals, the quantities $y$. The situation is shown in Table 12.5. According to this table, the regression equation now becomes:

$$y_i = \beta X_i + \beta \sum_{k=1}^{i} \varepsilon_k \qquad (12.72)$$

where $X_i$ are the nominal values 1, 2, 3, etc. We are now in a better position to evaluate our least squares analysis. It is seen at once that one of the basic assumptions of the "classical" case is violated: the errors corresponding to different points are not statistically independent, because the series of numbers

$$\varepsilon_1, \ \varepsilon_1 + \varepsilon_2, \ \varepsilon_1 + \varepsilon_2 + \varepsilon_3, \ \ldots$$

are not independent, each one including all previous ones.

**TABLE 12.5**

| Nominal | Volume True | Reciprocal pressure $y = 1000/p$ |
|---|---|---|
| $X_1 = 1$ | $x_1 = 1 + \varepsilon_1$ | $y_1 = \beta x_1 = \beta + (\beta \varepsilon_1)$ |
| $X_2 = 2$ | $x_2 = 2 + (\varepsilon_1 + \varepsilon_2)$ | $y_2 = \beta x_2 = 2\beta + (\beta \varepsilon_1 + \beta \varepsilon_2)$ |
| $X_3 = 3$ | $x_3 = 3 + (\varepsilon_1 + \varepsilon_2 + \varepsilon_3)$ | $y_3 = \beta x_3 = 3\beta + (\beta \varepsilon_1 + \beta \varepsilon_2 + \beta \varepsilon_3)$ |
| $\vdots$ | $\vdots$ | $\vdots$ |
| $X_i = i$ | $x_i = i + \sum_{k=1}^{i} \varepsilon_k$ | $y_i = \beta x_i = i\beta + \beta \sum_{k=1}^{i} \varepsilon_k$ |
| $\vdots$ | $\vdots$ | $\vdots$ |

A series of values in which each element is correlated with a neighboring element is said to have *serial correlation*. It is because of the serial correlation in the errors of the pressure values (or of their reciprocals) that the residuals of the least squares analysis show the pattern of strong non-randomness evident in Table 12.4.

We will denote the general model expressed by Eq. 12.72 as a straight-line model with *cumulative errors*. Evidently, cumulative errors constitute a special case of serial correlation.

We now see that the lack of randomness displayed by the residuals is not at all an indication of inadequacy of the assumed model, i.e., of Boyle's law, but simply a mathematical consequence of the cumulative nature of the errors.

The question that faces us now is that of devising a valid analysis of the data, an analysis that takes into account the cumulative nature of the errors. For our problem this question is readily solved (8). From Table 12.5 we immediately infer that

$$y_2 - y_1 = \beta + (\beta \varepsilon_2)$$

$$y_3 - y_2 = \beta + (\beta \varepsilon_3)$$

$$\cdot \quad \cdot \quad \cdot \quad \cdot \quad \cdot \quad \cdot$$

$$y_i - y_{i-1} = \beta + (\beta \varepsilon_i)$$

$$\cdot \quad \cdot \quad \cdot \quad \cdot \quad \cdot \quad \cdot$$

$$y_N - y_{N-1} = \beta + (\beta \varepsilon_N) \qquad (12.73)$$

Thus, the *increments* $y_2 - y_1$, $y_3 - y_2$, etc., are random variables with an expected value $\beta$ and *independent* errors $(\beta \varepsilon_1)$, $(\beta \varepsilon_2)$, etc.

In Boyle's experiment the series of $X_i$ values starts at $X_1 = 12$, and increases by steps of one or two units, to $X_N = 48$. To derive a formula applicable to the general case, let $L_i$ represent the difference between the $i$th and the $(i-1)$st $X$ value; thus

$$L_i = X_i - X_{i-1} \qquad (12.74)$$

Denote the corresponding increment of $y$ by $z$; i.e.,

$$z_i = y_i - y_{i-1} \qquad (12.75)$$

Then, instead of Eqs. 12.73 we have the general model:

$$z_i = \beta L_i + \beta \varepsilon_i \qquad (12.76)$$

where $i$ varies from 2 to $N$.

We may denote the error $\beta \varepsilon_i$ by $\eta_i$; thus:

$$\eta_i = \beta \varepsilon_i \qquad (12.77)$$

hence

$$z_i = \beta L_i + \eta_i \qquad (12.78)$$

The important point is that the $\varepsilon_i$ (and therefore also the $\eta_i$) corresponding to different $z$ values are statistically independent. Furthermore it can be proved that the variance of $\eta_i$ depends on the length of the $X$ interval and is in fact proportional to it:

$$V(\eta_i) = k L_i \qquad (12.79)$$

where $k$ is a constant. Dividing Eq. 12.78 by $\sqrt{L_i}$, we obtain

$$\frac{z_i}{\sqrt{L_i}} = \beta(\sqrt{L_i}) + \left( \frac{\eta_i}{\sqrt{L_i}} \right) \qquad (12.80)$$

This equation satisfies all the requirements of the classical case: the errors

$\eta_i/\sqrt{L_i}$ are statistically independent; their mean is zero; their variance is $(1/L_i)V(\eta_i)$ which, in view of Eq. 12.79, equals the constant $k$.

Applying the ordinary least squares calculations, we obtain:

$$\hat{\beta} = \frac{\sum_i [(z_i/\sqrt{L_i})\sqrt{L_i}]}{\sum_i (\sqrt{L_i})^2} = \frac{\sum_i z_i}{\sum_i L_i} \qquad (12.81)$$

From this equation we infer:

$$V(\hat{\beta}) = \frac{1}{(\sum L_i)^2} \sum_i V(z_i)$$

$$= \frac{1}{(\sum L_i)^2} \sum_i V(\eta_i) = \frac{1}{(\sum L_i)^2} k \sum_i L_i$$

hence

$$V(\hat{\beta}) = \frac{k}{\sum L_i} \qquad (12.82)$$

Since $k$ is the variance of $\eta_i/\sqrt{L_i}$ in Eq. 12.80, an estimate for $k$ is obtained from the residuals*:

$$\hat{k} = \frac{\sum_i [(z_i/\sqrt{L_i}) - \hat{\beta}\sqrt{L_i}]^2}{N - 2}$$

or

$$\hat{k} = \frac{\sum_i [(z_i - \hat{\beta}L_i)^2/L_i]}{N - 2}$$

Hence:

$$\hat{V}(\hat{\beta}) = \frac{\sum_i (d_i^2/L_i)}{(N - 2)\sum_i L_i} \qquad (12.83)$$

where the residual $d_i$ is defined as

$$d_i = z_i - \hat{\beta}L_i \qquad (12.84)$$

Summarizing, we have the results:

$$\hat{\beta} = \frac{\sum_i z_i}{\sum_i L_i}$$

$$\hat{V}(\hat{\beta}) = \frac{\sum_i (d_i^2/L_i)}{(N - 2)\sum_i L_i}$$

* The denominator $N - 2$ is explained by the fact that there are $(N - 1)$ $z$-values, and the number of fitted parameters is one.

where $d_i$ is defined by Eq. 12.84. Applying this theory to Boyle's data, Table 12.4, we obtain the results shown in Table 12.6.

We will now compare the results of this analysis with those of our original analysis, in which we ignored the cumulative nature of the errors. Such a comparison is not only instructive; it is also of great importance for a better understanding of the nature of cumulative data and for the avoidance of the very serious misinterpretations that result from the failure to recognize the presence of cumulative errors.

**TABLE 12.6**  Correct Least Squares Analysis of Boyle's Data

| $L$ | $z$ | $d = z - \hat{\beta}L$ | $L$ | $z$ | $d = z - \hat{\beta}L$ |
|-----|-----|------------------------|-----|-----|------------------------|
| 2 | 1.615 | 0.180 | 1 | 0.693 | −0.024 |
| 2 | 1.409 | −0.026 | 1 | 0.700 | −0.017 |
| 2 | 1.460 | 0.025 | 1 | 0.699 | −0.018 |
| 2 | 1.532 | 0.097 | 1 | 0.764 | 0.047 |
| 2 | 1.292 | −0.143 | 1 | 0.656 | −0.061 |
| 2 | 1.590 | 0.155 | 1 | 0.650 | −0.067 |
| 2 | 1.413 | −0.022 | 1 | 0.756 | 0.039 |
| 2 | 1.393 | −0.042 | 1 | 0.705 | −0.012 |
| 2 | 1.383 | −0.052 | 1 | 0.635 | −0.082 |
| 2 | 1.372 | −0.063 | 1 | 0.789 | 0.072 |
| 2 | 1.464 | 0.029 | 1 | 0.681 | −0.036 |
| 2 | 1.409 | −0.026 | 1 | 0.769 | 0.052 |

$$\hat{\beta} = \frac{\sum z}{\sum L} = 0.7175, \qquad \hat{V}(\hat{\beta}) = \frac{\sum [d_i^2/L_i]}{(N - 2) \sum L_i} = 94 \times 10^{-6}, \qquad \hat{\sigma}_{\hat{\beta}} = .0097$$

1. The estimates of the slope, while different by the two methods, are quite close to each other.  It can be shown that even the incorrect method of analysis provides an unbiased estimate of the slope, and that the true precision of this estimate is only slightly worse than that of the estimate by the correct method.

2. We have already observed the striking non-randomness of the residuals obtained by the incorrect method and explained them in terms of the serial correlation of the errors.  In contrast, the residuals obtained from the increments, by the correct method, show the usual random behavior.

3. By far the most important difference between the two analyses is in the estimate of the standard error of the slope.  We have already noted that the two slope-estimates are quite similar.  Yet, according to the incorrect method of analysis, the variance of the estimated slope is $\hat{V}(\hat{\beta}) = 29 \times 10^{-8}$ while the correct analysis gives a variance of $94 \times 10^{-6}$

for essentially the same quantity. This very large discrepancy is not an accident. It can be proved mathematically that the variance estimate of the slope given by the incorrect method is always badly in error, leading to a very serious overestimation of the precision of the data. This point deserves further comment. As is shown by the example of Boyle's data, only a careful conceptual analysis of the experiment will lead one to regard the data as cumulative. A superficial appraisal of the situation may well result in treating the data by the classical case. Such an erroneous analysis might lead to two misleading conclusions: it might cast doubt on the validity of the underlying straight-line model, because of the non-randomness of the residuals; and it might lead us to regard the data, and the estimate of the slope, as far more precise than they actually are. This brings us to the next point.

4. In many experimental situations it is by no means clear, *a priori*, whether one has a genuine case of cumulative errors. Consider, for example, the measurement of the birefringence of glass caused by compressive loading (10). A glass specimen is compressed between two horizontal plates. By applying increasing loads to the upper plate, by means of calibrated weights, the compression of the glass is gradually increased. At each load, a measurement of birefringence of the glass is made. Are successive measurements subject to serial correlation? It turned out, in this case, that they were, but this is by no means self-evident, *a priori*.

Further complication may arise when the measurement of the "response" variable, birefringence in this case, and pressure in Boyle's experiment, is itself subject to error. In such cases, the total error of the response is composed of two parts, one random and the other cumulative. It is possible to devise a valid statistical analysis for such cases provided that the ratio of the magnitudes of these two errors is known. We will, however, not deal here with this matter.

At any rate it is clear that situations involving cumulative errors are quite frequent in laboratory experimentation. Wherever they occur serious misinterpretation of the data is a distinct possibility. Boyle's data provided us with one example. Other instances, of more recent interest, are found in studies of chemical reaction rates.

## 12.8  CHEMICAL REACTION RATES

We will discuss this important matter in some detail, using data obtained by Daniels and Johnston (4), in a study of the decomposition of nitrogen pentoxide.

It is well known that a first order reaction is represented by the relation

$$\ln \frac{c}{a} = -kt \tag{12.85}$$

where $a$ is the initial concentration (at time zero), $c$ the concentration at time $t$, and $k$ is the rate constant of the reaction. The problem consists in determining $k$ and estimating its precision.

According to Daniels and Johnston, the decomposition of nitrogen pentoxide is a first order reaction. Their experimental procedure consisted in measuring, at chosen time-intervals, the pressure $p$ of the oxygen formed in the course of the decomposition:

$$2N_2O_5 \rightarrow O_2 + 2N_2O_4 \tag{12.86}$$

It can readily be shown that Eq. 12.85 can be written in terms of the pressure $p$, as follows:

$$\ln \left(1 - \frac{p}{p_\infty}\right) = -kt \tag{12.87}$$

where $p_\infty$ is the pressure at the completion of the reaction; $p_\infty$ is also obtained experimentally. Table 12.7 gives data of $t$ and $p$ for duplicate experiments carried out at the same temperature (55 °C).

**TABLE 12.7** Decomposition of Nitrogen Pentoxide at 55 °C

| Time ($t$) | Pressure ($p$) First set | Second set |
|:---:|:---:|:---:|
| 3 | 72.9 | — |
| 4 | 95.0 | — |
| 5 | 115.9 | — |
| 6 | 135.2 | 85.3 |
| 7 | 151.9 | 110.2 |
| 8 | 168.0 | 136.1 |
| 9 | 182.8 | 148.8 |
| 10 | 195.7 | 166.2 |
| 12 | 217.8 | 200.0 |
| 14 | 236.4 | 226.7 |
| 16 | 252.2 | 250.0 |
| 18 | 265.9 | 267.3 |
| 22 | 285.7 | 294.3 |
| 26 | 299.3 | 314.9 |
| 30 | 309.0 | 327.6 |
| 38 | 320.0 | 342.5 |
| 46 | — | 349.8 |
| 54 | — | 353.6 |
| $\infty$ | 331.3 | 357.8 |

Using common logarithms we have

$$\log \left( 1 - \frac{p}{p_\infty} \right) = \frac{-k}{2.30} \, t \qquad (12.88)$$

Writing:

$$X = t \qquad (12.89)$$

$$y = \log \left( 1 - \frac{p}{p_\infty} \right) \qquad (12.90)$$

$$\beta = \frac{-k}{2.30} \qquad (12.91)$$

we obtain from Eq. 12.88:

$$y = \beta X \qquad (12.92)$$

Table 12.8 shows the calculations for both sets of data, assuming a Berkson model, since the time $t$ is a controlled variable. To allow for a possible error in determining the starting point of the reaction, the equation actually fitted was:

$$y = \alpha + \beta X \qquad (12.93)$$

The data are shown graphically in Fig. 12.8. The slopes of the two lines are somewhat different. Yet the scatter about the lines, especially the lower one, is rather small. The analysis shows that the difference of the slopes equals $0.03910 - 0.03817 = 0.00093$. The estimate for the standard error of this difference is $\sqrt{[1.28^2 + 2.94^2]10^{-8}} = 0.00032$. Thus the difference equals about 3 times its estimated standard error. One might conclude that a systematic difference exists between the two sets of results. We will not make a formal test of significance, since the entire foregoing analysis ignores an important feature of the data: their cumulative nature.

By the nature of the experiment, the pressure measurements are subject to cumulative fluctuations. Indeed, if during any time interval, the rate of reaction changed as a result of slight changes in the environmental conditions (such as temperature), this would affect not only the pressure reading at the end of the interval, but also all subsequent pressure readings. Thus, while the *measurement errors* of the pressure readings are independent, they are nevertheless correlated as a result of what we may call *process fluctuations*. If the true (independent) measurement errors of the pressure readings are small as compared to the effects of process fluctuations, the data are essentially of a cumulative nature and should be analyzed in terms of their increments, in accordance with the theory of cumulative errors; such an analysis is presented in Table 12.9. The difference between the slopes of the two sets is very nearly the same as in

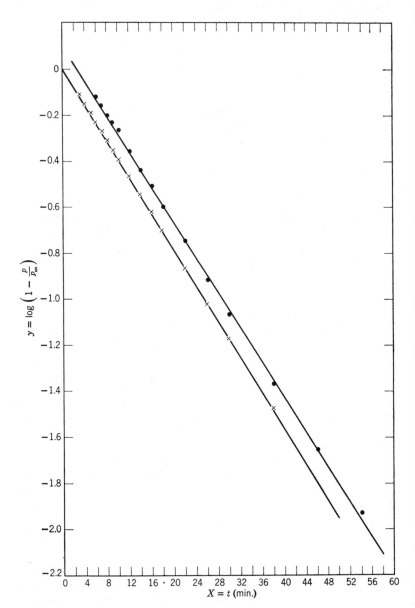

**Fig. 12.8**  Fit of $N_2O_5$—decomposition data by Berkson model.  (Data in Table 12.8.)

**TABLE 12.8**  Decomposition of Nitrogen Pentoxide at 55 °C.  Analysis According to Berkson Model

|  | $-y$ | | Residuals | |
| --- | --- | --- | --- | --- |
| $X$ | First set | Second set | First set | Second set |
| 3 | 0.1079 | — | 0.0041 | — |
| 4 | 0.1467 | — | 0.0044 | — |
| 5 | 0.1870 | — | 0.0032 | — |
| 6 | 0.2277 | 0.1183 | 0.0016 | 0.0131 |
| 7 | 0.2664 | 0.1599 | 0.0020 | 0.0096 |
| 8 | 0.3072 | 0.2079 | 0.0003 | −0.0002 |
| 9 | 0.3485 | 0.2335 | −0.0019 | 0.0124 |
| 10 | 0.3880 | 0.2712 | −0.0023 | 0.0128 |
| 12 | 0.4652 | 0.3556 | −0.0013 | 0.0048 |
| 14 | 0.5430 | 0.4360 | −0.0009 | 0.0007 |
| 16 | 0.6219 | 0.5210 | −0.0016 | −0.0080 |
| 18 | 0.7046 | 0.5971 | −0.0061 | 0.0077 |
| 22 | 0.8614 | 0.7508 | −0.0065 | 0.0088 |
| 26 | 1.0150 | 0.9211 | −0.0037 | −0.0264 |
| 30 | 1.1720 | 1.0737 | −0.0043 | −0.0263 |
| 38 | 1.4672 | 1.3686 | 0.0133 | −0.0159 |
| 46 | — | 1.6498 | — | 0.0083 |
| 54 | — | 1.9318 | — | 0.0316 |

|  | $\hat{\alpha}$ | $\hat{\beta}$ | $\hat{V}(\hat{\beta})$ | $\hat{\sigma}_{(\hat{\beta})}$ |
| --- | --- | --- | --- | --- |
| First set | 0.0053 | −0.03910 | $1.63 \times 10^{-8}$ | $1.28 \times 10^{-4}$ |
| Second set | 0.0976 | −0.03817 | $8.64 \times 10^{-8}$ | $2.94 \times 10^{-4}$ |

the Berkson analysis.  However, the estimates of the standard deviations are now considerably larger, and so is the estimate of the standard error of the difference between the slopes.  We now find that the ratio of this difference to its standard error is

$$\frac{0.00106}{\sqrt{0.0078^2 + 0.0038^2}} = 0.12$$

There is no reason, according to this analysis, to suspect a systematic error in either set.

We are not in a position to decide, on the basis of the data, which analysis—the Berkson type or the cumulative type—is the correct one. We have no reliable information on the relative magnitude of the measurement error and of the process fluctuations.  Had each set been based on a larger number of observations, say 25 or more, there might then have been some basis of inference in the behavior of the residuals obtained by

**TABLE 12.9** Decomposition of Nitrogen Pentoxide at 55 °C. Analysis by Cumulative Model

| | First set | | | Second set | |
|---|---|---|---|---|---|
| L | $-z$ | d | L | $-z$ | d |
| 1 | 0.0388 | 0 | 1 | 0.0416 | $-0.0038$ |
| 1 | 0.0403 | $-0.0015$ | 1 | 0.0480 | $-0.0102$ |
| 1 | 0.0407 | $-0.0019$ | 1 | 0.0256 | 0.0122 |
| 1 | 0.0387 | 0.0001 | 1 | 0.0377 | 0.0001 |
| 1 | 0.0408 | $-0.0020$ | 2 | 0.0844 | $-0.0088$ |
| 1 | 0.0413 | $-0.0025$ | 2 | 0.0804 | $-0.0048$ |
| 1 | 0.0395 | $-0.0007$ | 2 | 0.0850 | $-0.0094$ |
| 2 | 0.0772 | 0.0005 | 2 | 0.0761 | $-0.0005$ |
| 2 | 0.0778 | $-0.0001$ | 4 | 0.1537 | $-0.0404$ |
| 2 | 0.0789 | $-0.0012$ | 4 | 0.1703 | $-0.0570$ |
| 2 | 0.0827 | $-0.0050$ | 4 | 0.1526 | $-0.0393$ |
| 4 | 0.1568 | $-0.0015$ | 8 | 0.2949 | $-0.1438$ |
| 4 | 0.1536 | 0.0017 | 8 | 0.2812 | $-0.1301$ |
| 4 | 0.1570 | $-0.0017$ | 8 | 0.2820 | $-0.1309$ |
| 8 | 0.2952 | 0.0155 | | | |

| | $\hat{\beta}$ | $\hat{V}(\hat{\beta})$ | $\hat{\sigma}_{(\hat{\beta})}$ |
|---|---|---|---|
| First set | $-0.03884$ | $6200 \times 10^{-8}$ | $78 \times 10^{-4}$ |
| Second set | $-0.03778$ | $1412 \times 10^{-8}$ | $38 \times 10^{-4}$ |

the Berkson analysis; considerable non-randomness would be an indication of the preponderance of cumulative types of error. Even then, alternative explanations would exist: a systematic error in the value of $p_\infty$, or non-linearity, i.e., non-validity of the first-order reaction model.

The example we have just discussed raises an important question in regard to the determination of chemical rate constants. If the experiment is conducted in a way that can lead to cumulative process fluctuations, the possibility that these fluctuations may overshadow the independent measurement errors must be kept in mind. If they do, an analysis in terms of increments is the only correct one. Actually, this type of analysis is frequently used in reaction rate work. In particular, it was used by Daniels and Johnston in the study from which our example is taken. On the other hand, this method has been severely criticized by Livingston (7). Referring to this method of analysis as the "short-interval" method, Livingston observes: "In other words, when the time interval is constant, the use of the short-interval method of computing the average is equivalent to rejecting all but the first and the last measurements.... In spite of its apparent absurdity, this method of averaging has appeared a number of times in the chemical literature."

Now, it is true enough that the slope estimate obtained by the method of increments involves only the first and the last measurements. This is apparent from Eq. 12.81, since in the case of adjacent intervals, $\sum L_i$ is simply $X_N - X_1$, and $\sum z_i = y_N - y_1$ (see Eqs. 12.74 and 12.75). But the intermediate measurements are not "rejected." In the first place, each one of them is part of the final value $y_N$. Secondly, they are explicitly used in the estimation of the precision of the slope, as is seen from Eq. 12.83, in which the $d_i$ involve the individual $y$ measurements. Furthermore the method is by no means "absurd," as can be seen from the following example. If a motorist wanted to estimate his average speed of travel between two cities, say Washington and New York, he would obviously divide the *total* mileage by the *total* time of travel. He would use this method of calculation even if he had made a series of mileage observations at intermediate time instants, say each half-hour. Yet one would not accuse him of thus "rejecting" any information. On the other hand, if the motorist is a traveling salesman between Washington and New York and wished to use the information gathered during this trip for estimating the *variability* of his average speed of travel *from trip to trip*, he must make use of the intermediate mileage observations, in accordance with Eq. 12.83.

In conclusion, the determination of a rate constant, in the form of the slope of a straight line, involves a number of problems of a statistical nature. If a numerical value for the constant is all that is required, the selection of the correct model is generally not critical, since most methods will yield approximately the same value. If, on the other hand, an estimate of the precision of this numerical value is also required, the selection of the correct model is extremely important. In particular, the presence of cumulative errors would, if unrecognized, lead to a very serious underestimation of the true variability of the data, and to a much higher confidence in the value obtained for the rate constant than it really deserves.

## 12.9　SUMMARY

The proper method for fitting a straight line to a set of data depends on what is known, or assumed, about the errors affecting the two variables. In this chapter four cases are considered: (1) the classical case, in which the $x$-variable is not subject to error and the errors of $y$ are independent and have a common variance; (2) the case in which both variables are subject to error; (3) the Berkson case, in which the $x$-variable is controlled; and (4) the case in which the errors of the $y$ variable are cumulative.

For the classical case, detailed consideration is given to the uncertainties of the parameters, as well as of the fitted line. Of special interest is the problem of evaluating the precision of values read from a calibration line. The case in which both variables are subject to error can be treated by Deming's generalized use of the method of least squares. Formulas are given and the procedure is illustrated by means of a numerical example.

The Berkson case is of great importance in the physical sciences, because it deals with situations in which the value of the independent variable, although subject to error, can be treated as in the classical case. This occurs whenever the $x$-variable is *controlled*, i.e., when the values recorded for it (the *nominal* values) are target values, differing by unknown, but randomly distributed amounts from the *actual* values taken by this variable. The scatter observed by the $y$ variable is, in such cases, composed of two parts: an error of measurement of $y$, and an error *induced* in $y$ as a result of the difference between the nominal and the actual value of $x$.

A frequently occurring type of experimentation is that in which the changes of a physical or chemical system are studied as a function of time. A typical example is given by the determination of chemical reaction rates. In this type of situation, the fluctuations affecting consecutive measurements of the system are not independent, because any change in experimental conditions affects *all* measurements subsequent to its occurrence. In such cases, the error associated with any particular measurement $M_i$ is the algebraic sum of the error affecting the previous measurement, $M_{i-1}$, and of the effect of fluctuations that occurred in the time interval between $M_{i-1}$ and $M_i$. This case is referred to as that of *cumulative errors*. Special vigilance is needed in the analysis of data subject to cumulative errors, inasmuch as the scatter of the experimental points about the fitted line vastly underestimates the true error of the measurements. As a result, the uncertainty of the slope of the fitted line is also considerably underestimated, unless the analysis takes the cumulative nature of the errors into account. The matter is especially pertinent in the evaluation of reaction rates and their precisions.

## REFERENCES

1. Berkson, J., "Are There Two Regressions?" *J. Amer. Statistical Assn.*, **45**, 164–180 (1950).
2. Boyle, R., "A Defence of the Doctrine Touching the Spring and Weight of the Air," reprinted in *The Laws of Gases, Memoirs by Robert Boyle and E. H. Amagat*, by C. Barus, Harper, New York, 1899.
3. Brown, J. E., M. Tryon, and J. Mandel, "Determination of Propylene in Ethylene–Propylene Copolymers by Infrared Spectrophotometry," *Analytical Chemistry*, **35**, 2172–2176 (1963).

4. Daniels, F., and E. H. Johnston, "The Thermal Decomposition of Gaseous Nitrogen Pentoxide; A Monomolecular Reaction," *J. Amer. Chem. Soc.*, **43**, 53–71 (1921).

5. Deming, W. E., *Statistical Adjustment of Data*, Wiley, New York, 1943.

6. Finney, D. J., *Probit Analysis*, Cambridge University Press, Cambridge, 1952.

7. Livingston, R., "Evaluation and Interpretation of Rate Data," in *Technique of Organic Chemistry*, Vol. VIII, by A. Weissberger, ed., Interscience, New York, 1953.

8. Mandel, J., "Fitting a Straight Line to Certain Types of Cumulative Data," *J. Amer. Statistical Assn.*, **52**, 552–566 (1957).

9. Mandel, J., and F. J. Linnig, "Study of Accuracy in Chemical Analysis Using Linear Calibration Curves," *Analytical Chemistry*, **29**, 743–749 (May 1957).

10. Waxler, R. M., and A. Napolitano, "The Relative Stress-Optical Coefficients of Some NBS Optical Glasses," *J. Research Natl. Bur. Standards*, **59**, 121–125 (1957).

# chapter 13

# THE SYSTEMATIC EVALUATION OF MEASURING PROCESSES

## 13.1  INTRODUCTION

In Chapter 6 we have discussed the concepts of precision and accuracy in terms of a measuring technique, as applied to a particular system or portion of material.  We have noted, however, that the complete study of a measuring technique must include its application to all the systems or samples for which it is intended, or at least to a representative set of such systems or samples.

To be more specific, let us consider a method for determining the uranium content of an ore.  It is not enough to establish the merits of the method for a single sample of uranium ore, if it is to be used as a general method of determination.  What is necessary is a study of its merits when applied to all the various types of uranium ore for which it is claimed to be a valid method of analysis.  In this chapter we are concerned with statistical methods designed to aid us in evaluating a measuring process viewed in this more general way.

We will discuss this problem in rather considerable detail, not only because of its intrinsic importance, which is considerable, but also because it provides a good illustration of what the author believes to be the proper use of statistical methods in solving a scientific problem.  An important element in such a study is the setting up of a valid model.  But provision must also be made, in the course of the study, to check the validity of the model at various points.    Furthermore, a problem of this type involves an almost constant interlacing of statistical and non-statistical considerations. By "non-statistical" we mean, in this connection, considerations derived from the physical, chemical, or technological background information about the problem.

In this chapter we will be concerned more particularly with questions relating to precision.    Another important problem connected with the study of measuring processes is that of comparing two or more processes for the measurement of the same property.    This problem will be dealt with in the next chapter.

## 13.2    A MODEL FOR MEASURING PROCESSES

The fundamental idea in our statistical representation of a measuring process is precisely the "response curve" alluded to in Chapter 1.    We consider a property $Q$, susceptible of representation by a numerical quantity, such as the iron content of steel, the velocity of a bullet, the wavelength of an electro-magnetic wave, the density of a liquid, the breaking strength of yarn, the abrasion rate of rubber tires, the distance of a star, etc.    Excluded are such properties as the style of a piece of furniture, the elegance of a mathematical proof, the perspicacity of a human mind, because there exist no well-defined ways of representing these properties by numbers of an arithmetic scale.    There are borderline cases such as human intelligence as measured by IQ, but we will not be concerned with properties of this type.

A measuring process for the determination of $Q$ is an operation which, when carried out in a prescribed way, associates with each value of $Q$ a corresponding value of $M$, the measured value.    For example, the iron content of steel is determined by going through a series of well-defined steps constituting a chemical analysis; the end-result may be the weight of a precipitate, a number of milliliters of reagent, or an optical absorbance value.    In the case of the breaking strength of yarn, the process involves an instrument in which the yarn is pulled at both ends in a prescribed way, and the force measured at the instant the yarn breaks.

It is clear that $M$, the result of the measurement, must be a function of $Q$, the value of the property it is designed to measure.    Furthermore, this function should be "monotonic," i.e., steadily increasing or steadily decreasing, but not both increasing and decreasing within the range of $Q$ over which it applies.

If both $Q$ and $M$ are entirely free of fluctuations the entire measuring process would be completely represented by the $Q$, $M$ curve (Fig. 13.1). However, it is well known that no measuring process is perfect.    Specifically, the response $M$ depends, in addition to $Q$, also on other factors,

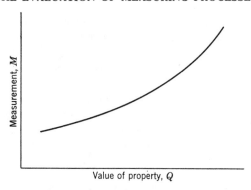

**Fig. 13.1**   The measurement is a monotonic function of the property.

such as the surrounding temperature and relative humidity, the levelling of the instrument, the impurities in the chemical reagents, the correctness of a balance, the skill of the operator.   We will refer to these as "environmental factors."   It is also well-known that attempts are always made to "exercise control" over such environmental factors, i.e., to keep them from changing from some fixed state.   For example, through the use of a "conditioning-room," the temperature and relative humidity may be kept within stated "tolerances."   Reagents of certified purity may be used, meaning that upper limits are given for the amounts of impurities of a given type, and other precautions of "control" of a similar type may be taken.

It is, however, clear that it is humanly impossible to achieve *total* control; for not only is it impossible to keep given environmental conditions *absolutely* constant, but it is impossible even to be aware of *all* factors that could affect the measuring process.   In our model, environmental conditions are taken into account in the following manner.

We conceive of an "input" $Q$, varying in a continuous manner, under *absolutely constant* environmental conditions.   (This is of course only a conceptual experiment).   At each value of $Q$, we observe the "output" $M$, and plot $M$ versus $Q$.   Upon completion of this first phase of the conceptual experiment, we allow the environmental conditions to change from their initial state, but then immediately "freeze" them in their new state, and repeat the experiment.   The $M$, $Q$ curve obtained this time cannot, in general, be expected to coincide with the first one, since it was obtained under different conditions.   The experiment is continued in the same fashion, each time for a different set of values for the environmental factors.   Clearly, we will then generate a "bundle" of curves, each curve corresponding to a particular set of values for the environmental conditions. This bundle of curves is our conceptual model of the measuring process (Fig. 13.2). In defining this bundle we must of course impose limits on the amount of variation allowed in the environmental conditions. Fortunately every well-defined measuring process specifies tolerances within which the factors that are known to affect the result must be kept. Thus the bundle is automatically limited by these tolerances. Within these limits, a measurement could conceivably belong to any curve in the bundle, depending on the state of environmental conditions that prevailed during the taking of the measurement.

Consider a particular value of $Q$. All measurements corresponding to it must lie on the vertical line going through $Q$ (Fig. 13.2). The range $M$ to $M'$ covered is the total amount of variability of the measured value for the value of $Q$. However, the curves will in general not be equally spaced within this range. If the factors generating the curves are subject to purely random fluctuations, there will generally be a tendency for greater compactness of the curves in the central part of the $MM'$ range so that the probability is greater to find a measured value in that central region than in the vicinity of either $M$ or $M'$. The ordinates corresponding to the chosen value of $Q$, i.e., the measured values $M$, will form a statistical population, with a definite mean value and a definite standard deviation. There is generally no *a priori* reason that such a population will necessarily be Gaussian, nor even that it will be symmetrical. But in any case the variability exhibited by this population is the statistical representation of what is usually called "experimental error" of the measured value.

A number of facts emerge at once from our model. In the first place,

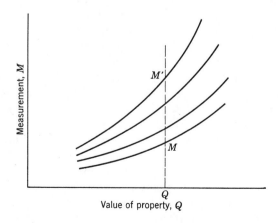

**Fig. 13.2**  Bundle of curves representing a measuring process.

the standard deviation of experimental error is obviously not necessarily the same for different levels of $Q$, since there is no assurance that the lines in the bundle will necessarily be parallel, or stay within parallel limiting lines. It is a well known fact that the magnitude of experimental error often increases as the level of the measured property increases; the law of increase is by no means necessarily the same for different measuring processes.

Secondly, since the multiplicity of response curves in the bundle is due to the variation of many factors, a more rigid control of some of these may result in an appreciable narrowing of the total width of the bundle. Now, the notion of control of environmental conditions really comprises two types. In the first type, control involves verbal specifications, aimed at limiting the variation of important environmental factors within specified tolerances. For example, in the spectrophotometric determination of a particular antioxidant in rubber the wavelength is specified to be $309 \pm 1$ millimicrons.

In determining the breaking strength of cotton cord, the rate of the pulling clamp is specified to be $12 \pm 0.5$ inches per minute. Examples of this type abound in all areas of physical and chemical measurement.

The second type of control is more rigid: it is the control exercised by a good experimenter in his own laboratory. Thus, the wavelength used by a particular experimenter, using a particular instrument, may be far more constant than is required by the specified tolerance of 1 millimicron. It is evident that this second type of control, which we may call "local control," is generally superior to that due to the mere observance of specified tolerances.

It follows from what has just been said that among the curves in a bundle, there will generally be found much narrower subsets of curves, each subset corresponding to conditions of particular local control. In such cases, the values of $M$ for a particular value of $Q$ can also be partitioned into subpopulations, each subpopulation corresponding to a situation in which the environmental factors are held under conditions of local control. A partitioning that suggests itself at once is one made according to laboratories. It would seem that within a laboratory local control is always possible, but this is not necessarily true when measurements are spaced with respect to time. For some types of measurements, the pertinent environmental conditions in a laboratory can be considered to remain stable for weeks or even many months. For other types, an interval of hours, or even minutes between measurements may cause detectable changes in some of the environmental factors. Of course, even local control is not perfect and fluctuations are to be expected even under very stable conditions.

## 13.3  THE EXPERIMENTAL STUDY OF MEASURING PROCESSES

In attempting to design an actual experiment for the study of the bundle of response curves representing a given measuring process, one is faced with several practical limitations.

In the first place, it is generally impossible to vary $Q$ in a continuous way. What is generally done is to carry out the study on a set of different materials chosen in such a way that the $Q$ values corresponding to them cover the range of interest at roughly equal intervals. For example, in studying a chemical method of analysis of propylene in ethylene–propylene copolymers, the materials used in the study may range from zero per cent propylene (polyethylene) to one hundred per cent propylene (poly-propylene) in steps of approximately 20 per cent increase. This would require six materials, containing 0, 20, 40, 60, 80, and 100 per cent propylene, respectively, where the six concentrations need only be approximately equal to these round values. Here the range is the largest possible (0 to 100 per cent) but in many situations the range is restricted in a number of ways.

The $Q$ values are not the only ones for which a selection must be made. Obviously it is impossible to translate into action the concept of holding all environmental factors rigorously constant. Suppose that a particular experimenter makes a series of measurements consisting of a single deter-mination for each of the six copolymers mentioned above. Since even under these conditions of local control, some environmental factors will have changed between measurements, the six points obtained will not fall on a single curve. The best we can do is to draw a smooth line close to the points and attribute the scatter of the points about this line to the unavoid-able fluctuations present under the best conditions of environmental control.

If the same set of six copolymers is now measured by a second experi-menter, in a different laboratory, a second line will result, with its own scatter due to within-laboratory fluctuations. By performing this experiment in a larger number of laboratories, we may hope to obtain some idea of the bundle of curves in our theoretical model. Methods for extracting this information from the data will be discussed shortly.

## 13.4  INTERFERING PROPERTIES

A careful examination of the experimental plan described in the preced-ing section reveals a major flaw. The theoretical model calls for a single, continuously varying, property $Q$. In the experimental plan, this is approximated by a series of samples or materials for which $Q$ has different values. But it is seldom possible to find a series of samples differing *in no other respect* than in the value of one single property.   In chemical analysis one can sometimes approach this ideal situation by making up a series of solutions of increasing concentration in one of the constituents (the one whose concentration is measured by the analysis).   But in most cases, and particularly for physical properties, a change in one property is generally accompanied by changes in other properties.   If the experiment is performed with such a set of samples, the failure of the experimental points to lie on a smooth curve will be due not only to changes in environmental conditions from one measurement to the next, but also to the fact that it is not only the level of $Q$ that changes from one sample to the next but also one or more other properties.   An ideal method of measurement would be so "*specific*" that it responds only to the change in $Q$ and is completely unaffected by changes in other properties.   Most methods of measurement fall far short of this ideal specificity.   In those cases, the other properties are said to be "interfering" with the measurement of $Q$. In chemical analysis this phenomenon is well-known as that of "interfering substances."

It has already been pointed out that when interfering properties are present, the points yielded by the experiment show greater scatter about a smooth curve than can be accounted for by changes in environmental conditions.   Is it possible to disentangle these two sources of scatter? The answer to this question is the subject of the following section.

## 13.5   REPLICATION ERROR

Suppose that a particular experimenter, working under conditions of good local control, makes three determinations on each of the six copolymers mentioned in Section 13.3.   Then, any disagreement between the three results obtained for the same sample is due *solely* to changes in environmental conditions, since both the property $Q$ and *all other properties* are the same for the three determinations.   Such a set of repeated determinations on the same sample are called *replicate determinations* or *replicates*.   (In the case of two repeated determinations one speaks of *duplicates*.)

It follows that by making replicate determinations, one can isolate the effect of environmental changes from that due to interfering properties. It is for this reason that the *replication error*, i.e., the variability observed among replicates, is considered as a *measure of the effect of environmental conditions under local control*.   When this effect is subtracted from the scatter of the experimental points corresponding to a particular experimenter, the remainder is a measure of interfering properties.

There is an important qualification to the above statements.   In order for the variability between replicates to be representative of the changes in environmental conditions *as they usually occur between measurements made on different samples*, one must allow these changes to take place as if the sample had been changed between measurements. For example, if three successive measurements were made on sample *A*, then three on sample *B*, and so on, the average time elapsed between replicates would be smaller than that occurring between measurements on different samples. Similarly, if the three replicates on sample *A* were made using one hot-plate, the three replicates on *B* on another hot-plate, and similarly for the other samples, differences in temperature between hot-plates would not operate between replicates. It is therefore important to spread the replicate determinations with respect to time, equipment, and other environmental factors in the same manner as measurements made on different samples. Statistically this is achieved through the process of *randomization*.

To illustrate randomization, consider again the three replicates run on samples *A* through *F*, a total of 18 determinations. Table 13.1 lists the 18 determinations in systematic order first and identifies them in that order by the numerals 1 through 18 (column 3).

**TABLE 13.1** Randomization

| Sample | Replicate | Identification number | Randomized order |
|--------|-----------|-----------------------|------------------|
| *A* | 1 | 1 | 8 |
| | 2 | 2 | 14 |
| | 3 | 3 | 9 |
| *B* | 1 | 4 | 5 |
| | 2 | 5 | 13 |
| | 3 | 6 | 7 |
| *C* | 1 | 7 | 11 |
| | 2 | 8 | 16 |
| | 3 | 9 | 18 |
| *D* | 1 | 10 | 17 |
| | 2 | 11 | 4 |
| | 3 | 12 | 15 |
| *E* | 1 | 13 | 10 |
| | 2 | 14 | 6 |
| | 3 | 15 | 2 |
| *F* | 1 | 16 | 3 |
| | 2 | 17 | 1 |
| | 3 | 18 | 12 |

Column 4 is a purely random rearrangement of the numbers 1 through 18. In practice, such a random order can be read from a "table of random numbers." (A brief table of random numbers is given in the Appendix, as Table VII.) Now the order of *running* the 18 determinations is that of this last column. Thus, the first determination to be made is the one whose identification number is 17, i.e., $F2$, the second is $E3$, etc. It may seem that the numerals 2 and 3, in $F2$ and $E3$, are now meaningless. Actually, however, each replicate may correspond to an individually prepared specimen; in that case the numerals in question serve to identify the specimens. From our discussion of the separation of replication error from the effect of interfering properties, it is clear that randomization is not the silly game it may appear to be, but rather a means of validating this separation.

## 13.6   THE SELECTION OF MATERIALS

We have seen that interfering properties will tend to increase the scatter of the data. Actually, such increased scatter appears only when the interfering properties differ from one material to another, and to the extent that the interfering properties are not functionally related to the property $Q$. To clarify this point, consider the determination of the atomic weight of an element. If this element occurs as the mixture of two isotopes in a constant ratio, then an increase, from one material to another, in the concentration of one of the isotopes will always be accompanied by a proportional increase in the concentration of the other isotope. Chemical methods are unable to differentiate between the two isotopes. Therefore, and because of the constant ratio of the isotopes in all materials, the presence of a mixture of isotopes will never appear in the form of undue scatter in any chemical determination. However, if the ratio of isotopes had varied from one material to another, then an experiment involving different materials might have revealed their presence as inconsistencies (or excessive scatter), exactly as the discrepancy between density measurements of nitrogen extracted from air and nitrogen extracted from ammonia revealed the presence of inert gases in the atmosphere.

It follows from these considerations that the more heterogeneous are the materials selected for the study of a measuring process, the more scatter may be expected in the data. The word heterogeneous, as used here, does *not* refer to drastic differences in the property $Q$ (a systematic change in the $Q$ value is in fact required for the experiment) but rather to the presence of different types of interfering properties. For example, if one wanted to study the determination of barium as the insoluble barium sulfate precipitate, either one of the following two experiments could be run:

(a) A series of solutions of barium chloride are prepared, containing for example, 5, 10, 15, 20, 25 grams of barium per liter of solution.

(b) Samples taken from a series of rocks containing barium in different amounts are pulverized and prepared for analysis.

In the case a the amount of precipitate obtained should be strictly proportional to the amount of barium present. In the case b, on the other hand, the precipitate may be contaminated by various amounts of other materials, whose behavior in regard to the analysis is similar to barium. Since the materials vary widely in regard to the presence of these interfering substances, the amount of precipitate, when plotted against the true quantity of barium present in each sample, may show appreciable scatter from the strictly proportional relationship.

The avoidance of this type of heterogeneity is not necessarily always a virtue: by restricting the experiment to a homogeneous set of materials one fails to learn anything about the disturbances created by interfering substances or properties. In the routine application of the test it may not always be possible to make corrections for these effects and a study of the test which ignores them will fail to give a realistic estimate of the amount of scatter encountered in practice. Thus, the selection of the materials for the study of a test method is a complex problem that should be solved on the basis of an analysis of the objectives of the study. It is not a statistical problem. What is required from a statistical point of view is that enough materials be present to cover the range of interest, so that the statistical analysis be of sufficient power to estimate the required parameters with satisfactory precision.

## 13.7   INTERLABORATORY STUDIES OF TEST METHODS

The best way of obtaining a realistic appraisal of the degree of control of the environment that can be achieved through the specification of tolerances is to run an interlaboratory study. Essentially such a study is made by sending a number of materials, covering the desired range of $Q$ values, to a number of laboratories, each of which submits its values for the measurement of these materials. Sometimes several properties are studied simultaneously. In other cases, two or more alternate measuring techniques pertaining to the same property are covered in a single study. For clarity of presentation we will first deal with interlaboratory studies of a single measuring technique.

The basic design for an interlaboratory study of a measuring process (often referred to as an "interlaboratory test") is shown in Table 13.2. The study involves $q$ materials, each of which is measured by the $p$ laboratories participating in the study. Each of these laboratories makes $n$

replicate measurements on each of the $q$ materials. We will refer to the $n$ replicates made by laboratory $i$ on material $j$ as the "$i, j$ cell." Thus, the study comprises $p \times q$ cells of $n$ measurements each, a total of $p \times q \times n$ measurements.

The selection of the materials has been discussed in Section 13.6. We now turn our attention to the participating laboratories. It is evident that unless the participating laboratories are entirely qualified to run the test, both in terms of equipment and personnel, the study will be a waste of time. It is also evident that the greater the number of participating laboratories, the more informative will be the study. Exactly how many laboratories are required depends on a number of factors, such as the objectives of the study (precisely stated), the availability of qualified laboratories, the funds available for the study, and other factors. The important point here is not to delude oneself by expecting a great deal of information from a meager set of data. The point made in Chapter 10 concerning the relatively large sample sizes required to obtain reliable estimates of variability is highly relevant for interlaboratory studies, since their general objective is the study of variability.

The number of replicates per cell, $n$, need not, in general, be large. Under certain circumstances, $n = 2$ will give excellent results. Some physical and dimensional types of measurements are quite rapid and inexpensive, but also quite variable. For these types it is advisable to run a number of replicates considerably larger than 2, for example, $n = 8, 10$, or even 15.

Consider now any particular column of Table 13.2. From the material represented by this column, a total of $p \times n$ test specimens will be tested. It is good policy to entrust the preparation and distribution of these specimens to a central group (generally a technical committee or a laboratory selected by the committee). For reasons analogous to those stated in Section 13.6, a thorough randomization of the $p \times n$ specimens is indispensable for a valid study.

Each participating laboratory contributes $n \times q$ measurements. Whenever possible, the laboratories should be instructed to run these $n \times q$ measurements in a completely randomized order.* It is also advisable to provide each laboratory with one or two additional specimens from each material, to be used in case of loss or damage to any test specimen. But apart from this possibility, the number of specimens of each material, to be measured by each laboratory should be exactly $n$, the number specified in the accepted design. The laboratories should be instructed to report their results as obtained, preferably on a standardized

---

* See, however, the discussion in Section 13.19, describing a situation in which only partial randomization is required.

**TABLE 13.2**   Basic Design for Interlaboratory Test

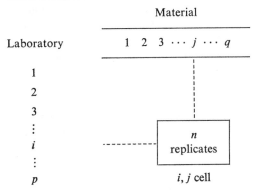

form sent to them by the committee. While the participants should be requested to adhere strictly to the prescribed procedure for the measuring technique—identical copies of this procedure should be made available to all participants—they should be encouraged to submit observations and remarks that might prove helpful in the interpretation of the data.

Occasionally, a participating laboratory is generally well qualified, but lacks previous experience in the use of the measuring process that is being studied. When the method is entirely new, most laboratories will fall in this category. In such cases, it is advisable to precede the interlaboratory study with a preliminary test, for the purpose of allowing the participants to familiarize themselves with the new process. All this requires is the sending out, to all participants, of identical copies of the procedure, described with as much detail as necessary, and of test specimens for experimentation. This preliminary step may well save considerable time and money; failure to carry it out may cause a great deal of unnecessary frustration.

## 13.8   STATISTICAL ANALYSIS: BASIC OBJECTIVES

The basic objective of the interlaboratory study is to obtain precise and usable information about the variability of results yielded by the measuring process. In terms of the discussion given in the beginning section of this chapter, the object is to derive from the data information about the structure of the bundle of curves representing the measuring process, including estimates for the replication error, the effect of interfering properties and of systematic differences between laboratories. We will show in the sections that follow how these objectives can be attained by a proper statistical analysis of the data.

A further objective of many, if not all interlaboratory studies, is to discover previously unidentified sources of variation in the test results and to obtain clues on their elimination through possible modifications of the specified procedure. Often this takes the form of better control (closer tolerances) of certain environmental conditions, but it may also appear as a method for eliminating or neutralizing the effects of certain interfering properties in the materials.

Finally, if any modification of the procedure is undesirable or if no reasonable modifications have been suggested by the study, it is often still possible to improve the precision of the measuring technique through the use of standard samples. We will examine these objectives in succession.

## 13.9   THE DETERMINATION OF REPLICATION ERROR

The error of replication is obtained without difficulty from the data provided by an experiment run in accordance with the design of Table 13.2. Each cell provides its own estimate of replication error with $(n - 1)$ degrees of freedom. The question is whether and how these various estimates should be combined. This matter, as well as further steps in the analysis are best discussed in terms of a specific set of data.

Table 13.3 shows a portion of the results obtained in an interlaboratory study of the Bekk method for the measurement of the smoothness of paper. The complete study involved 14 materials and 14 laboratories and is available in published form (4). The reason for presenting here only a portion of the data is merely to facilitate the discussion and to allow us to concentrate on the principles of the analysis rather than on computational details.

The number of replicates per cell in this study is $n = 8$; the test is rapid, inexpensive, and subject to a great deal of variability from specimen to specimen.

In the present section we are concerned only with the replication error, but a meaningful analysis of this source of variability requires a preliminary step in the over-all analysis of the data. This step consists in computing for each cell of the original table: (a) the average of the $n$ replicates in the cell; and (b) the standard deviation of these $n$ replicates. This computation leads to a new tabulation, shown for our data in Table 13.4, in which the 8 measurements of each cell are now replaced by two numbers, their average, and their standard deviation. We denote these two quantities by $\bar{y}_{ij}$ and $s_{ij}$, respectively. For reasons that will soon become apparent, the materials are always arranged in order of increasing magnitude of the measured quantity. Since this order may vary from laboratory to laboratory the order adopted is that of the average results of all

**TABLE 13.3**   Bekk Smoothness Data.   Original Values in Each Cell

| Laboratory Code number | Material Code Number 10 | 3 | 5 | 8 | 6 |
|---|---|---|---|---|---|
| 1 | 6.5 | 12.9 | 38.7 | 166.3 | 125.9 |
|  | 6.5 | 12.5 | 47.7 | 151.8 | 113.5 |
|  | 6.6 | 12.8 | 44.5 | 141.0 | 123.4 |
|  | 8.1 | 11.0 | 45.5 | 149.4 | 174.0 |
|  | 6.5 | 9.3 | 41.5 | 151.7 | 130.1 |
|  | 6.5 | 13.7 | 46.0 | 166.1 | 158.2 |
|  | 6.8 | 13.1 | 43.8 | 148.6 | 144.5 |
|  | 6.5 | 14.0 | 58.0 | 158.3 | 180.0 |
| Average | 6.750 | 12.41 | 45.71 | 154.2 | 143.7 |
| 5 | 6.0 | 13.2 | 42.9 | 178.5 | 196.7 |
|  | 6.0 | 14.2 | 39.3 | 183.3 | 230.8 |
|  | 6.1 | 12.2 | 39.7 | 150.8 | 146.0 |
|  | 5.8 | 13.1 | 45.8 | 145.0 | 247.3 |
|  | 5.7 | 9.9 | 43.3 | 162.9 | 183.7 |
|  | 5.7 | 9.6 | 41.4 | 184.1 | 237.2 |
|  | 5.7 | 12.6 | 41.8 | 160.5 | 229.3 |
|  | 6.0 | 9.0 | 39.2 | 170.8 | 185.9 |
| Average | 5.875 | 11.73 | 41.68 | 167.0 | 207.1 |
| 9 | 6.0 | 11.0 | 39.0 | 150.0 | 150.0 |
|  | 5.0 | 10.0 | 44.0 | 150.0 | 165.0 |
|  | 6.0 | 12.0 | 43.0 | 150.0 | 172.0 |
|  | 6.0 | 13.0 | 43.0 | 160.0 | 162.0 |
|  | 6.0 | 11.0 | 40.0 | 150.0 | 170.0 |
|  | 6.0 | 10.0 | 35.0 | 160.0 | 150.0 |
|  | 6.0 | 11.0 | 34.0 | 182.0 | 158.0 |
|  | 6.0 | 12.0 | 45.0 | 140.0 | 198.0 |
| Average | 5.875 | 11.25 | 40.38 | 155.3 | 165.6 |
| 13 | 5.0 | 11.4 | 35.4 | 121.8 | 123.8 |
|  | 5.0 | 10.0 | 37.2 | 127.4 | 162.0 |
|  | 4.9 | 8.8 | 34.8 | 145.0 | 128.4 |
|  | 4.8 | 8.2 | 41.2 | 162.4 | 153.0 |
|  | 4.6 | 10.0 | 42.6 | 122.2 | 164.4 |
|  | 4.5 | 8.4 | 37.8 | 124.0 | 140.0 |
|  | 4.8 | 10.0 | 34.8 | 110.2 | 130.2 |
|  | 4.3 | 12.6 | 34.0 | 141.2 | 198.8 |
| Average | 4.738 | 9.925 | 37.23 | 131.8 | 150.1 |

laboratories for each of the materials.   Thus, material No. 10 occurs at the extreme left because its over-all average, 5.810, is the smallest; it is followed by material No. 3, whose average, 11.329, is the next-to-smallest, and so on.

**TABLE 13.4** Bekk Smoothness Data: Averages and Standard Deviations*

| Laboratory Code number | Material code number | | | | |
| | 10 | 3 | 5 | 8 | 6 |
| --- | --- | --- | --- | --- | --- |
| 1 | 6.750 | 12.41 | 45.71 | 154.2 | 143.7 |
| | 0.556 | 1.55 | 5.70 | 8.83 | 24.7 |
| 5 | 5.875 | 11.73 | 41.68 | 167.0 | 207.1 |
| | 0.167 | 1.94 | 2.30 | 14.7 | 34.7 |
| 9 | 5.875 | 11.25 | 40.38 | 155.3 | 165.6 |
| | 0.354 | 1.04 | 4.14 | 12.6 | 15.4 |
| 13 | 4.738 | 9.925 | 37.23 | 131.8 | 150.1 |
| | 0.250 | 1.51 | 3.17 | 16.6 | 25.0 |
| Avg. of averages, $\bar{y}_j$ | 5.810 | 11.329 | 41.25 | 152.1 | 166.6 |
| Avg. of std. devs., $\bar{s}_j$ | 0.332 | 1.51 | 3.83 | 13.2 | 25.0 |
| $(\bar{s}_j/\bar{y}_j)$ | 0.0571 | 0.1333 | 0.0928 | 0.0868 | 0.1501 Avg. $=0.1040$ |

* Of 8 replicate measurements per laboratory and material; the top number in each cell is the average, and the lower number, the standard deviation of 8 measurements.

In addition to the column-averages, denoted $\bar{y}_j$ (which are really the averages of cell-averages for each individual material), there are also listed in Table 13.4 the averages of the standard deviations of each column. These latter averages are denoted $\bar{s}_j$.

We now return to our basic model (Fig. 13.2) for the interpretation of these quantities. An individual $s_{ij}$-value represents the variability among the curves of the subset corresponding to laboratory $i$ at a $Q$ level represented by material $j$. If the participating laboratories are all competent (as we have assumed) the spread measured by $s_{ij}$ should not vary appreciably from one laboratory to another, when measured at the same $Q$ level. Therefore, the average $\bar{s}_j$ can be accepted as a measure of this variability *within* subsets at level $j$. On the other hand, there is no justification for assuming that this spread is necessarily the same at all $Q$ levels (all $j$). And in fact, the results in Table 13.4 show a very marked increase in $\bar{s}_j$ as the level of the property increases. It is for ease of detection of this effect that the materials have been arranged in increasing order of magnitude of the measured characteristic.

It is instructive to examine the relationship between $\bar{s}_j$ and $\bar{y}_j$, by plotting the first quantity versus the second. The plot shown in Fig. 13.3 suggests a proportional relation, and this is confirmed by the last row of Table 13.4, which lists the ratio $\bar{s}_j/\bar{y}_j$ for each $j$. The average value of

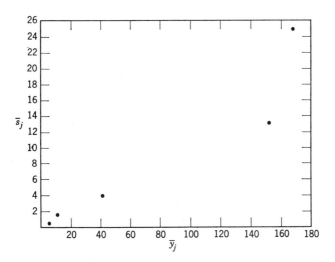

**Fig. 13.3**  Bekk smoothness data; relation between replication error and measured value.

this ratio, 0.104, represents of course the slope of the line in the $\bar{s}_j$ versus $\bar{y}_j$ plot.    Thus, the standard deviation of replication error is approximately 10 per cent of the measured value; in other words, the coefficient of variation of replication error is close to 10 per cent, regardless of the level of the smoothness measurement.    While the *coefficient of variation* here is a constant, the data in different cells of Table 13.3 are, of course, not homoscedastic, since the *standard deviation* varies appreciably from material to material.    For many reasons, most of which are of a statistical technical nature, it is desirable to deal with homoscedastic data wherever this is possible.    This can be accomplished in the present case by a *transformation of scale*.

Let $y$ represent a smoothness measurement expressed in Bekk seconds and let $\sigma_y$ represent the standard deviation of replication error.    Then we have just found that approximately

$$\sigma_y = 0.10y \tag{13.1}$$

Now let $z$ represent $y$ expressed in a transformed scale defined by

$$z = f(y) \tag{13.2}$$

Then, according to Eq. 4.43 the standard deviation of $z$ is given by

$$\sigma_z = \left[\frac{df(y)}{d(y)}\right]\sigma_y \tag{13.3}$$

We wish to select the function $f(y)$ in such a way that $\sigma_z$ is a constant (homoscedasticity in the $z$ scale). Thus we wish to have:

$$\left[\frac{df(y)}{dy}\right]\sigma_y = K \tag{13.4}$$

where $K$ is independent of $z$.

Replacing $\sigma_y$ by its value as given by Eq. 13.1 we obtain:

$$\left[\frac{df(y)}{dy}\right](0.10)y = K$$

or

$$df(y) = \frac{K}{0.10}\frac{dy}{y} \tag{13.5}$$

Integration of this relation gives

$$f(y) = K' \ln y \tag{13.6}$$

where $\ln y$ denotes the natural logarithm of $y$. The value of $K'$ is arbitrary.

We have thus established that the transformation of scale that will accomplish homoscedasticity for our present data is a simple logarithmic transformation. We can of course use logarithms of base 10 rather than natural logarithms, since this means nothing more than a change in the constant $K'$.

More generally, if the relation between measurement and standard deviation is

$$\sigma_y = \Phi(y) \tag{13.7}$$

the proper transformation of scale (Eq. 13.2), to achieve homoscedasticity is given by the differential Eq. 13.4 which becomes:

$$\frac{df(y)}{dy}\Phi(y) = K$$

or

$$df(y) = K\frac{dy}{\Phi(y)} \tag{13.8}$$

which, upon integration gives:

$$z = f(y) = K\int\frac{dy}{\Phi(y)} + \text{arbitrary constant} \tag{13.9}$$

We have now extracted from the data all the information we require about the replication error. This information can be summarized as follows:

(a) The standard deviation of replication error is approximately equal to 10 per cent of the measured value.

(b) To achieve homoscedasticity, express the data in a new scale, $z$, by applying a simple logarithmic transformation; for convenience the arbitrary factor in the transformation may be taken to be 1000. Thus the transformation is:

$$z = 1000 \log y \qquad (13.10)$$

Having reached this point, the individual values within each cell are no longer required and all further steps in the analysis of the data are made using only the cell averages (upper value in each cell of Table 13.4).

## 13.10    ANALYSIS OF THE TRANSFORMED CELL-AVERAGES

Had the previous analysis not indicated the desirability of a transformation of scale, then the cell-averages, taken as they appear in Table 13.4, would form the starting point of the subsequent analysis. Because of the desirability of a transformation of scale for the present data, all cell-averages were transformed by Eq. 13.10 prior to further analysis. Table 13.5 shows the transformed cell-averages. Thus, the value 829 is equal to 1000 times the logarithm (base 10) of 6.750, in accordance with Eq. 13.10, and similarly for all other entries.

In analyzing Table 13.5 we follow the general procedure explained in Section 9.4. The analysis of variance is shown in Table 13.6; the sums of squares were calculated in accordance with Eqs. 9.42 through 9.46.

**TABLE 13.5**    Bekk Smoothness Data—Cell Averages in Transformed Scale[a]

| Laboratory Code number | Material code number | | | | | Average, $\hat{\mu}_i$ |
|---|---|---|---|---|---|---|
| | 10 | 3 | 5 | 8 | 6 | |
| 1 | 829 | 1093 | 1660 | 2188 | 2157 | 1585.4 |
| 5 | 769 | 1069 | 1620 | 2223 | 2316 | 1599.4 |
| 9 | 769 | 1051 | 1606 | 2191 | 2219 | 1567.2 |
| 13 | 676 | 997 | 1571 | 2120 | 2176 | 1508.0 |
| Average $x_j$ | 760.75 | 1052.50 | 1614.25 | 2180.50 | 2217.00 | 1565.0 |

[a] Transformation: $z = 1000 \log y$.

From Table 13.6 it may be inferred that the linear model is significantly better than the additive model, as shown by the $F$-ratio of $3562/756 = 4.71$ which, for 3 and 9 degrees of freedom, is significant at the 5 per cent level. The values of the parameter estimates for the linear model are given in Table 13.7, and the residuals in Table 13.8.

**TABLE 13.6**   Bekk Smoothness Data—Analysis of Variance

| Source of variation[a] | DF | SS | MS |
|---|---|---|---|
| Laboratories | 3 | 24,267 | 8089 |
| Materials | 4 | 6,863,376 | 1,715,844 |
| Interaction (add. model) | 12 | 17,488 | 1457 |
| Slopes | 3 | 10,686 | 3562 |
| Residual (lin. model) | 9 | 6802 | 756 |

[a] The "Total" sum of squares is omitted since it is of no interest.

**TABLE 13.7**   Bekk Smoothness Data—Estimates of Parameters, Linear Model

| Laboratory Code number | $\hat{\mu}$ | $\hat{\beta}$ | Material Code number | $\gamma_j$ |
|---|---|---|---|---|
| 1 | 1585.4 | 0.937 | 10 | −804.25 |
| 5 | 1599.4 | 1.044 | 3 | −512.50 |
| 9 | 1567.2 | 1.001 | 5 | 49.25 |
| 13 | 1508.0 | 1.018 | 8 | 615.50 |
|  |  |  | 6 | 652.00 |

**TABLE 13.8**   Bekk Smoothness Data—Residuals, Linear Model

| Laboratory Code number | Material code number | | | | | Sum of squares of residuals |
|---|---|---|---|---|---|---|
|  | 10 | 3 | 5 | 8 | 6 |  |
| 1 | −2.8 | −12.2 | 28.5 | 25.9 | −39.3 | 3184 |
| 5 | 9.2 | 4.6 | −30.8 | −19.0 | 35.9 | 2704 |
| 9 | 6.9 | −3.2 | −10.5 | 7.7 | −0.8 | 228 |
| 13 | −13.3 | 10.7 | 12.9 | −14.6 | 4.3 | 689 |
|  |  |  |  |  |  | 6805 |

We must now justify the analysis in the light of the basic model proposed in Section 13.2 for the statistical representation of a measuring process.

The equation underlying the analysis by the linear model is:

$$z_{ij} = \mu_i + \beta_i \gamma_j + \eta_{ij} \qquad (13.11)$$

The symbol $z$ is used instead of $y$ to indicate that a transformation of scale has been made. Furthermore, the error term is expressed by the symbol $\eta$, rather than $\varepsilon$, because we wish to reserve the latter to denote replication error.   We also know that

$$\gamma_j = x_j - \bar{x} \tag{13.12}$$

where the $x_j$ are the column averages shown in the bottom row of Table 13.5.   Combining Eqs. 13.11 and 13.12 we obtain

$$z_{ij} = (\mu_i - \bar{x}\beta_i) + \beta_i x_i + \eta_{ij} \tag{13.13}$$

which shows that *for any given i, $z_{ij}$* is, apart from random error, a linear function of $x_j$.   This means that if we plot $x_j$ on the abscissa of rectangular co-ordinates, and $z_{ij}$ on the ordinate, we obtain, *for each laboratory* (fixed $i$), a set of points scattering (with errors $\eta_{ij}$) about a straight line.   The intercept and the slope of the line depend on $i$, i.e., on the laboratory. Thus there results a bundle of straight lines, each representing a particular laboratory.   This is exactly the model we have assumed for the measuring process (Section 13.2) except for two circumstances:

(*a*) The curves corresponding to the various laboratories are now specified to be straight lines.

(*b*) The values plotted on the abscissa are the laboratory averages for each of the materials (as provided by the data themselves), rather than the theoretical $Q$-values.

Let us first consider *b*.   This apparent restriction is really not a fundamental one.   For the property $Q$ could be expressed in any desired scale. The substitution of the $x_j$ for the $Q$-values is really nothing more than the expression of $Q$ in a special scale, that of the measured values transformed in accordance with Eq. 13.10.   It is true that the $x_j$, as averages of values affected by random error, are themselves subject to random errors. However, if the number of laboratories is not too small, there will be appreciable compensation of the errors through the averaging process. Besides, the $x_j$ are generally the best measure we have for the unknown $Q$. Furthermore, since our analysis is concerned with the *internal structure* of the data (the question of accuracy will be dealt with in Chapter 14), small fluctuations affecting the $x_j$ values will have little effect on our conclusions.

As to point *a*, this is really a consequence of *b*.   Indeed, the curvedness suggested by Figs. 13.1 and 13.2 for the response curves under controlled conditions is due in most cases to the fact that $M$ and $Q$ are often expressed in different scales.   For example, $Q$ might represent the concentration of a solution and $M$ the optical "transmittance" in a spectrophotometric measurement made on the solution.   While transmittance is a function of concentration, it is not a linear function.   But by expressing $Q$ in the scale in which the measurement itself is expressed, one is very likely to obtain, to a good approximation at least, a straight line relationship.   Experience confirms this: in the few cases in which a laboratory showed systematic curvature for the plot of $z_{ij}$ versus $x_j$, there was always serious departure, on the part of this laboratory, from the specified procedure for carrying out the measuring process.

Of course, the possibility of representing the bundle of response curves by a bundle of straight lines has the important advantage of greatly facilitating the statistical analysis of the data.

Returning now to our data, and to Table 13.8, the last column lists the sum of squares of residuals computed separately for each laboratory. It would appear that laboratories 1 and 5 show greater scatter than laboratories 9 and 13. However, the four sums of squares are not independent, and each one is based on only 3 degrees of freedom. With such a limited set of data, one should be extremely cautious in drawing conclusions regarding the relative precision of the participating laboratories. The full study, involving 14 laboratories and 14 materials showed much less variation in the sums of squares for the 4 laboratories in question. If a larger number of laboratories is available, it is advisable to compute the individual sums of squares, as in Table 13.8, and then derive from them separate estimates of variance, by dividing each sum of squares by the number of materials minus 2. These estimates may be denoted $\hat{V}_i(\eta)$, where the index $i$ denotes the laboratory. For the data of our example, the four values of $\hat{V}_i(\eta)$ would be as follows:

| Laboratory Code number | 1 | 5 | 9 | 13 |
|---|---|---|---|---|
| $\hat{V}_i(\eta)$ | 1061 | 901 | 76 | 230 |

Note that the pooled estimate of $V(\eta)$, $\hat{V}(\eta) = 756$, shown in Table 13.6 is not the average of these four values. Indeed, the total sum of squares of residuals in Table 13.8 is 6805, which is identical, except for a slight rounding error, with the value 6802 in Table 13.6. Now, the degrees of freedom corresponding to this sum of squares is 9, whereas the divisor based on the average of the four individual $\hat{V}_i(\eta)$ would be $3 \times 4 = 12$. The reason for this discrepancy is that those four individual estimates are correlated with each other. This correlation arises from the use of the column averages in the regression analyses: the $z_{ij}$ for any value of $j$ are of course correlated with their average, $x_j$. In general, the pooled value of $\hat{V}(\eta)$ is obtained by dividing the sum of the individual $\hat{V}_i(\eta)$ by $p - 1$, where $p$ is the number of laboratories:

$$\hat{V}(\eta) = \frac{\sum_i \hat{V}_i(\eta)}{p - 1} \qquad (13.14)$$

If a satisfactory number of laboratories have participated in the study, an examination of the individual $\hat{V}_i(\eta)$ sometimes reveals one or more laboratories for which this estimate is considerably greater than for all others. This might be indicative of either a lower level of control of environmental factors in these laboratories, or of failure to follow exactly the instructions for the measuring process. In such cases, a better appraisal of the measuring process is obtained by omitting such laboratories from the analysis. Of course, a situation of this type may offer a valuable opportunity for the discovery and correction of omissions or unsatisfactory statements in the specification of the measuring method.

Turning now to the $\mu$ and $\beta$ values for the participating laboratories, here again it is advisable to scrutinize these values for the purpose of detecting "outlying" laboratories. What interpretation should be given to a markedly discrepant $\mu$ or $\beta$ value? We will discuss this matter in detail in the following section. At this point we merely point to the extreme difficulty of judging whether any one of *four* values of $\mu$ and $\beta$ can be considered to be discrepant. This again points to the paucity of information provided by an interlaboratory study of insufficient scope. The selected data we have been using for illustration (Table 13.3) provide an example of such an unsatisfactory state of affairs. We will therefore continue our discussion on the basis of the complete study, by listing the parameter estimates obtained in this study.

## 13.11  INTERPRETATION OF THE PARAMETERS

The first three columns in Table 13.9 show the estimates of $\mu$, $\beta$, and $V(\eta)$ obtained from the complete study of the Bekk method. The laboratories have been reordered according to increasing $\mu$-value, to facilitate the examination of the data.

It is apparent that laboratory 6 shows very low values both for $\hat{\mu}$ and $\hat{\beta}$, and that laboratory 4 shows a very low value for $\hat{\beta}$. These two laboratories made their measurements at 65 per cent relative humidity while all other laboratories operated at 50 per cent relative humidity. It is therefore legitimate to eliminate the data from laboratories 4 and 6 from the analysis. The two columns on the right of Table 13.9 list the values of $\hat{\beta}$ and $\hat{V}(\eta)$ recalculated after omission of the two discrepant laboratories. (The values for $\hat{\mu}$ are unaffected by this omission, since each $\hat{\mu}_i$ is the average of the values occurring in row $i$ only.) The analysis of variance for the data from the 12 laboratories is shown in Table 13.10.

From the viewpoint of significance testing, Tables 13.6 and 13.10 are practically equivalent; both show the need for the linear model and the significance of the "laboratories" effect, i.e., the existence of significant

**TABLE 13.9**  Parameter Estimates for Bekk Data, Complete Study

| Laboratory Code number | $\hat{\mu}$ | 14 Laboratories $\hat{\beta}$ | $\hat{V}(\eta)^a$ | 12 Laboratories $\hat{\beta}$ | $\hat{V}(\eta)^a$ |
|---|---|---|---|---|---|
| 6 | 1403.2 | 0.881 | 575 | — | — |
| 4 | 1481.8 | 0.913 | 309 | — | — |
| 10 | 1488.6 | 1.021 | 286 | 1.003 | 254 |
| 14 | 1516.2 | 1.048 | 578 | 1.030 | 620 |
| 13 | 1521.3 | 1.035 | 235 | 1.017 | 255 |
| 7 | 1537.4 | 1.028 | 760 | 1.010 | 758 |
| 8 | 1542.4 | 1.028 | 569 | 1.010 | 589 |
| 3 | 1561.1 | 0.991 | 1190 | 0.974 | 1081 |
| 9 | 1568.0 | 1.028 | 524 | 1.010 | 443 |
| 5 | 1584.1 | 1.056 | 869 | 1.038 | 940 |
| 11 | 1591.4 | 1.015 | 810 | 0.998 | 808 |
| 1 | 1597.9 | 0.942 | 928 | 0.926 | 898 |
| 2 | 1598.6 | 0.960 | 219 | 0.944 | 217 |
| 12 | 1607.6 | 1.056 | 194 | 1.038 | 213 |

$^a$ The values of $\hat{V}(\eta)$ given in the original paper differ slightly from those listed here: the formula used in that paper is $\hat{V}_i(\eta) = p \sum_j (\text{res})_{ij}^2/(p-1)(q-2)$ where the factor $p/(p-1)$ is a correction for the bias introduced by the correlation between different $\hat{V}_i(\eta)$. Since the values of $\hat{V}_i(\eta)$ are needed only for internal comparison, the adjustment is not indispensable.

**TABLE 13.10**  Bekk Smoothness Data, Complete Study—Analysis of Variance

| Source of variation | DF | SS | MS |
|---|---|---|---|
| Laboratories | 11 | 226,170 | 20,561 |
| Materials | 13 | 53,076,850 | 4,082,835 |
| Interaction (add. model) | 143 | 144,845 | 1013 |
| Slopes | 11 | 60,410 | 5492 |
| Residual (lin. model) | 132 | 84,435 | 640 |

differences in the $\mu_i$. Furthermore the estimates of $V(\eta)$ provided by the two tables are in good agreement (756 and 640). Whether the two tables also agree quantitatively in their evaluation of the difference between laboratories cannot be decided without extracting "components of variance," a subject to be dealt with in Section 13.12.

Two points need to be emphasized. The first concerns the desirability of correlating the values of $\hat{\mu}_i$, $\hat{\beta}_i$, and $\hat{V}_i(\eta)$ with all that is known about differences in environmental factors in the laboratories. Thus, the low values obtained by laboratories 4 and 6 for both $\mu$ and $\beta$ could be correlated with the different relative humidity in these laboratories. It is here that the notes and remarks made by the participants of an interlaboratory study acquire real importance. For although an experimenter usually makes many side observations that might become valuable clues, he generally has little opportunity to perceive their true significance until he compares his results with those obtained by other experimenters. This is exactly what an interlaboratory study allows one to do.

The second observation to be made at this point is that there sometimes exists a correlation between the $\hat{\mu}$ and the $\hat{\beta}$, and that such a correlation is a valuable feature in the interpretation of the data. If one ranked the $\mu$ and $\beta$ estimates in Table 13.9 one would find no significant correlation between the ranks. Table 13.11, on the other hand, which is taken from an interlaboratory study of a method of analysis for butyl rubber (7), shows a marked correlation between the $\hat{\mu}$ and $\hat{\beta}$ values.

**TABLE 13.11**     Analysis of Butyl Rubber—Parameter Estimates

| Laboratory[a] | $\hat{\mu}$ | $\hat{\beta}$ | Laboratory[a] | $\hat{\mu}$ | $\hat{\beta}$ |
|---|---|---|---|---|---|
| 3A | 31.86 | 0.902 | 5A | 35.02 | 1.003 |
| 3B | 33.42 | 0.979 | 5B | 35.11 | 0.991 |
| 6B | 34.59 | 0.982 | 2  | 35.78 | 1.020 |
| 6A | 34.70 | 0.995 | 1B | 36.32 | 1.045 |
| 4B | 34.73 | 0.998 | 1A | 36.54 | 1.056 |
| 4A | 34.93 | 1.029 |    |       |       |

[a] All laboratories, except 2, submitted results from two operators, denoted A and B.

The laboratories in Table 13.11 are ordered in accordance with increasing $\hat{\mu}$. It is easily seen that the order of the $\hat{\beta}$ values is far from random, the order of their ranks being 1, 2, 3, 5, 6, 9, 7, 4, 8, 10, 11. Thus, with some exceptions, a laboratory for which $\hat{\mu}$ is high also shows a high $\hat{\beta}$ value. This suggests that $\hat{\mu}$ and $\hat{\beta}$ are functionally related. Figure 13.4 is a plot of this relationship and reveals that $\hat{\beta}$ is essentially a linear function of $\hat{\mu}$. In Section 11.12 we have seen, using an algebraic argument, that such a linear relationship implies that the bundle of straight lines postulated by the model all pass through a common point. This can also be made plausible by a geometric representation. As is seen in Fig. 13.5, $\hat{\mu}$ is the ordinate corresponding to $\bar{x}$, i.e., the "height" of the line at its centroid; $\hat{\beta}$ is of course the slope of the line. If we now imagine a sequence of lines of increasing $\hat{\mu}$ values and observe that their $\hat{\beta}$ values also increase, this indicates a tendency of the lines to "fan out" from left to right. In other

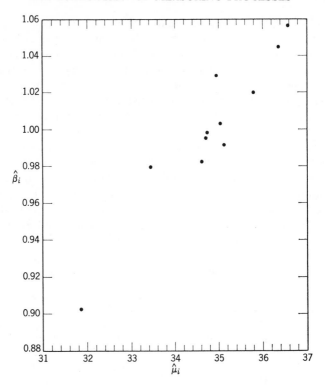

**Fig. 13.4**  Butyl rubber data; relation between $\hat{\mu}$ and $\hat{\beta}$.

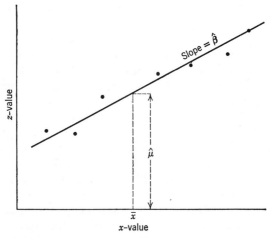

**Fig. 13.5**  Geometric interpretation of $\hat{\mu}$ and $\hat{\beta}$.

words, one could then consider the lines as representing different positions of a single line rotating about a fixed point. Therefore a significantly positive correlation between $\hat{\mu}_i$ and $\hat{\beta}_i$ indicates a tendency for "concurrence" of the lines at a point on the left; a significantly negative correlation indicates a tendency for concurrence at a point situated on the right of the graph. To test the significance of the correlation, use is made of the correlation coefficient between $\hat{\mu}_i$ and $\hat{\beta}_i$, defined by the relation

$$r = \frac{\sum_i (\hat{\mu}_i - \bar{\mu})(\hat{\beta}_i - \bar{\beta})}{\sqrt{\sum_i (\hat{\mu}_i - \bar{\mu})^2 \sum_i (\hat{\beta}_i - \bar{\beta})^2}} \tag{13.15}$$

For the Bekk study (Table 13.9, 12 laboratories), the correlation coefficient between $\hat{\mu}_i$ and $\hat{\beta}_i$ is low: $r = 0.33$. For the butyl rubber data (Table 13.11), we obtain $r = 0.9474$. This value is close to unity and poir.ts therefore to a bundle of approximately concurrent lines. If the lines were *exactly* concurrent, there would exist a particular $Q$-value—the point of concurrence—for which all laboratories obtain exactly the same result. In that case, materials would be subject to laboratory-to-laboratory variability to the extent that the value of the measured property differed from this $Q$-value. If the concurrence were only approximate, these conclusions would only be approximately true. The existence and the exactness of concurrence can be established by tests of significance. We will illustrate the method in terms of the butyl data, for which the analysis of variance is given in Table 13.12.

**TABLE 13.12** Analysis of Butyl Rubber—Analysis of Variance

| Source of variation | DF | SS | MS |
|---|---|---|---|
| Laboratories | 10 | 68.23 | 6.823 |
| Materials | 3 | 20,425.82 | 6808.607 |
| Interaction (add. model) | 30 | 45.56 | 1.519 |
| Slopes | 10 | 39.88 | 3.988 |
| Concurrence | 1 | 35.79 | 35.79 |
| Non-concurrence | 9 | 4.09 | 0.454 |
| Residual (lin. model) | 20 | 5.68 | 0.284 |

The sum of squares for "slopes" has been partitioned into two parts, obtained by multiplying it successively by $r^2$ and by $1 - r^2$:

$$r^2(39.88) = (0.9474)^2(39.88) = 35.79$$

$$(1 - r^2)(39.88) = 39.88 - r^2(39.88) = 39.88 - 35.79 = 4.09$$

The mean squares corresponding to these two quantities are measures for, respectively, the degree of concurrence and the extent to which the lines fail to concur.

To test for the existence of concurrence, compute the ratio of these two mean squares. A significant $F$ ratio indicates a tendency for concurrence. The exactness of the concurrence is tested by comparing the mean square for non-concurrence with the mean square for "residual." If the mean square for non-concurrence is not significantly larger than that of residual error (the comparison is made by the appropriate $F$ ratio), the concurrence can be considered to be exact to within experimental error. If, on the other hand, the mean square for non-concurrence is significantly larger than that of the residuals, the concurrence is not absolute and there remains a significant amount of laboratory-to-laboratory variability even at the point at which the lines for all laboratories tend to be closest to each other.

For the butyl data, we have

$$F_{\text{conc./non-conc.}} = \frac{35.79}{0.454} = 78.8$$

and

$$F_{\text{non-conc./resid.}} = \frac{0.454}{0.284} = 1.60$$

The first $F$, with 1 and 9 degrees of freedom, is very highly significant. The second $F$, with 9 and 20 degrees of freedom, shows no significance even at the 10 per cent level. We conclude that there is concurrence to within experimental error. In such a case, we can go a step further and compute the location, on the $z$-scale, of the point of concurrence. Denoting by $z_0$ the point of concurrence,* we derive from Eqs. 11.36 and 11.39:

$$z_0 = \bar{x} - \frac{\sum_i (\hat{\mu}_i - \bar{\mu})^2}{\sum_i (\hat{\mu}_i - \bar{\mu})(\hat{\beta}_i - \bar{\beta})} \tag{13.16}$$

For the butyl data this formula gives the value $z_0 = 1.27$. No transformation of scale was used for these data. Therefore, this value is expressed in the original scale, per cent butyl. On the other hand, the materials used in this study had butyl contents ranging from 3 to 60 per cent. The concurrence point, 1.27, is therefore quite close to the origin, zero per cent butyl. The analysis has thus established that the bundle of lines for this test is a concurrent one, the point of concurrence being, for all practical purposes, the origin. Further discussion of this matter will be found in Section 13.14 and its sequel.

* The ordinate of the point of concurrence is equal to its abscissa.

## 13.12   COMPONENTS OF VARIANCE

In Section 9.2 we have seen that the mean squares in an analysis of variance cannot always be interpreted as estimates of single variances. This is however not true for the mean square corresponding to experimental error.   Consider the analysis given in Table 13.10.   If the linear model is an adequate representation of the data, the mean square labeled "Residual" is a valid measure of the variance of experimental error. According to Eq. 13.11 this is simply the variance $V(\eta)$.   More specifically, denoting as usual an expected value by the symbol $E$, we have

$$E(\text{MS}_{\text{residual}}) = V(\eta) \tag{13.17}$$

The operational interpretation of this equation is as follows.   If the entire experiment involving 12 laboratories and 14 materials were repeated an unlimited number of times, the sequence of values obtained for the mean square of residuals would constitute a definite statistical population and the mean of this population would be exactly equal to $V(\eta)$.   This formulation raises the question: is the infinite replication of the experiment made by the *same* 12 laboratories and on the *same* set of 14 materials? Strictly speaking the answer depends on the way in which the selection of the materials and the selection of laboratories were made, and on the precise way in which the analysis is to be interpreted.   Thus, if the inferences made from the analysis are to cover a given population of laboratories, the "sample" of laboratories included in the experiment should be a random selection from the population.   Then each repetition of the experiment would be made on a new sample of laboratories, each sample being selected at random from the population of laboratories. On the other hand, one might in certain cases wish to restrict all inferences to a particular set of laboratories, in which case all successive repetitions of the experiment would involve the same set of laboratories.   A random selection of laboratories is quite often beyond practical possibilities.   It is then not unreasonable to extend the inferences to a *hypothetical* population of laboratories "similar" to those selected.   Under this assumption the hypothetical replication of the experiment would indeed involve a new set of laboratories for each repetition, all sets being random selections from this hypothetical population.   While this interpretation may seem artificial, it should be observed that most inferences, those of our every day life as well as those made in scientific work, involve similar modes of reasoning.   The important point is to be aware of the degree and type of generalization involved in the inferences, and to formulate them in precise language.

In regard to the materials, it is of course absurd to consider those used in the study as a random sample from a single population, since they were purposely selected to cover a wide range of $Q$ values. It is however entirely reasonable to consider them as random selections from *several* populations. More specifically, we may consider material $j$ as a random selection from a population of materials, *having all the same Q value* (as material $j$), but differing from each other in other properties. The extent to which they differ in these properties (other than $Q$) is exactly the same as that to which the materials actually selected for the interlaboratory study differ from each other in these properties. In this manner the inferences made from the data acquire a precise meaning.

The degree of homogeneity of the materials, discussed in Section 13.6, is now directly linked to the $q$ conceptual populations from which the $q$ materials are considered to be taken.

Turning now to the other mean squares in the analysis of variance we first note that they involve estimates of the following parameters: the experimental error $V(\eta)$ already mentioned, the differences between laboratories in terms of $\mu_i$, and the differences between laboratories in terms of $\beta_i$. Referring to the hypothetical population of laboratories discussed above, let us denote by $V(\mu)$ the variance of $\mu_i$ in this population. Similarly, let $V(\beta)$ represent the variance of $\beta_i$ in the same hypothetical population of laboratories. It can then be shown that each mean square in the analysis of variance is a random variable, the expected value of which is a linear combination of $V(\eta)$, $V(\mu)$, and $V(\beta)$. The term "expected value" refers here, as before, to the conceptual endless sequence of repetitions of the experiment. Table 13.13 contains the formulas relating the expected values of the mean squares to the parameters $V(\eta)$, $V(\mu)$, and $V(\beta)$.

**TABLE 13.13**  Expected Values of Mean Squares

| Source of variation | Mean square | Expected value of mean square |
|---|---|---|
| Laboratories | $MS_L$ | $V(\eta) + qV(\mu)$ |
| Materials | $MS_M$ | $V(\eta) + p\left(\dfrac{\sum_j (x_j - \bar{x})^2}{q - 1}\right)$ |
| Interaction (add. model) | $MS_{LM}$ | $V(\eta) + \left(\dfrac{\sum_j (x_j - \bar{x})^2}{q - 1}\right)V(\beta)$ |
| Slopes | $MS_\beta$ | $V(\eta) + \left[\sum_j (x_j - \bar{x})^2\right]V(\beta)$ |
| Residual (lin. model) | $MS_\eta$ | $V(\eta)$ |

It is customary to compute estimates of unknown variances, such as $V(\mu)$ and $V(\beta)$, by equating the mean squares with their expected values. Thus, instead of writing

$$E[MS_L] = V(\eta) + qV(\mu)$$

one writes

$$MS_L \approx V(\eta) + qV(\mu)$$

This latter equation is of course only approximately true. This is indicated by substituting the symbol $\approx$ for the equality sign. Proceeding similarly for all rows in Table 13.13, one obtains*

$$MS_L \approx V(\eta) + qV(\mu) \tag{13.18}$$

$$MS_M \approx V(\eta) + p\frac{\sum (x_j - \bar{x})^2}{q - 1} \tag{13.19}$$

$$MS_{LM} \approx V(\eta) + \left(\frac{\sum (x_j - \bar{x})^2}{q - 1}\right)V(\beta) \tag{13.20}$$

$$MS_\eta \approx V(\eta) \tag{13.21}$$

From Eqs. 13.21 and 13.19 one first derives the estimates†:

$$\hat{V}(\eta) = MS_\eta \tag{13.22}$$

$$\text{Estimate of } \left[\frac{\sum (x_j - \bar{x})^2}{q - 1}\right] = \frac{MS_M - \hat{V}(\eta)}{p} = \frac{MS_M - MS_\eta}{p} \tag{13.23}$$

Using these estimates in the remaining equations, one obtains:

$$\hat{V}(\mu) = \frac{MS_L - MS_\eta}{q} \tag{13.24}$$

$$\hat{V}(\beta) = \frac{MS_{LM} - MS_\eta}{(MS_M - MS_\eta)/p} \tag{13.25}$$

The three quantities $\hat{V}(\eta)$, $\hat{V}(\mu)$, and $\hat{V}(\beta)$, obtained from Eqs. 13.22, 13.24, and 13.25, are basic for the evaluation of the precision of the test method. Applying these equations to the Bekk data (Table 13.10), we obtain:

$$\hat{V}(\eta) = 640$$
$$\hat{V}(\mu) = 1423$$
$$\hat{V}(\beta) = 0.001095 \tag{13.26}$$

* The row for "slopes" is omitted because it contains no information that is not already provided by the other rows.

† The quantity $\sum (x_j - \bar{x})^2/(q - 1)$ can be computed directly, since each $x_j$ is known. The two methods of computation do not yield identical results, but the difference is practically negligible.

To see how these quantities are used in the evaluation of reproducibility of the test method under various conditions, let us refer again to the basic equation (13.11), using it this time for a fixed value of the property $Q$, but, for different laboratories.  Since $Q$ is fixed, so is $x_j$, and $\gamma_j$.  We are interested in the variability of $z$ due to a change in laboratory.  From Eq. 13.11 it follows that for fixed $\gamma_j$,

$$V(z) = V(\mu) + \gamma_j{}^2 V(\beta) + V(\eta)$$

provided that the three quantities $\eta$, $\mu$, and $\beta$ are statistically independent.  Now, the error $\eta$ measures the scatter of the $z$-values of any particular laboratories *about its own response line*; it is therefore a measure of *within laboratory variability*, whereas both $\mu$ and $\beta$ refer to the variability *between laboratories*.  Therefore $\eta$ is always independent of $\mu$ and $\beta$, but it is not always true that the latter two are mutually independent.  We have seen that for the butyl rubber data $\mu$ and $\beta$ showed a high degree of correlation.  Since $\eta$ is always independent of $\mu$ and $\beta$, the following equation is always true:

$$V_j(z) = \underbrace{V[\mu + \gamma_j\beta]}_{\substack{\text{Between} \\ \text{laboratories}}} + \underbrace{V(\eta)}_{\substack{\text{Within} \\ \text{laboratories}}} \qquad (13.27)$$

## 13.13  WITHIN-LABORATORY VARIABILITY

In Sections 13.4 and 13.5 we have seen how the scatter of the data obtained by a single laboratory about the $Q$, $M$ line is composed of at least two parts: the replication error, and effects due to interfering properties.  We have also seen that the first of these two sources of error, the replication error, can be measured independently as the variance among within-cell replicates.

The quantity we have referred to as $\eta$ represents the total scatter about the response line.  It contains the replication error, $\varepsilon$, reduced by a factor related to $n$, the number of replicates per cell.  Indeed, the analysis leading to $\eta$ is carried out on cell averages.  Therefore that portion of the variance $\eta$ that is due to replication is actually equal to $V(\varepsilon)/n$ where $\varepsilon$ is the replication error (in the transformed scale $z$).  It follows that $V(\eta)$ must be at least equal to $V(\varepsilon)/n$.  For the Bekk data we have found that $\hat{V}(\eta) = 640$ (see Eq. 13.26).  The estimate for $V(\varepsilon)$ for these data is obtained from the replication error of the original data, using the transformation Eq. 13.10.  From the latter we obtain

$$\sigma_z = 1000 \frac{d \log y}{dy} \sigma_y$$

$$= 1000 \frac{1}{2.30y} \sigma_y$$

Using the complete set of Bekk data for 12 laboratories and 14 materials (4), and fitting a straight line through the origin to the $\bar{y}_j \bar{s}_j$ values, we obtain a more exact relation than Eq. 13.1.   This calculation gives:

$$\hat{\sigma}_y = 0.128y$$

$$\hat{\sigma}_z = \frac{1000}{2.30y}(0.128y) = 55.6$$

Actually this is the standard deviation of $z$ due to replication error.   Hence

$$\hat{V}(\varepsilon) = (55.6)^2 = 3091$$

Each cell contained 8 replicates.   Therefore the contribution of replication error to $V(\eta)$ is $\hat{V}(\varepsilon)/8$ or $3091/8 = 386$.

It is seen that the replication error does not account for all of the variance of $\eta$.   We will denote by the symbol $\lambda$ that part of $\eta$ which is not accounted for by replication error.   Thus

$$V(\lambda) = V(\eta) - \frac{V(\varepsilon)}{n} \tag{13.28}$$

For the Bekk data, this formula gives an estimate for $V(\lambda)$ equal to $\hat{V}(\lambda) = 640 - 3091/8 = 254$.   According to our discussion of Section 13.4, $\lambda$ may be due to interfering properties.   We must now qualify this statement.

Let us first note that in our analysis the results of each laboratory are fitted, by linear least squares, against the $x_j$, the latter being averages over all laboratories of the results for each material $j$.   Let us now suppose that material $j$ has a property $Q'$, different from $Q$, which affects the measurement $M$.   If the way in which $Q'$ affects $M$ is the same for all laboratories, it will also affect the average $x_j$ in the same way.   Therefore the interference due to $Q'$ will result essentially in a shift of all the points corresponding to material $j$ to either a slightly higher or a slightly lower position on their respective response lines.*   No additional scatter will result from such a property and the study will not detect its presence unless and until its results are compared with those obtained by a different method of measurement or with theoretical values.

On the other hand, if the effect of the interfering property $Q'$ on the measurement $M$ differs from laboratory to laboratory, it will obviously cause additional scatter of the points representing the measurements about their fitted lines.   In that case it will become part of $V(\eta)$ and since it is unrelated to $V(\varepsilon)$, it is in fact part of $V(\lambda)$.   Thus, $V(\lambda)$ is due in part at least to interfering properties.

---

* In a good method of measurement, the effects of interfering properties are generally small compared to that of the measured property $Q$.

There are other causes that may contribute to $V(\lambda)$, for example, calibration errors. This can be seen from the following sample. Suppose that several chemists are asked to weigh three objects $A$, $B$, and $C$, weighing respectively, a little over 1 gram, a little over 5 grams, and a little over 10 grams, and that the results are plotted, separately for each chemist, against the true weights of $A$, $B$, and $C$. If, for each of the chemists, there are small errors in his standard weights of 1, 5, and 10 grams, and if these errors are independent of each other, then the points for each single chemist will not fall exactly on a straight line, even if no other error is made in the weighing process. If we now plot the results of each chemist against the average result of all chemists for each of the three objects, the points will still show scatter about their response lines, as a result of the miscalibration of the weights. (In the absence of other errors, the true response lines for the different laboratories will coincide, but if systematic differences exist between the operating techniques of the chemists, for example parallax reading errors, the lines will differ. This in no way affects the scatter due to miscalibration.) We see from this example that $V(\lambda)$ may be the result of a number of factors inherent in the equipment, in addition to those inherent in the materials. The statistical analysis can do no more than point to the existence of $\lambda$-factors. Their explanation requires knowledge of the physico-chemical phenomena involved in the measuring process.

### 13.14 BETWEEN-LABORATORY VARIABILITY

According to Eq. 13.27 the between laboratory variability is measured by the quantity

$$V[\mu + \gamma_j \beta]$$

Let us first consider the case in which $\mu$ and $\beta$ are mutually independent. (The Bekk data fall in this category.) In this case we have:

$$V_j(z) = \overbrace{V(\mu) + \gamma_j{}^2 V(\beta)}^{\text{Between laboratories}} + \overbrace{V(\eta)}^{\text{Within laboratories}} \tag{13.29}$$

The index $j$ is attached to the variance symbol to indicate its dependence on $\gamma_j$. The reason for this dependence is of course the non-parallelism of the lines corresponding to different laboratories. For if the lines were all parallel, their variability at a fixed value of $\gamma_j$ (or $Q$), i.e., their variability along a vertical line drawn through $x_j$, would be the same, regardless of the value of $x_j$. This is also shown by Eq. 13.29, since a bundle of parallel lines would correspond to a constant slope. Thus $V(\beta)$ would be zero and $V_j(z)$ would no longer depend on $\gamma_j$.

Considering now the case in which $\hat{\mu}_i$ and $\hat{\beta}_i$ are not independent, there then exists at least an approximate relation between these two quantities. Figure 13.4 is a plot of $\hat{\beta}_i$ versus $\hat{\mu}_i$ for the butyl rubber data (Table 13.11).

In this case the relation is very adequately represented by a straight line. Denoting the average value of the $\mu_i$ over all laboratories by $E(\mu)$ and that of the $\beta_i$ by $E(\beta)$, the equation of this straight line is

$$\beta_i - E(\beta) = \alpha[\mu_i - E(\mu)] + \delta_i \tag{13.30}$$

where $\alpha$ is the slope of the line, and $\delta_i$ the scatter of the $i$th point about the line.

From Eq. 13.30 it follows that

$$V(\beta) = \alpha^2 V(\mu) + V(\delta) \tag{13.31}$$

and also (see Eq. 4.12):

$$\text{Cov}\,(\beta, \mu) = \text{Cov}\,\{\alpha[\mu_i - E(\mu_i)] + \delta_i, \mu_i\} = \alpha V(\mu) \tag{13.32}$$

since $\delta_i$ and $\mu_i$ are independent.

Applying now Eq. 4.10 to the expression $V(\mu + \gamma_j\beta)$ we obtain

$$V(\mu + \gamma_j\beta) = V(\mu) + 2\gamma_j\,\text{Cov}\,(\mu, \beta) + \gamma_j{}^2 V(\beta)$$

which, because of Eqs. 13.31 and 13.32, becomes:

$$V(\mu + \gamma_j\beta) = V(\mu) + 2\gamma_j[\alpha V(\mu)] + \gamma_j{}^2[\alpha^2 V(\mu) + V(\delta)]$$

or

$$V(\mu + \gamma_j\beta) = [1 + \alpha\gamma_j]^2 V(\mu) + \gamma_j{}^2 V(\delta) \tag{13.33}$$

Thus, for the case in which the relation between $\mu_i$ and $\beta_i$ can be represented by a straight line, Eq. 13.27 becomes:

$$V_j(z) = \underbrace{(1 + \alpha\gamma_j)^2 V(\mu) + \gamma_j{}^2 V(\delta)}_{\substack{\text{Between} \\ \text{laboratories}}} + \underbrace{V(\eta)}_{\substack{\text{Within} \\ \text{laboratories}}} \tag{13.34}$$

This equation contains two new parameters, $\alpha$ and $V(\delta)$. Estimates for these quantities are obtained by applying the usual formulas of linear regression to the numerical values of $\hat{\mu}_i$ and $\hat{\beta}_i$: an estimate for $\alpha$ is given by the slope of the line ($\hat{\beta}_i$ versus $\hat{\mu}_i$) and an estimate for $V(\delta)$ by the residual variance about this line:

$$\hat{\alpha} = \frac{\sum_i (\hat{\beta}_i - 1)(\hat{\mu}_i - \bar{x})}{\sum (\hat{\mu}_i - \bar{x})^2} \tag{13.35}$$

$$\hat{V}(\delta) = \frac{\sum_i (\text{residuals})^2}{p - 2} \tag{13.36}$$

where the $i$th residual is given by $(\hat{\beta}_i - 1) - \hat{\alpha}(\hat{\mu}_i - \bar{x})$.

We may summarize the results of this section as follows:

The between laboratory variability at a point $x_j$ is given by the expression $V(\mu + \gamma_j\beta)$ where $\gamma_j = x_j - \bar{x}$.

(a) If the plot of $\hat{\beta}_i$ versus $\hat{\mu}_i$ reveals no relationship between these two quantities, this expression becomes

$$V(\mu) + \gamma_j{}^2 V(\beta)$$

and it is estimated by

$$\hat{V}(\mu) + \gamma_j{}^2 \hat{V}(\beta) \tag{13.37}$$

where the variance estimates are derived from the mean squares of the analysis of variance, in accordance with Eqs. 13.24 and 13.25.

(b) If a plot of $\hat{\beta}_i$ versus $\hat{\mu}_i$ reveals an approximate straight line relationship between $\beta_i$ and $\mu_i$, of slope $\alpha$ and scatter $V(\delta)$, the between laboratory variability is given by:

$$[1 + \alpha\gamma_j]^2 V(\mu) + \gamma_j{}^2 V(\delta)$$

and it is estimated by

$$[1 + \hat{\alpha}\gamma_j]^2 \hat{V}(\mu) + \gamma_j{}^2 \hat{V}(\delta) \tag{13.38}$$

where $\hat{\alpha}$ and $\hat{V}(\delta)$ are given by Eqs. 13.35 and 13.36 and $\hat{V}(\mu)$ by Eq. 13.24.

In terms of the tests of significance discussed in Section 13.11, regarding possible concurrence of all laboratory lines, an exact concurrence (to within $V(\eta)$) means that $V(\delta)$ is negligible. The point of concurrence is then given by Eq. 13.16, which is identical with:

$$z_0 = \bar{x} - \frac{1}{\hat{\alpha}} \tag{13.39}$$

**TABLE 13.14** Components of Variance

| Component | Bekk data | Butyl rubber data |
|---|---|---|
| *Between laboratories* | | |
| $\hat{V}(\mu)$ | 1423 | 1.635 |
| $\hat{V}(\beta)$ | 0.001095 | 0.001995 |
| $\hat{\alpha}$ | — | 0.02981 |
| $\hat{V}(\delta)$ | — | 0.0001922 |
| $z_0$ | — | 1.27 |
| *Within laboratories* | | |
| $\hat{V}(\lambda)$ | 254 | 0.04 |
| $\hat{V}(\varepsilon)$ | 3091 | 0.48[a] |
| $\hat{V}(\eta) = \hat{V}(\lambda) + \dfrac{\hat{V}(\varepsilon)}{n}$ | 640 | 0.284 |

[a] This estimate was derived from the within-cell variation. The value of $n$ was 2. The detailed data are not reproduced here.

Let us apply this theory to the two sets of data we have used as illustrations, the Bekk data and the butyl rubber data (Tables 13.9 and 13.11 for the parameters and Tables 13.10 and 13.12 for the analyses of variance). Table 13.14 summarizes the results, including those on within laboratory variability.

The following section deals with the use of information of the type of Table 13.14 for the computation of variability.

## 13.15   VARIABILITY OF MEASURED VALUES

It cannot be emphasized too much that the uncertainty of a measured value is not an absolute. It acquires meaning only in the context of a well-specified situation. Among the many possibilities, the following are of particular importance.

(*a*) *Comparisons involving replication error only.* This situation arises when an experimenter makes a number of replicate measurements of the same quantity and wishes to know whether the agreement he obtains is satisfactory. The standard of comparison is then $V(\varepsilon)$, or more specifically, the estimate $\hat{V}(\varepsilon)$ available from the interlaboratory study. Using this reference value, an experimenter using the test method for the first time can evaluate his technique. Another application is in testing for compliance with specifications: if for example 5 replicate measurements are required by the specification, the agreement between them should be consistent with $\hat{V}(\varepsilon)$ and a check for such consistency should be made *before they are used* to pass judgment on the acceptability of the lot of merchandise.

(*b*) *Within laboratory comparisons of two materials.* Suppose that two manufacturers submit samples of the same nominal product to a testing laboratory. For example, the material is paper and both samples are claimed to have a smoothness of 150 Bekk seconds. The laboratory tests 20 specimens from each product. If the claims of both manufacturers are justified, how closely should the two average values agree?

Since the two samples originated from different manufacturers, it is likely that they will differ in a number of ways, some of which may affect the smoothness measurement. We now make an important assumption: namely, that the type of heterogeneity between the two samples with respect to such interfering properties is of the same nature and degree as the heterogeneity displayed by the various materials with which the interlaboratory study was made. This means that the estimate $\hat{V}(\lambda)$ obtained in that study is an appropriate measure for the possible heterogeneity of the two samples with respect to interfering properties. Since the two materials are tested by the same laboratory, no between laboratory components enter the comparison. The variance of each average of 20 specimens is therefore

$$\hat{V}(\lambda) + \frac{\hat{V}(\varepsilon)}{20}$$

and the variance of the difference between the two averages is

$$2\left[\hat{V}(\lambda) + \frac{\hat{V}(\varepsilon)}{20}\right]$$

According to Table 13.14, this value, *in the z-scale*, is equal to

$$2\left[254 + \frac{3091}{20}\right] = 817$$

The standard deviation in the $z$-scale is $\sqrt{817} = 28.5$.

To convert this value to the original scale, we remember that

$$z = 1000 \log y$$

and consequently

$$\sigma_z = 1000 \frac{\sigma_y}{2.30y}$$

Hence

$$\frac{\sigma_y}{y} = 2.3 \times 10^{-3}\sigma_z = 2.3 \times 10^{-3}(28.5) = 0.0656$$

Since we have assumed that $y = 150$, we obtain:

$$\sigma_y = 150 \times 0.0656 = 9.8$$

Thus, the uncertainty of the difference between the averages for the two samples is measured by a standard deviation of 9.8. If upon completion of the experiment, we find that the two averages are 156.5 and 148.2 we have no reason to believe that there is a real difference in smoothness between the two materials, since $156.5 - 148.2 = 8.3$ is only about 0.84 standard deviations. A difference equal to 0.84 standard deviations or more will occur by chance in about 40 per cent of all cases. It is interesting to observe that an increase in the number of specimens tested from each sample, beyond the 20 already tested, would hardly improve the precision of the comparison: this precision is limited by the $\lambda$-variance which must be included in the comparison.

(c) *Between laboratory comparisons.* It is readily understood that when a comparison is made between results obtained in different laboratories, the variance $V(\lambda)$ is always included in the comparison, whether the latter involves a single material or different materials. It is true that the interfering properties for a single material are constant, but the response of different laboratories to the same interfering property is not necessarily the same, and the variability of this response is precisely what is measured by $\lambda$.

The case of between-laboratory comparisons is further complicated by the dependence of between-laboratory variability on the level of the measured property, i.e., on $\gamma_j$ (or $x_j$). It is here that the nature of this dependence, as expressed by Eqs. 13.37 or 13.38, whichever applies, acquires its true significance. We will illustrate the use of these equations by two examples.

Consider first the measurement of smoothness by the Bekk method. Suppose that samples *from the same material* are sent to two laboratories of comparable competence to that of the laboratories included in the inter-laboratory study. Each of the two laboratories makes 20 replicate measurements. The averages are $\bar{y}_A$ for laboratory $A$ and $\bar{y}_B$ for laboratory $B$. What is the variance of $\bar{y}_A - \bar{y}_B$? Applying Eq. 13.37 and the estimates in Table 13.14 we have for the between-laboratory variability in the $z$-scale:

$$1423 + \gamma_j^2(0.001095)$$

Adding the within-laboratory components $\hat{V}(\lambda)$ and $\hat{V}(\varepsilon)/20$ we obtain for each of the two averages, in the $z$-scale,

$$1423 + \gamma_j^2(0.001095) + 254 + 3091/20$$

For the difference between the two averages, the variance is twice this quantity:

$$2\left[1423 + \gamma_j^2(0.001095) + 254 + \frac{3091}{20}\right]$$

or

$$3664 + 0.00219\gamma_j^2$$

The variance depends on $\gamma_j$. Let us suppose that the Bekk smoothness of the material is known to be about 150 seconds. The corresponding $z$ value is $1000 \log (150)$, or 2176. Since $\bar{x}$ in the interlaboratory test was 1559 (the average of all $\hat{\mu}_i$ in Table 13.9, deleting the first two laboratories, No. 6 and No. 4), we have for $\gamma_j = x_j - \bar{x}$:

$$\gamma_j = 2176 - 1559 = 617$$

Consequently the variance in question is

$$3664 + (0.00219)(617)^2 = 4498$$

The corresponding standard deviation is $\sqrt{4498} = 67.1$.

Converting, as before, to the original scale, we obtain

$$\frac{\sigma_y}{y} = 2.3 \times 10^{-3}\sigma_z = 2.3 \times 10^{-3} \times 67.1 = 0.154$$

and for $\sigma_y$:

$$\sigma_y = (0.154)(150) = 23.1$$

Thus, the difference $\bar{y}_A - \bar{y}_B$ has a standard deviation of approximately 23 Bekk seconds.

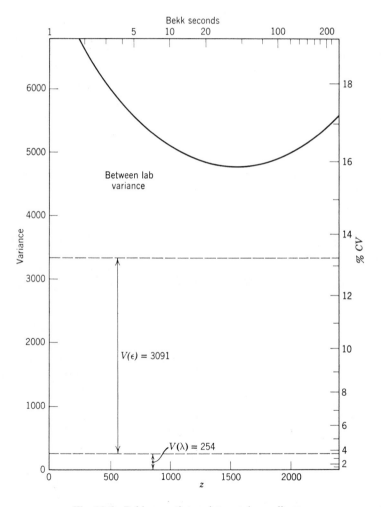

**Fig. 13.6**  Bekk smoothness data; variance diagram.

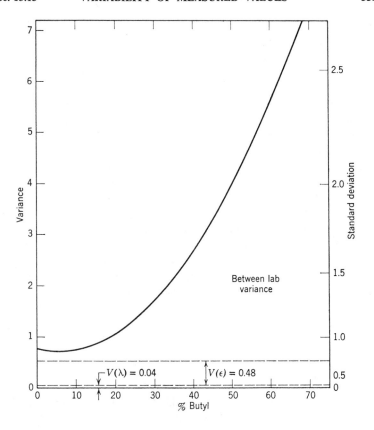

**Fig. 13.7**   Butyl rubber data; variance diagram.

Detailed calculations of this type may be avoided through the use of a diagram the idea of which is due to T. W. Lashof (4). In Fig. 13.6, the horizontal scale is laid out in accordance with the transformed scale $z$, but labeled in both scales. The ordinate represents the variances $\hat{V}(\lambda)$, $\hat{V}(\varepsilon)$, and $\hat{V}(\mu) + \gamma_j^2 \hat{V}(\beta)$ added to each other as indicated. Reading the ordinate corresponding to $y = 150$ we obtain at once* the total variance 5180. Since, however, we require $\hat{V}(\varepsilon)/20$, we have to reduce the ordinate by 19/20 of the height corresponding to $\hat{V}(\varepsilon)$. This is easily accomplished by means of a ruler (or a simple calculation). The result obtained for the total variance, after this modification, is 2250. For convenience, a scale

---

* Reading the diagram to this degree of precision implies of course that it be drawn on a large enough scale with a sufficiently fine grid.

of $\%CV$ has been added and from it we obtain readily $\%CV = 100\sigma_y/y$ $= 10.8$ . For $y = 150$ this yields $\sigma_y = 16.2$. Multiplying by $\sqrt{2}$ (since we are interested in the standard deviation of a *difference* of two averages) we obtain the standard deviation 22.9 which is in good agreement with our directly calculated value.

In addition to allowing us to compute rapidly the variance of test results for different cases and at any level of the property, a variance-diagram of the type of Fig. 13.6 gives the best over-all picture of the precision of the test method. At a glance we see the relation between within- and between-laboratory variability, the way both depend on the level of the property, and the magnitude of the $\lambda$-effects with respect to replication error.

The variance diagram for the butyl rubber data is shown in Fig. 13.7. No transformation of scale was made for the data. The values for "per cent butyl" extended from about 3 to 60 per cent. The marked difference in behavior, with regard to precision, between the two sets of data, Bekk smoothness and butyl rubber, is apparent from a comparison of Figs. 13.6 and 13.7. The following two sections deal with the practical conclusions that can be drawn from the variance-diagram, using these two sets of data as illustrations.

### 13.16  THE IMPROVEMENT OF PRECISION THROUGH REPLICATION

An important question arising in the use of any measuring process concerns the number of replicates that it is useful to run. Superficially, one might be tempted to answer: "the more the better," by referring to the theorem on the standard error of the mean (Eq. 4.14), to show that the standard deviation can be depressed to any desired quantity, no matter how small, through sufficient replication. The theory proposed in this chapter shows the fallacy of such reasoning. For replication error is only a portion, sometimes quite small, of the total error.

If a test is used as a means of controlling quality in a production process, agreement of the test results with those obtained in other laboratories may be irrelevant. This happens when the test is made merely to insure *uniformity of production* from batch-to-batch, or from shift-to-shift. Assuming that the testing process itself is stable with time, the only component of variance is $\varepsilon$, and the necessary number of replicates, $n$, is obtained by applying the method of Section 10.6 to $V(\varepsilon)$.

On the other hand, the test may be made for the purpose of determining, not uniformity of production, but the actual level of the measured property. The value obtained must show agreement with that obtained in any other competent laboratory. In such cases, it is the interlaboratory component that matters, unless it is overshadowed by within-laboratory variability. The latter is composed of two parts: $V(\lambda)$ and $V(\varepsilon)$. If $n$ replicates are made, the component $V(\varepsilon)$ is reduced to $V(\varepsilon)/n$, *but $V(\lambda)$ is left intact.* Consequently, there is no justification in replicating beyond the point where $V(\varepsilon)/n$ is small with respect to $V(\lambda)$. If the sum $V(\lambda) + V(\varepsilon)$ is small to begin with, in comparison with the between-laboratory component, no replication whatsoever is necessary for the improvement of precision. However, to guard against gross errors it is always advisable to make at least duplicate measurements.

Applying these considerations to the Bekk data, we see at once from Fig. 13.6 that there is good justification for making 12 to 20 replicates (since for averages of 12 replicates, the quantity $V(\varepsilon)/n$ is approximately equal to $V(\lambda)$, and for $n = 20$, it becomes small with respect to $V(\lambda)$). For the butyl data, Fig. 13.7, the situation is somewhat different. At low values of "per cent butyl," say 3 per cent, the between-laboratory component is small, and threefold replication results in a marked improvement in precision. For appreciably larger levels of butyl-content, the between-laboratory component becomes predominant; replication then fails to achieve any appreciable improvement in precision.

## 13.17   THE IMPROVEMENT OF
## BETWEEN-LABORATORY PRECISION

Precision can be improved by reducing either the within- or the between-laboratory variability. The former cannot be reduced to less than $V(\lambda)$ without modifying the measuring process itself. How can one reduce the variability between laboratories?

It is not uncommon to find, in the analysis of interlaboratory test results, a segregation of the laboratories into two groups: the "acceptable" and the "outlying" laboratories. Generally this procedure is carried out on the basis of the "laboratory main effects," which are none other than the $\mu_i$-values in our analysis. Laboratories whose $\mu_i$-values are far from the average are considered to be unreliable. Sometimes statistical "tests for outliers" are applied for this purpose. Obviously the removal of laboratories whose $\mu_i$ are at the extremes of the range results in a sizable reduction in $\hat{V}(\mu_i)$. If, as is commonly done, this component alone is taken as a measure of between laboratory variability, the latter is appreciably reduced by the removal of the "outliers." The question is of course whether the lower estimate thus obtained is a realistic representation of the variability that will be encountered in the practical use of the measuring process.  Should this not be the case, then the removal of the "outlying" laboratories is nothing but dangerous self-deception.

Now, the selection of the laboratories for participation in the inter-laboratory study was, or should have been, based on their recognized competence.  If the analysis of the data is carried out in accordance with the method presented in this chapter, an incompetent laboratory is likely to be detected in several ways.  In the first place, the $\varepsilon$-variability may differ from that of other laboratories.  Careless work tends to result in an excessive variability among replicates (the comparison can be made for each material, thus providing cumulative evidence); dishonest reporting of data will generally result in abnormally low replication errors.  But a far more important index of competence is the value of $\hat{V}_i(\eta)$ for each laboratory.  For while $\varepsilon$-variability can be calculated by each operator for his own data, the estimate of $V_i(\eta)$ is obtainable only after analysis of the results from all laboratories.  Dishonesty is of no avail here, while carelessness or incompetence are bound to result in excessive scatter, i.e., in an abnormally large value of $\hat{V}_i(\eta)$.

It happens occasionally that a participating laboratory that shows no discrepancy in its $\hat{V}_i(\eta)$ has a visibly discrepant $\mu_i$ or $\beta_i$.  For the reasons that have just been discussed, it is unlikely that such a laboratory is guilty of careless work or of incompetence.  A far more likely interpretation is that the laboratory departed in some important respect from the procedure used by the other participants.  This may be accidental, such as the use of the wrong load in a physical test, or a miscalibration of an instrument or a standard solution; or it may be due to a different interpretation of the prescribed procedure.  In either case the matter is generally cleared up by further inquiry into the procedure used by the discrepant laboratory.  Such follow-up steps in the examination of interlaboratory test results are invaluable: they often result in the elimination of unsuspected ambiguities in the procedure or in greater emphasis on the control of important environmental factors.  Our discussion of the Bekk data is a case in point: two laboratories were discrepant in $\mu_i$ and/or $\beta_i$ because of the use of a different relative humidity.  The elimination of these laboratories from the analysis is justified by the assumption that henceforth the prescribed procedure will specify without ambiguity the relative humidity under which the test is to be performed.

Let us now suppose that even though no discrepant laboratories have been found, the between-laboratory component of variability is appreciably larger than the within-laboratory components (assuming a reasonable number of replicates).  There still remains, in such cases, a well-known method for reducing the between-laboratory component: the use of standard samples.  This will be the subject of the next section.

## 13.18   THE USE OF STANDARD SAMPLES

A standard sample consists of a quantity of material, as homogeneous as possible, to which values have been assigned for one or more properties. In many cases, the value or values assigned are the result of painstaking measurements made by an institution of recognized competence. Such standards have many scientific uses. In connection with our present discussion, however, the accuracy of the assigned value or values is not an important issue. We are concerned with the reduction of between laboratory variability for a particular method of measurement. Suppose for the moment that the response lines for all laboratories constitute a parallel bundle.

If the lines can be moved parallel to themselves until they agree in one point, they will of course coincide completely. This can apparently be accomplished in the following way. Let $S$ represent a standard sample and let $S_0$ be a value assigned to $S$ for the property investigated. If a particular laboratory measures $S$ and obtains for it a value $S'$ (see Fig. 13.8), the difference $S' - S_0$ represents, *except for experimental error* in the measurement $S'$, the "bias" of that laboratory.

Let us now move the response line of the laboratory to a parallel position by a distance equal to $S' - S_0$, downward if $S' - S_0$ is positive and

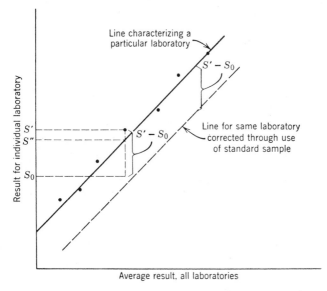

**Fig. 13.8**   Use of a standard sample; geometric interpretation.

upward if this quantity is negative. If this process is carried out for each laboratory, the lines will now differ from each other only by *the experimental errors in the measurements S' of the standard*. If these differences are smaller than the original distances between the lines, an improvement in interlaboratory precision will have been achieved. It is clear that the success of this operation depends in no way upon the assigned value $S_0$ as long as the same value $S_0$ is used for the adjustment of all the laboratories.

The parallel translation of the lines is accomplished in practice by *subtracting* from any measured value the bias $S' - S_0$. We have just seen that the agreement of the laboratories, *after adjustment by means of the standard*, is still not complete. It is only to within experimental error. What is the precise meaning of this expression in this context?

It is readily seen that the failure of the adjusted lines to coincide is due to the amounts by which the measurements $S'$ depart from their respective response lines, i.e., by distances such as $S'S''$ (see Fig. 13.8). Assuming that $S'$ is the average of $m$ replicates, $S'S''$ has a variance equal to

$$V(\lambda) + \frac{V(\varepsilon)}{m}$$

Now, for parallel lines, the interlaboratory variance is equal to $V(\mu)$. It results therefore from the preceding argument that the improvement in precision due to the standard sample is measured by a *reduction* in variance equal to

$$V(\mu) - \left[ V(\lambda) + \frac{V(\varepsilon)}{m} \right]$$

This quantity is not necessarily positive, and therefore the use of a standard sample may occasionally do more harm than good.

Let us assume that a material $M$ is measured by a number of laboratories for which the response lines are parallel. Suppose that each laboratory makes $n$ replicate determinations on $M$ and $m$ replicates on the standard sample. Then the variance among the values obtained for $M$, *after adjustment by the standard*, is obtained by subtracting $V(\mu)$ from the original variance and adding $V(\lambda) + V(\varepsilon)/m$: Thus

$$V_a(z) = \left[ V(\mu) + V(\lambda) + \frac{V(\varepsilon)}{n} \right] - [V(\mu)] + \left[ V(\lambda) + \frac{V(\varepsilon)}{m} \right]$$

where the suffix $a$ refers to the adjustment process. Thus:

$$V_a(z) = 2V(\lambda) + \left[ \frac{1}{n} + \frac{1}{m} \right] V(\varepsilon) \tag{13.40}$$

An estimate of this quantity can be read off from the variance diagram, or it can be computed from the estimates $\hat{V}(\lambda)$ and $\hat{V}(\varepsilon)$. By proper choice of the values $n$ and $m$ the term in $V(\varepsilon)$ can be made as small as desired, but the term $2V(\lambda)$ limits the precision that can be achieved. This conclusion is merely a mathematically precise statement for the intuitively obvious fact that the use of a standard improves precision only to the extent that the effects of interfering properties and similar factors are small with respect to between laboratory variability.

If the lines representing the laboratories are not parallel, the effectiveness of a standard sample depends on the level of the property for the standard sample and on the way it is used.   Let $x_s$ denote the level of the standard sample (on the $x$-scale; see Section 13.10).   If it is used for parallel translations of all response lines by amounts $S' - S_0$, as in the previous case, the ordinate of the adjusted line for laboratory $i$, at a level $x_j$ (ignoring for the moment $\lambda$ and $\varepsilon$ errors) will now be:

$$\mu_i + \beta_i x_j - [\mu_i + \beta_i x_s - S_0]$$

or

$$\beta_i(x_j - x_s) + S_0$$

The variance of this quantity with respect to laboratories is

$$(x_j - x_s)^2 V(\beta)$$

Adding the variance due to $\lambda$ and $\varepsilon$ effects, we then have for the total variance of adjusted values:

$$V_a(z) = (x_j - x_s)^2 V(\beta) + 2V(\lambda) + \left[\frac{1}{n} + \frac{1}{m}\right] V(\varepsilon) \qquad (13.41)$$

Thus the gain in precision is maximum when $x_j = x_s$, i.e., for a material having the same level for the property as the standard.   As $x_j$ departs from $x_s$, either to the left or to the right, the variance of the adjusted values increases, with a corresponding decrease in the effectiveness of the standard sample.

To illustrate this formula, consider the Bekk data, Table 13.14.   Suppose that both $n$ and $m$ are chosen to be 10, and let the smoothness of the standard sample be 50 Bekk seconds.   Then $x_s = 1000 \log (50) = 1699$. We have:

$$V_a(z) = (x_j - 1699)^2(0.001095) + 2(254) + \frac{2}{10}(3091)$$

hence

$$V_a(z) = (x_j - 1699)^2(0.001095) + 1126$$

The unadjusted variance at $x_j$ would be

$$V(z) = 1423 + (x_j - 1559)^2(0.001095) + 254 + \frac{3091}{10}$$

or

$$V(z) = (x_j - 1559)^2(0.001095) + 1986$$

Thus the reduction in variance due to the standard sample is

$$V(z) - V_a(z) = 0.307x_j + 361$$

A study of this function for values of $x_j$ ranging from 600 to 2000, using the variance diagram for reference, shows that the use of a standard sample may well be justified in this case.

Since the object of the standard sample is to move the response lines for all laboratories as close to each other as possible, it is clear that in the general case where the lines criss-cross at random, *two* standard samples are required to achieve this goal to the largest possible extent. For parallel lines one standard sample is all that is required. This is also the case when the lines are strictly concurrent: the standard sample should then be selected as far as possible from the point of concurrence. For example, for the butyl rubber data, the point of concurrence is for all practical purposes at the origin. A standard sample with a high butyl content will be very effective in reducing the interlaboratory variance. Adjustments are made in this case by *multiplying* the results from each laboratory by a correction factor.

When two standards are used, the best way of adjusting the data is by linear interpolation between the standards. Suppose, for example, that for the Bekk data two standards, $S$ and $T$, are used, with assigned Bekk values equal to 5 seconds and 140 seconds, respectively. The corresponding transformed values are 699 and 2146. Suppose now that a particular laboratory measures both standards in addition to a material $M$ and obtains the following values (in the original scale):

$$\frac{S}{4.2} \quad \frac{M}{38.8} \quad \frac{T}{165.4}$$

Transforming to the $z$-scale and writing the assigned values under the observed, we obtain:

| $S$ | $M$ | $T$ |
|-----|-----|-----|
| 623 | 1589 | 2219 |
| 699 | ? | 2146 |

By linear interpolation this gives the adjusted value for $M$:

$$699 + (1589 - 623)\frac{2146 - 699}{2219 - 623} = 1575$$

Transformed back to the original scale, this yields the value: 37.6 seconds.

The interpolation can be made graphically by means of a nomogram, but we will not discuss this matter.

Further information on the general subject of interlaboratory testing may be found in references 1, 5, 6, 8, 9, and 10.

## 13.19   NESTED DESIGNS

In the preceding sections we have dealt with the problem of estimating the contributions of various factors to the over-all variability of a testing process.   The general design we have discussed lends itself to the estimation of essentially two components of variability; that occurring *within* laboratories and that due to differences *among* laboratories.   A finer analysis of the problem has led us to the partitioning of each of these two components into two parts: $\lambda$ and $\varepsilon$ variance for within-laboratory variability, and position and slope effects for between-laboratory variability.   Now the components of variance that are of practical interest in any particular situation are determined by the manner in which the test method is to be used in practice.   Occasionally, one is interested in differentiating between the reproducibility of results made simultaneously (or almost simultaneously) and that of results obtained either on different days or separated by even longer time intervals.   It also happens that interest is centered on the differences in results obtained by different operators or on different instruments in the same laboratory.   A particular type of design, known as *nested* or *hierarchical*, has been found useful in investigations of this type.   An example of a nested design is represented by the diagram shown in Fig. 13.9.   In each of three laboratories, two operators made duplicate measurements on each of three days.   The diagram justifies the names " nested " and " hierarchical."   In this example only one material is involved, but the design can of course be enlarged to include two or more materials.

In contrast to nested classifications are the designs known as *crossed classifications*, in which *each level* of one factor occurs in combination with *each level* of every other factor.   For example, in extending the design of Fig. 13.9 to four materials, one must choose between a number of alternatives.   Thus, operator $A_1$ could test all four materials on each of the three days $A_{1a}$, $A_{1b}$, $A_{1c}$, and similarly for all other operators.   In that case, the design contains a crossed classification between the factor

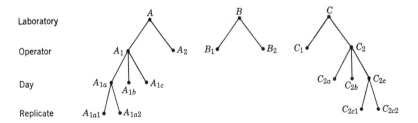

**Fig. 13.9**   A nested design.

"material" (4 levels) and the factor "day of test" (18 levels). On the other hand, each operator could devote a different set of three days to each of the four materials. In that case, the factor "material" does not cross the factor "day of test"; in fact this latter factor is then nested within the factor "material."

From the viewpoint of the statistical analysis it is of the utmost importance to ascertain whether any two factors are crossed or whether one is nested within the other. Generally speaking, when a factor $B$ crosses a factor $A$, the analysis involves: $a$, the main effect of $A$; $b$, the main effect of $B$; $c$, the interaction between $A$ and $B$. On the other hand, if $B$ is nested within $A$, the analysis involves: $a'$, the effect of $A$; $b'$, the effect of $B$ within $A$, but no interaction, since no level of $B$ occurs for more than one level of $A$.

The absence of interactions in a purely nested design provides the key to the analysis of variance. Suppose that factor $B$ is nested within factor $A$. Perform first an analysis of variance as though $B$ and $A$ were crossed (items $a$, $b$, and $c$ above). Then $b'$ is obtained by summing the sums of squares, as well as the degrees of freedom, for $b$ and $c$. The effect of $A$, $a'$, is identical with $a$.

We can make use of the concept of nesting in the design of interlaboratory studies. For example, the $n$ replicates within each cell of Table 13.2 could be partitioned into between- and within-day replications. Thus, if $n = 6$, we could consider two days of test, with three replicates on each day. We must then decide whether all materials will be tested on the same two days or whether several sets of two days each will be required by each laboratory. Depending on the type of information that is desired, the randomization of the test specimens and of the order of testing then will have to be carried out in different ways. Thus, if in the above-cited example one is interested in within-day replication error, the set of $6 \times q$ specimens sent to a given laboratory should not be completely randomized, but rather divided into groups such that each group contains three specimens of the same material. The $2q$ groups of three specimens each may then be randomized.

Further information on nested design, including the calculation of components of variance, may be found in references 2, 3, and 9; examples of the use of nested classifications in the design and interpretation of interlaboratory studies will be found in references 1 and 9.

## 13.20   SUMMARY

A meaningful approach to the problem of evaluating measuring processes must be based on a plausible conceptual model for the characterization of variability in measurements.

Underlying the present chapter is a model of the measuring process based on the consideration of variable environmental factors, of calibration errors, and of the effect of interfering properties of the materials that are subjected to the measuring process. A distinction is made between local control of environmental conditions, i.e., the ability to limit fluctuations of these conditions in a given laboratory at a given time, and their control in terms of specified tolerances, the aim of which is to achieve uniformity of test results among different laboratories or at different times. A systematic study of the effectiveness of both types of control is best made by conducting an interlaboratory study of the measuring process. In planning such a study, careful consideration must be given to the selection of the materials to which the measuring process is to be applied and to the choice of laboratories. Proper randomization procedures in the allocation of test specimens to the participating laboratories, and in the order of tests in each laboratory are essential for a meaningful interpretation of the results. The statistical analysis of the data obtained in an interlaboratory study must be made in terms of the model adopted for the measuring process, but this model must be kept flexible, to allow for plausible explanations of the peculiar features of each specific interlaboratory study. The method of analysis proposed in this chapter allows for the isolation of sources of error arising under conditions of local control, from variability due to between-laboratory differences. Both types of variability are further partitioned into component parts, each of which corresponds to an experimentally meaningful aspect of the measuring process. Thus, consideration is given to the improvement of precision that can be achieved through replication and through the use of standard samples. The analysis leads to prediction of the amount of variability that may be expected under various uses of the testing method. A brief discussion is given of a type of experimental design, called nested, which is often used in the study of test methods.

## REFERENCES

1. *ASTM Manual for Conducting an Interlaboratory Study of a Test Method*, ASTM Special Technical Publication No. 335, American Society for Testing and Materials, Philadelphia, 1963.
2. Bennett, C. A., and N. L. Franklin, *Statistical Analysis in Chemistry and the Chemical Industry*, Wiley, New York, 1954.
3. Brownlee, K. A., *Statistical Theory and Methodology in Science and Engineering*, Wiley, New York, 1960.
4. Lashof, T. W., and J. Mandel, "Measurement of the Smoothness of Paper," *Tappi*, **43**, 385–399 (1960).
5. Mandel, J., and T. W. Lashof, "The Interlaboratory Evaluation of Testing Methods," *ASTM Bulletin*, No. **239**, 53–61 (July 1959).

6. Pierson, R. H., and E. A. Fay, "Guidelines for Interlaboratory Testing Programs," *Analytical Chemistry*, **31**, 25A–33A (1959).
7. Tyler, W. P., Private communication of data obtained by Subcommittee XI of ASTM Committee D-11.
8. Wernimont, G., "The Design and Interpretation of Interlaboratory Test Programs," *ASTM Bulletin*, **No. 166**, 45–48 (1950).
9. Wernimont, G., "Design and Interpretation of Interlaboratory Studies of Test Methods," *Analytical Chemistry*, **23**, 1572–1576 (1951).
10. Youden, W. J., "Graphic Diagnosis of Interlaboratory Test Results," *Industrial Quality Control*, **15**, No. 11, 24–28 (May 1959).

*chapter 14*

# THE COMPARISON OF METHODS OF MEASUREMENT

## 14.1 INTRODUCTION

There exists hardly a physical or chemical property that can not be measured in more than one way. To cite but two typical examples: numerous chemical analyses, such as the determination of iron in a solution, can be performed gravimetrically, volumetrically, spectrophotometrically, and in a number of other ways; a physical property such as the resistance to tear of certain materials (for example: plastic film) can be measured by a number of instruments based on different physical principles.

Whenever two or more alternative methods exist for the measurement of the same property, there arises at once the problem of comparing their relative merits. But first a more fundamental question must be answered: do the different methods *really* measure the same property?

Despite its simplicity, this question is far from trivial. It could be argued on philosophical grounds that *different* methods can never measure exactly the same thing. We do not intend to discuss this philosophical question, but we do wish to recall that no measuring technique is completely specific, i.e., no measuring technique responds to only one single property. Consequently the relationship between two measuring methods may well depend on the choice of the materials selected for their comparison.

To illustrate this point, suppose that two methods are compared for the determination of nickel in an alloy. The first method, by precipitation with dimethyl–glyoxime, is not affected by the presence of small amounts of cobalt. The second method, by electrolytic deposition, is affected by the presence of cobalt.. Thus, if we performed two separate experiments, one involving a series of samples of varying nickel-content, but all free of cobalt, and the second a series of similar samples in which, however, cobalt is present, and if in each experiment, the electrolytic method were compared with the gravimetric method, we would find a different relationship between the two methods for the samples that are free of cobalt than for the samples containing cobalt. Suppose now that a third experiment is conducted, this time on a series of samples some of which contain cobalt, other of which do not. Then the over-all relationship between the two methods, as revealed by this third experiment, would show peculiarities which would depend on which samples (i.e., which nickel concentrations) contain cobalt.

We will assume in this chapter that the materials selected for the study are representative of the class of materials for which the alternative methods are to be compared.

## 14.2  FUNCTIONAL RELATIONSHIPS BETWEEN ALTERNATIVE METHODS

In the preceding chapter we have stressed the distinction between the property to be measured and the actual measurement made for that purpose. We have, however, emphasized that the two must be related in a functional way, in fact by a monotonic function. If we now consider more than one type of measurement for the same property, we must require, in an analogous way, that the various quantities considered be functionally related to each other. For example, bound styrene in rubber can be determined by measuring either a density or a refractive index. If both quantities are functionally related to bound styrene, they must be functionally related to each other.

This simple fact is of the greatest importance, since it allows us to compare alternative methods of measurement of a given property without making specific use of this property. Thus, in the bound styrene example, if we have measured both the density and the refractive index of a given series of rubber samples, we can plot the densities versus the refractive index values. Such a plot can be made whether or not we know the bound styrene content of each sample. If both measurements are indeed valid methods for determining bound styrene, our plot should represent a functional relationship. This can of course be judged in the absence of any knowledge of the true bound styrene values for the rubber samples.

It follows that the first step in the comparison of alternative methods of measurement consists in verifying that a functional relationship exists between the quantities representing the two measurements. While this step is necessary, it is by no means sufficient.

## 14.3   TECHNICAL MERIT AND ECONOMIC MERIT

The choice between alternative methods of test is only partly a technical question. Suppose that a density measurement requires a half hour of the operator's time while a refractive index measurement can be made in five minutes. Clearly such information cannot be ignored in choosing between the two methods. It is also possible that one of two alternative methods requires only a limited amount of skill on the part of the operator while the other is of a delicate nature and demands considerable experience to be successful. Nor can we ignore the cost of the equipment required by either method, or the cost and difficulty of calibrating it. In brief, the choice between alternative methods is dictated by many factors of an economic nature, in addition to the scientific merit of either method.

We will return briefly to a consideration of these economic factors after discussing the more basic question concerned with the purely scientific, or "technical" merit of alternative methods of test.

## 14.4   THE NEED FOR A CRITERION OF TECHNICAL MERIT

When asked which of two methods is better, one is inclined to answer: the one that has the better precision and accuracy. Reflection shows, however, that the requirement of accuracy, understood in the light of our discussion of Chapter 6, is not pertinent to our present problem. We have seen that accuracy implies a comparison with a reference value. If we consider a *range* of values, rather than an isolated point, the determination of accuracy would require the availability of a reference value at each point in the range. In the absence of a set of reference values, the problem of accuracy cannot be discussed. But even if a reference value were available for each point, the problem of accuracy would be trivial: by plotting the measured values versus the reference values, a calibration curve would be obtained. The only requirement for this calibration curve to be valid is that the measured values be sufficiently free of *random errors* to obtain a smooth relationship. The calibration curve itself defines the accuracy of the method. There remains of course the problem of *factors of accuracy*, but the discussion of this problem belongs to the *development* of a method of measurement, not to its comparison with an alternative method.

The real problem in the comparison of methods from the viewpoint of technical merit is therefore the comparison of their *precisions*. We have encountered many measures of precision: standard deviation, coefficient of variation, and others. None of these can be used *directly* for the comparison of alternative methods of measurement.

Consider again the determination of bound styrene in rubber. It is a relatively simple task to determine the reproducibility, either within a laboratory, or on an interlaboratory scale, of either of the two methods, the density method and the refractive index method. The reproducibility will be measured in terms of a standard deviation, or more generally, by a function giving the standard deviation of reproducibility (either within or between laboratories) for each magnitude of the measured value. But unless calibration curves are used to convert each density value, and each refractive index, into a bound styrene value, the standard deviations will be expressed in density units and refractive index units respectively. How does one compare a standard deviation of density with one of refractive index?

## 14.5 A CRITERION FOR TECHNICAL MERIT

The preceding discussion itself suggests the answer to this question. If density and refractive index were both first converted into bound styrene values, the comparison of the precisions would be simple: the method yielding the lower standard deviation *in bound styrene units* is more precise. However, this approach necessitates a knowledge of two

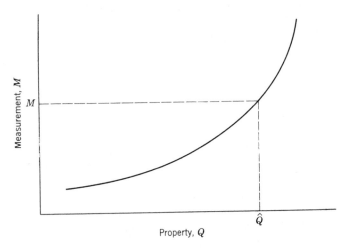

**Fig. 14.1** Relation between measurement and property.

calibration curves.   It is easy to see that the same result can be achieved without any recourse to bound styrene values.   To show this, let us first examine how a standard deviation is converted by means of a calibration curve.

Let $M$ represent the measured value and $Q$ the value of the property that is being measured.   In our example, $Q$ is per cent bound styrene and $M$ might be density.   A relation must exist between $M$ and $Q$.

$$M = f(Q) \tag{14.1}$$

Equation 14.1 represents of course the calibration curve of $M$ in terms of $Q$.   Now, given a measured value $M$, we use the calibration curve by reading off the corresponding value of $Q$ (see Fig. 14.1).   Let us represent this value by $\hat{Q}$.   Then $\hat{Q}$ is obtained by "inverting" the function (Eq. 14.1).   For example if we had

$$M = \frac{k}{Q}$$

we would obtain

$$\hat{Q} = \frac{k}{M}$$

Or if we had

$$M = a + b \log Q$$

we would infer

$$\hat{Q} = 10^{(M-a)/b}$$

In general we write

$$\hat{Q} = g(M) \tag{14.2}$$

where $g$ is the "inverse function" of $f$.

In Eq. 14.2, $M$ must be considered as a measured value, and $\hat{Q}$ as a value derived from $M$.   Representing the error of measurement of $M$ by $\varepsilon$, and applying the law of propagation of errors (Chapter 4), we obtain,* for the standard deviation of $\hat{Q}$:

$$\sigma_{\hat{Q}} = \left| \frac{dg}{dM} \right| \sigma_\varepsilon \tag{14.3}$$

Since $g$ is the inverse function of $f$, we have

$$\frac{dg}{dM} = g_M' = \frac{1}{f_Q'} = \frac{1}{df/dQ} \tag{14.4}$$

where the prime denotes a derivative.

Hence

$$\sigma_{\hat{Q}} = \frac{\sigma_\varepsilon}{|f_Q'|} \tag{14.5}$$

* The two vertical bars in Eq. 14.3 and other subsequent equations denote an absolute value.   This is necessary since the derivative $dg/dM$ may be negative and a standard deviation is of necessity a positive quantity.

This equation permits us to convert the standard deviation of the measured quantity, $\sigma_\varepsilon$, into a standard deviation of the property of interest. It requires a knowledge of the derivative of the function $f$, that is, a knowledge of the tangent to the calibration curve.

If we now consider two measuring processes, $M$ and $N$, for the determination of the same property $Q$, we will have two calibration functions:

$$M = f(Q) \tag{14.6}$$

and

$$N = h(Q) \tag{14.7}$$

In our example, $Q$ would represent "per cent bound styrene," $M$ density, and $N$ refractive index. Let us represent by $\varepsilon$ and $\delta$ the random errors associated with $M$ and $N$, respectively. Furthermore, let the $Q$-estimates derived from these two functions be denoted $\hat{Q}_M$ and $\hat{Q}_N$. Then:

$$\sigma_{\hat{Q}_M} = \frac{\sigma_\varepsilon}{|f_Q'|} \tag{14.8}$$

$$\sigma_{\hat{Q}_N} = \frac{\sigma_\delta}{|h_Q'|} \tag{14.9}$$

According to our previous argument, the method with the greater technical merit is the one for which $\sigma_{\hat{Q}}$ is smaller. Let us therefore consider the ratio of these two standard deviations; we have:

$$\frac{\sigma_{\hat{Q}_M}}{\sigma_{\hat{Q}_N}} = \left|\frac{\sigma_\varepsilon/f_Q'}{\sigma_\delta/h_Q'}\right| = \frac{\sigma_\varepsilon/\sigma_\delta}{|f_Q'/h_Q'|} \tag{14.10}$$

Now, the quantity $f_Q'/h_Q'$, which is the ratio of two derivatives taken with respect to the same quantity $Q$, has a simple interpretation. We have already observed that since both $M$ and $N$ are functions of $Q$, they must be functionally related to each other. Suppose that we plot $M$ versus $N$. Then the derivative $dM/dN$ can be written:

$$\frac{dM}{dN} = \frac{dM}{dQ}\frac{dQ}{dN} = \frac{dM/dQ}{dN/dQ} = \frac{f_Q'}{h_Q'} \tag{14.11}$$

Thus, Eq. 14.10 becomes:

$$\frac{\sigma_{\hat{Q}_M}}{\sigma_{\hat{Q}_N}} = \frac{\sigma_\varepsilon/\sigma_\delta}{|dM/dN|} \tag{14.12}$$

This equation expresses a remarkable fact: the ratio of the standard deviations of the two methods, *expressed in the same unit* (that of $Q$), can be determined from a knowledge of the relation between $M$ and $N$, *without any recourse to the calibration curves of $M$ and $N$ in terms of $Q$.* (The standard deviations $\sigma_\varepsilon$ and $\sigma_\delta$, which express merely the reproducibility of $M$ and $N$ (in their respective units) can of course be determined without recourse to $Q$.) For example, the comparison of the technical merits of the density method and the refractive index method can be made without recourse to the calibration of either method in terms of bound styrene.

Let us now examine the problem from an experimental viewpoint. The comparison expressed by Eq. 14.12 involves three quantities: $\sigma_c$, $\sigma_0$, and $dM/dN$. The first two of these can be obtained experimentally by the methods discussed in Chapter 13. To obtain the third quantity, $dM/dN$, no further experimental work is required, provided that the studies of methods $M$ and $N$ have been made on the same series of samples. Simply plot for each sample, the average of $M$ versus the corresponding average of $N$; draw a smooth curve through these points, and determine the derivative $dM/dN$ of this curve.

## 14.6  THE SENSITIVITY RATIO

With regard to Eq. 14.12 we note that as the ratio $\sigma_{\hat{Q}_M}/\sigma_{\hat{Q}_N}$ *increases*, the technical merit of $M$ *decreases* with respect to that of $N$. It is therefore reasonable to consider the reciprocal of this ratio. This reciprocal has been called the *relative sensitivity of method M with respect to method N* (1).

$$\text{Relative sensitivity } (M/N) = \frac{|dM/dN|}{\sigma_\varepsilon/\sigma_\delta}$$

To simplify notation, we will represent "relative sensitivity" by the symbol $RS$. Thus we write:

$$RS(M/N) = \frac{|dM/dN|}{\sigma_\varepsilon/\sigma_\delta} \qquad (14.13)$$

If this ratio appreciably exceeds unity, method $M$ is "technically" superior to method $N$, i.e., of the two methods, method $M$ has the greater ability to detect a real difference in the property $Q$. This criterion automatically takes account of the uncertainty due to the experimental errors of both methods. The opposite conclusion holds when the ratio given by Eq. 14.13 is appreciably smaller than unity, since the relative sensitivity of $N$ with respect to $M$ is simply the reciprocal of the relative sensitivity of $M$ with respect to $N$.

Thus, the sensitivity ratio provides a simple measure, which can be determined experimentally, for deciding which of two methods has greater technical merit in the sense of Section 14.4. Evidently the method is still applicable when more than two methods are compared. To solve this problem, assign the value "one" to the absolute sensitivity of one of the methods (the choice is arbitrary) and compute the relative sensitivities of all other methods with respect to the selected one. These values can then be used to rank all the methods with respect to their technical merit.

## 14.7 THE RELATIVE SENSITIVITY CURVE

It is evident from the discussion so far that the sensitivity of a method is not necessarily a constant, but will, in general, vary with the value of the property $Q$. Indeed, the sensitivity ratio of $M$ with respect to $N$, as given by Eq. 14.13, depends on three quantities: the slope $dM/dN$, and the two standard deviations $\sigma_\varepsilon$ and $\sigma_\delta$. Unless the relation between $M$ and $N$ is linear, the slope will vary along the curve of $M$ versus $N$, i.e., with the value of $Q$. Also, the standard deviations $\sigma_\varepsilon$ and $\sigma_\delta$ are not necessarily constant throughout the ranges of variation of $M$ and $N$. Thus, the sensitivity ratio is really a function of $Q$. It is however not necessary to express it in terms of $Q$ (which is generally unknown), since either $M$ or $N$ are functionally (and monotonically) related to $Q$. By plotting the sensitivity ratio versus either $M$ or $N$ one obtains a complete picture of the relative merit of the two methods over their entire range of applicability. The curve obtained in such a plot may be denoted as the *relative sensitivity curve* of $M$ with respect to $N$.

## 14.8 TRANSFORMATIONS OF SCALE

In a subsequent section we will apply the sensitivity criterion to an actual problem. Before doing this, however, it is appropriate to demonstrate a remarkable property of the sensitivity ratio.

Suppose that in comparing two methods, $M$ and $N$, for the measurement of a property $Q$, we have found a certain value for the sensitivity ratio of $M$ with respect to $N$, say the value 2.7. Being a ratio of two similar quantities, this number is dimensionless. It is certainly pertinent to ask whether the same numerical value would have been obtained, had we converted all $M$ readings to a different scale, for example by affixing a logarithmic dial to the instrument on which $M$ is read. The answer is that the value of the sensitivity ratio is not changed by *any* transformation of scale. To prove this, let the transformation be given by the equation:

$$M^* = f(M) \tag{14.14}$$

Then, the relative sensitivity of $M^*$ with respect to $N$ is, by definition:

$$RS(M^*/N) = \frac{|dM^*/dN|}{\sigma_{M^*}/\sigma_N} \tag{14.15}$$

Now, since $M$ is functionally related to $N$, so is $M^*$; hence by differentiating Eq. 14.14 with respect to $N$, we obtain:

$$\frac{dM^*}{dN} = \frac{df(M)}{dN} = \frac{df(M)}{dM}\frac{dM}{dN} \tag{14.16}$$

Furthermore, by the law of propagation of errors we have, from Eq. 14.14,

$$\sigma_{M^*} = \left|\frac{df(M)}{dM}\right|\sigma_M \tag{14.17}$$

Introducing Eqs. 14.16 and 14.17 into Eq. 14.15, we obtain,

$$RS(M^*/N) = \left|\frac{\dfrac{df(M)}{dM}(dM/dN)}{\dfrac{df(M)}{dM}(\sigma_M/\sigma_N)}\right|$$

$$= \frac{|dM/dN|}{\sigma_M/\sigma_N} \tag{14.18}$$

But this last ratio is precisely equal to $RS(M/N)$, by Eq. 14.13. Thus, the transformation of scale of $M$ has not changed the value of its relative sensitivity with respect to $N$. A similar situation holds of course for any transformation of scale for $N$. In mathematical terminology, we would state that the sensitivity ratio is *invariant* with respect to any transformation of scale. Clearly this property is a *sine qua non* if the numerical, dimensionless value of the sensitivity ratio is to be a reliable measure for the relative merits of two methods. Indeed, as has been pointed out in Chapter 1, a transformation of scale can no more change the intrinsic merit of a measuring technique than a change of map can change the topological properties of a terrain. It is interesting to note that of all the measures of precision we have considered in this book, including the standard deviation, the variance, the range, confidence intervals, and even the coefficient of variation, the sensitivity ratio is the only one that enjoys the property of invariance with respect to transformations of scale. Let us consider, for example, the coefficient of variation, which is also a dimensionless quantity. Consider again two methods $M$ and $N$, for the measurement of property $Q$. The coefficients of variation are given by:

$$CV_M = \frac{\sigma_M}{M} \qquad CV_N = \frac{\sigma_N}{N} \tag{14.19}$$

The ratio of these quantities, taken with the positive sign, is:

$$\frac{CV_M}{CV_N} = \frac{\sigma_M/\sigma_N}{|M/N|} \tag{14.20}$$

Reasoning as before, a transformation of scale given by Eq. 14.14 will lead to the new ratio:

$$\frac{CV_{M^*}}{CV_N} = \left| \frac{\dfrac{df(M)}{dM} (\sigma_M/\sigma_N)}{M^*/N} \right| \tag{14.21}$$

In order that the ratio of coefficients of variation be unaffected by the transformation of scale, we must have:

$$\frac{CV_{M^*}}{CV_N} = \frac{CV_M}{CV_N}$$

or

$$\left| \frac{\dfrac{df(M)}{dM} \sigma_M/\sigma_N}{M^*/N} \right| = \frac{\sigma_M/\sigma_N}{|M/N|}$$

This equality requires that

$$\left| \frac{df(M)}{dM} \right| = \left| \frac{M^*}{M} \right| = \left| \frac{f(M)}{M} \right|$$

or

$$\left| \frac{df(M)}{f(M)} \right| = \left| \frac{dM}{M} \right| \tag{14.22}$$

Integrating this simple differential equation, we obtain:

$$\log f(M) = \text{constant} \pm \log M$$

Writing $\log K$ for the constant, this relation is equivalent to one or the other of the following transformations:

$$f(M) = KM \quad \text{or} \quad f(M) = \frac{K}{M} \tag{14.23}$$

Thus, the only transformations of scale that leave the coefficient of variation unaffected are the *proportional* one, such as a change of grams into pounds, or of inches into centimeters, and the reciprocal one. Any *non-linear* transformation of scale other than the reciprocal, or even a linear one in which the intercept is not zero, will, in general, change the numerical value of the ratio of coefficients of variation. In the following section we will further examine the coefficient of variation as a possible criterion for the comparison of methods of test.

### 14.9 RELATION BETWEEN SENSITIVITY AND THE COEFFICIENT OF VARIATION

A question that naturally arises is whether the coefficient of variation might not be considered as a criterion for comparing methods of test. In the preceding section we have already encountered one shortcoming of the coefficient of variation when used for this purpose: its dependence on the scales used for the expression of the measurements. We will now compare more directly the ratio of the coefficients of variation of two methods with the ratio of their sensitivities.

From Eq. 14.20 we obtain:

$$\frac{CV_N}{CV_M} = \frac{|M/N|}{\sigma_M/\sigma_N} \tag{14.24}$$

Comparing this expression with the relative sensitivity of $M$ with respect to $N$, Eq. 14.13, we obtain:

$$RS(M/N) = \frac{|dM/dN|}{\sigma_M/\sigma_N} = \frac{|M/N|}{\sigma_M/\sigma_N} \cdot \left|\frac{dM/dN}{M/N}\right|$$

$$= \frac{CV_N}{CV_M}\left|\frac{dM/M}{dN/N}\right| \tag{14.25}$$

In order that the ratio of the coefficients of variation be equivalent to the ratio of sensitivities, it is necessary that

$$\left|\frac{dM/M}{dN/N}\right| = 1$$

or

$$\left|\frac{dM}{M}\right| = \left|\frac{dN}{N}\right| \tag{14.26}$$

This differential equation is of the same type as Eq. 14.22. Its solution is given by one or the other of the following relations:

$$M = KN \quad \text{or} \quad M = \frac{K}{N} \tag{14.27}$$

Thus we see that *only* in the case in which the two methods $M$ and $N$ are *proportionally* or *reciprocally* related to each other, is the ratio of the coefficients of variation equivalent to the sensitivity ratio. In that case, the ratio of the coefficient of variation of $N$ to that of $M$ is equal to the relative sensitivity of $M$ with respect to $N$. It follows that the coefficient of variation cannot be considered as a valid criterion for the comparison of test methods except in the special case in which the measurements obtained by the different methods are proportional or reciprocal to each other.

To illustrate this point, consider two methods, $M$ and $N$, that are *linearly*, but *not proportionally*, related to each other. For example, $M$ and $N$ might represent the amounts of reagent used in two different volumetric methods of analysis. If $M$ is subject to a "blank titration," $B$, while $N$ does not require a blank, we have:

$$M - B = kN$$

or

$$M = B + kN$$

If we use the coefficient of variation as a criterion, we obtain:

$$\frac{CV_N}{CV_M} = \left|\frac{\sigma_N/N}{\sigma_M/M}\right| = \frac{|M/N|}{\sigma_M/\sigma_N} = \frac{|(B + kN)/N|}{\sigma_M/\sigma_N} = \frac{|k + (B/N)|}{\sigma_M/\sigma_N}$$

From this equation it follows that the larger the blank $B$, the more advantageous will method $M$ appear to be in comparison with method $N$. This conclusion is of course absurd, since a large systematic error (such as a blank) cannot possibly result in increased precision. The reason for the paradox is that a large blank *increases* the measurement and thereby *decreases* the coefficient of variation.

No difficulty is encountered when we use the sensitivity ratio as a criterion:

$$RS(M/N) = \frac{dM/dN}{\sigma_M/\sigma_N} = \frac{k}{\sigma_M/\sigma_N}$$

This quantity does not contain the blank $B$, which is as it should be. Indeed, a plot of $M$ versus $N$ yields a straight line with slope $k$ and intercept $B$; this line can be experimentally determined and it may be used as a calibration line of the method $M$ in terms of the method $N$. Thus, $B$ is not a factor of precision for $M$ and should not affect its relative technical merit.

## 14.10   GEOMETRIC INTERPRETATION OF THE SENSITIVITY RATIO

Consider Fig. 14.2 in which two methods, $M$ and $N$, are represented by the curve relating them and by their respective standard deviations of error, $\sigma_M$ and $\sigma_N$. We consider two materials, $A$ and $B$, for which the $M$ and $N$ values are represented by the two corresponding points on the curve. We assume that the points $A$ and $B$ are fairly close to each other. If we wish to differentiate between the two materials $A$ and $B$ on the basis of measurements made by either method, we see that the difference is greater by method $M$ than by method $N$, since $\Delta M > \Delta N$. Thus one would tend to prefer method $M$ on that basis. However, the actual measurements will not fall exactly on the theoretical curve; $\Delta M$ is subject to two errors: one in measuring $A$ and the other in measuring $B$. These errors belong to a population of standard deviation $\sigma_M$. A similar situation holds for $\Delta N$, which is subject to two errors characterized by $\sigma_N$. Evidently, the advantage of the larger $\Delta M$ may be offset if $\sigma_M$ happens to be appreciably larger than $\sigma_N$. It is intuitively clear that a valid comparison is obtained by comparing the *ratios* $\Delta M/\sigma_M$ and $\Delta N/\sigma_N$. Whichever of these is larger corresponds to the more sensitive method for differentiating

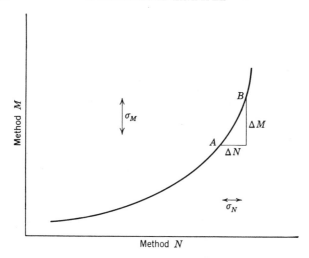

**Fig. 14.2**  Graphical interpretation of the sensitivity ratio.

between materials $A$ and $B$.    One is therefore led to an examination of the quantity

$$\frac{\Delta M}{\sigma_M} \bigg/ \frac{\Delta N}{\sigma_N}$$

which may be written

$$\frac{\Delta M/\Delta N}{\sigma_M/\sigma_N}$$

Since the curve of $M$ versus $N$ may be decreasing, which would entail a negative value for $\Delta M/\Delta N$, the sensitivity ratio is defined in terms of the absolute value of this ratio:

$$RS(M/N) = \frac{|\Delta M/\Delta N|}{\sigma_M/\sigma_N}$$

This relation is precisely the definition we arrived at previously on the basis of an algebraic argument (Eq. 14.13).

## 14.11   A NUMERICAL EXAMPLE

The data of Table 14.1 were obtained in a study of various physical tests for the characterization of vulcanized natural rubber (3).    They represent the experimental values for two tests, strain at 5 kg/cm², and stress at 100 per cent elongation, for six typical natural rubbers at each of three cures, 10 minutes, 20 minutes, and 40 minutes.    For each rubber and curing-time, four independent measurements were made by each of the two methods. The individual values are given in Table 14.1. Averages, standard deviations, and coefficients of variation of all sets of four are listed in Table 14.2.

**TABLE 14.1** Strain and Stress of Vulcanized Natural Rubber—Individual Measurements

| Rubber | Length of cure (min.) | Strain at 5 Kg/cm² (%) | | | | Stress at 100% elongation (Kg/cm²) | | | |
|--------|------|-------|-------|-------|-------|-------|-------|-------|-------|
| A | 10 | 83.0 | 82.5 | 81.5 | 83.0 | 5.445 | 5.515 | 5.520 | 5.530 |
|   | 20 | 63.0 | 64.0 | 63.5 | 63.5 | 6.435 | 6.365 | 6.395 | 6.545 |
|   | 40 | 56.5 | 56.5 | 57.0 | 57.0 | 6.855 | 6.885 | 6.875 | 6.820 |
| C | 10 | 285.5 | 275.5 | 295.0 | 281.5 | 2.740 | 2.820 | 2.715 | 2.765 |
|   | 20 | 180.5 | 165.5 | 178.5 | 167.5 | 3.685 | 3.745 | 3.590 | 3.715 |
|   | 40 | 130.5 | 123.5 | 127.5 | 123.0 | 4.250 | 4.395 | 4.345 | 4.390 |
| D | 10 | 108.5 | 110.0 | 113.5 | 112.5 | 4.770 | 4.635 | 4.605 | 4.625 |
|   | 20 | 75.5 | 75.0 | 76.5 | 75.0 | 5.850 | 5.730 | 5.630 | 5.755 |
|   | 40 | 63.0 | 63.0 | 63.5 | 63.0 | 6.470 | 6.370 | 6.295 | 6.435 |
| E | 10 | 141.5 | 139.5 | 139.5 | 139.5 | 4.115 | 4.180 | 4.170 | 4.165 |
|   | 20 | 108.0 | 105.5 | 106.0 | 106.0 | 4.750 | 4.765 | 4.790 | 4.885 |
|   | 40 | 93.5 | 93.0 | 92.0 | 92.5 | 5.495 | 5.125 | 5.135 | 5.120 |
| F | 10 | 246.5 | 236.0 | 253.5 | 251.0 | 3.085 | 3.135 | 3.060 | 3.040 |
|   | 20 | 146.0 | 139.5 | 150.5 | 152.5 | 4.070 | 4.130 | 3.995 | 3.940 |
|   | 40 | 115.0 | 115.0 | 119.0 | 119.5 | 4.570 | 4.575 | 4.525 | 4.470 |
| H | 10 | 202.5 | 194.0 | 201.5 | 200.0 | 3.485 | 3.545 | 3.585 | 3.485 |
|   | 20 | 109.0 | 110.5 | 107.5 | 109.5 | 4.695 | 4.620 | 4.740 | 4.675 |
|   | 40 | 81.0 | 81.0 | 80.5 | 81.5 | 5.485 | 5.550 | 5.525 | 5.500 |

It is known that the two tests in question are functionally related to each other and that the relationship is independent of the length of cure. This fact is confirmed by the present data, as can be seen in Fig. 14.3, which shows the hyperbolic type of relation obtained by plotting the 18 averages of stress (6 rubbers × 3 cures) versus the corresponding averages of strain. A considerable simplification is achieved by converting both the stress and the strain averages to logarithms. Figure 14.4 exhibits the relation between the two tests in the converted scales. It is seen that a straight line provides a very satisfactory fit to the logarithmic data.

The problem we wish to solve is the evaluation of the relative technical merits of the two tests, since either one could be used for a laboratory evaluation of the rubber. We have already established that the first desideratum—the existence of a functional relationship between the two tests—is satisfactorily fulfilled. Our next objective is to study the experimental errors affecting both methods.

**TABLE 14.2** Strain and Stress of Vulcanized Natural Rubber. Averages, Standard Deviations, and Coefficients of Variation

| Rubber | Length of cure | Strain Avg. | Strain Std. dev.[a] | Strain %CV | Stress Avg. | Stress Std. dev.[a] | Stress %CV |
|---|---|---|---|---|---|---|---|
| A | 10 | 82.5 | 0.73 | 0.88 | 5.502 | 0.041 | 0.75 |
|   | 20 | 63.5 | 0.49 | 0.77 | 6.435 | 0.088 | 1.36 |
|   | 40 | 56.8 | 0.24 | 0.42 | 6.859 | 0.032 | 0.46 |
| C | 10 | 284.4 | 9.47 | 3.22 | 2.760 | 0.051 | 1.85 |
|   | 20 | 173.0 | 7.28 | 4.21 | 3.684 | 0.075 | 2.04 |
|   | 40 | 126.1 | 3.64 | 2.89 | 4.345 | 0.070 | 1.62 |
| D | 10 | 111.1 | 2.43 | 2.19 | 4.659 | 0.080 | 1.72 |
|   | 20 | 75.5 | 0.73 | 0.97 | 5.741 | 0.107 | 1.86 |
|   | 40 | 63.1 | 0.24 | 0.38 | 6.392 | 0.085 | 1.33 |
| E | 10 | 140.0 | 0.97 | 0.69 | 4.158 | 0.032 | 0.76 |
|   | 20 | 106.4 | 1.21 | 1.14 | 4.798 | 0.066 | 1.37 |
|   | 40 | 93.0 | 0.73 | 0.78 | 5.219 | 0.182 | 3.49 |
| F | 10 | 246.8 | 8.50 | 3.44 | 3.080 | 0.046 | 1.50 |
|   | 20 | 147.1 | 6.31 | 4.29 | 4.034 | 0.092 | 2.29 |
|   | 40 | 117.1 | 2.19 | 1.87 | 4.535 | 0.051 | 1.12 |
| H | 10 | 199.5 | 4.13 | 2.07 | 3.525 | 0.049 | 1.38 |
|   | 20 | 109.1 | 1.46 | 1.34 | 4.682 | 0.058 | 1.25 |
|   | 40 | 80.9 | 0.49 | 0.60 | 5.515 | 0.032 | 0.57 |

[a] Estimated from range of four, using Eq. 6.2.

Since a logarithmic transformation has been found advantageous for expressing the relation between the two methods, we will carry out the entire sensitivity study in terms of these transformed data. In view of the results of Section 14.8, this is entirely legitimate: the ratio of the sensitivities of the two methods is invariant with respect to any transformation of scale.

Let us represent strain and stress by the symbols $E$ and $F$ ("elongation" and "force") respectively. Denote by $M$ and $N$ the logarithm of $E$ and $F$; thus:

$$M = \log E \qquad N = \log F \qquad (14.28)$$

Then Fig. 14.4 is a plot of $M$ versus $N$. We require three quantities: $dM/dN$, $\sigma_M$, and $\sigma_N$. The first of these is simply the slope of the straight

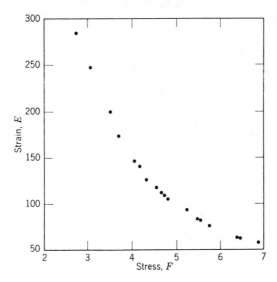

**Fig. 14.3**  Relation between strain and stress measurements for natural rubber vulcanizates.

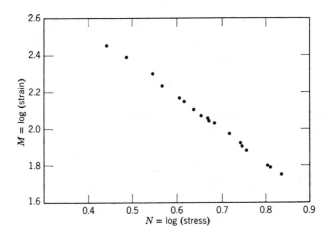

**Fig. 14.4**  Relation between strain and stress measurements, both in logarithmic scale, for natural rubber vulcanizates.

line fitted to the points in Fig. 14.4. In view of the small amount of scatter shown by these data, even a visual fit would give a very satisfactory slope value. However, a least squares fit was made, using the procedure of Section 12.5 and allowing equal error for both variables. That this is an acceptable assumption will soon be shown. The estimate of the slope obtained by these calculations is:

$$dM/dN = -1.838 \qquad (14.29)$$

Next we are to determine $\sigma_M$ and $\sigma_N$. Using Eq. 4.41 we obtain

$$V(M) = V(\log E) = \left[\frac{d \log E}{dE}\right]^2 \sigma_E{}^2 = \left(\frac{1}{2.3}\right)^2 \left(\frac{1}{E}\right)^2 \sigma_E{}^2$$

Hence

$$\sigma_M = \frac{1}{2.3} \frac{\sigma_E}{E} \qquad (14.30)$$

and similarly

$$\sigma_N = \frac{1}{2.3} \frac{\sigma_F}{F} \qquad (14.31)$$

The four replicate determinations available for each rubber and cure have provided us with estimates for the standard deviations $\sigma_E$ and $\sigma_F$ and for the averages $E$ and $F$ (Table 14.2). From these quantities we derive estimates for $\sigma_M$ and $\sigma_N$, using Eqs. 14.30 and 14.31. Table 14.3 lists the values of $M$, $N$, $\hat{\sigma}_M$ and $\hat{\sigma}_N$, as well as the ratio $\hat{\sigma}_M/\hat{\sigma}_N$. Evidently

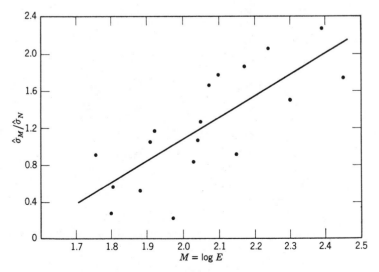

**Fig. 14.5** Relation between ratio of standard deviations and property value for natural rubber vulcanizates.

the ratios $\hat{\sigma}_M/\hat{\sigma}_N$ are subject to considerable sampling fluctuations. Nevertheless, it may be anticipated that if this ratio changes as a function of the magnitude of the measurement ($M$ or $N$; either may be used, since they are functionally related), a plot of $\hat{\sigma}_M/\hat{\sigma}_N$ versus either $M$ or $N$ will reveal such a relationship. Figure 14.5 is a plot of $\hat{\sigma}_M/\hat{\sigma}_N$ versus $M$; it reveals a definite trend which may be approximated by the least squares straight line shown on the graph. The equation of this line is

$$\sigma_M/\sigma_N = -3.595 + 2.335M \tag{14.32}$$

**TABLE 14.3** Natural Rubber Data.  Averages and Standard Deviations in Logarithmic Scale

| Rubber | Length of cure | $M = \log(\text{strain})$ | | $N = \log(\text{stress})$ | | $\hat{\sigma}_M/\hat{\sigma}_N$ |
|--------|------|---------|---------|---------|---------|------|
| | | Average | $\hat{\sigma}_M$ | Average | $\hat{\sigma}_N$ | |
| $A$ | 10 | 1.916 | $0.38 \times 10^{-2}$ | 0.741 | $0.32 \times 10^{-2}$ | 1.17 |
| | 20 | 1.803 | 0.33 | 0.808 | 0.58 | 0.57 |
| | 40 | 1.754 | 0.17 | 0.836 | 0.19 | 0.91 |
| $C$ | 10 | 2.454 | 1.44 | 0.441 | 0.80 | 1.74 |
| | 20 | 2.238 | 1.83 | 0.566 | 0.89 | 2.06 |
| | 40 | 2.101 | 1.25 | 0.638 | 0.70 | 1.78 |
| $D$ | 10 | 2.046 | 0.95 | 0.668 | 0.74 | 1.27 |
| | 20 | 1.878 | 0.42 | 0.759 | 0.80 | 0.52 |
| | 40 | 1.800 | 0.16 | 0.806 | 0.57 | 0.28 |
| $E$ | 10 | 2.146 | 0.30 | 0.619 | 0.33 | 0.91 |
| | 20 | 2.027 | 0.49 | 0.681 | 0.59 | 0.83 |
| | 40 | 1.968 | 0.34 | 0.718 | 1.49 | 0.22 |
| $F$ | 10 | 2.392 | 1.52 | 0.489 | 0.65 | 2.29 |
| | 20 | 2.168 | 1.88 | 0.606 | 0.99 | 1.87 |
| | 40 | 2.069 | 0.80 | 0.657 | 0.48 | 1.67 |
| $H$ | 10 | 2.300 | 0.90 | 0.547 | 0.59 | 1.50 |
| | 20 | 2.038 | 0.58 | 0.670 | 0.54 | 1.07 |
| | 40 | 1.908 | 0.26 | 0.742 | 0.25 | 1.05 |

Combining Eqs. 14.29 and 14.32, we obtain the sensitivity ratio of $M$ with respect to $N$:

$$RS(M/N) = \frac{|dM/dN|}{\sigma_M/\sigma_N} = \frac{1.838}{-3.595 + 2.335M}$$

In accordance with the results of Section 14.8, this quantity also represents the sensitivity ratio of $E$ with respect to $F$:

$$RS(E/F) = \frac{1.838}{-3.595 + 2.335M} = \frac{1.838}{-3.595 + 2.335 \log E} \quad (14.33)$$

Thus the sensitivity ratio of $E$ with respect to $F$ is a decreasing monotonic function of $E$; it is largest when $E$ is at the lower end of the range, and smallest when $E$ is at the upper end. Taking the extreme values of $E$ in the present experiment, we obtain:

$$\text{for } E = 70 \qquad RS(E/F) = 2.58$$

$$\text{for } E = 270 \qquad RS(E/F) = 0.88$$

We conclude that the strain test is, over most of the range, somewhat superior to the stress test, particularly for rubbers for which the strain value lies near the lower $E$ values listed in Table 14.1.

We make two further observations.

1. Since the ratio $\sigma_M/\sigma_N$ is, on the whole, not far from unity, the assumption of equal errors in $M$ and $N$, which was used in fitting the straight line to the $M$, $N$ relationship (Fig. 14.4) is practically justified. Indeed, the estimate of the slope would have been changed very little, had the correct ratio of standard deviations been used for the fit.

2. Our example demonstrates the fallacy of using the coefficient of variation as a criterion for comparing two methods. In view of Eqs. 14.30 and 14.31 the ratio of the coefficients of variation of $E$ and $F$ is simply the ratio of the standard deviations of $M$ and $N$, i.e., the quantity plotted in Fig. 14.5. Thus, the stress method would have been considered *superior* to the strain method for all rubbers and cures for which $M$ is greater than 1.97 (the value at which the line in Fig. 14.5 crosses the horizontal of ordinate 1.0), i.e., for the range $E > 93.3$. This range includes the majority of the vulcanizates in the experiment. This conclusion would have been seriously in error.

## 14.12  SENSITIVITY AND SAMPLE SIZE

What is the practical interpretation of a sensitivity ratio?

Suppose that in a particular study the relative sensitivity of method $M$ with respect to method $N$ has been found to be 2.5. Thus

$$\frac{|dM/dN|}{\sigma_M/\sigma_N} = 2.5$$

Let $Q$ represent the property of which both $M$ and $N$ are measures. It then follows from Eq. 14.12 that

$$\sigma_{\hat{Q}_M}/\sigma_{\hat{Q}_N} = \frac{1}{2.5}$$

where $\hat{Q}_M$ and $\hat{Q}_N$ represent the values of $Q$ as estimated from $M$ and $N$ respectively. It then follows from the results of Chapter 10 that for equal precision in $\hat{Q}$ the sample sizes required by $M$ and $N$ are in the ratio $(1/2.5)^2$, or approximately $1/6$. Thus a single measurement by method $M$ is about equivalent (in precision) to the average of 6 measurements by method $N$.

More generally, if $RS(M/N) = k$, a single measurement by method $M$ has the same precision as the average of $k^2$ measurements by method $N$.

This important theorem provides the link between technical merit and economic merit. For if we have

$$RS(M/N) = k$$

and the cost of a single measurement is $C_M$ for method $M$, and $C_N$ for method $N$, the ratio of costs to achieve equal precision by $M$ and $N$ respectively is

$$\frac{C_M}{k^2 \cdot C_N} \quad \text{or} \quad \frac{1}{k^2} \cdot \frac{C_M}{C_N}$$

For example, if $k^2 = 6$ and $C_M/C_N = 10$, the ratio in question is $10/6$. Thus method $N$ would, under such conditions, be economically preferable to $M$, in spite of its poorer precision. The use of $N$ would result in a cost of only $6/10$ of that of $M$, a saving of 40 per cent. It follows from the preceding considerations that the problem of selecting the most advantageous test method under given conditions can generally be solved by combining a knowledge of the relative sensitivities of the alternative methods with a knowledge of their relative costs per measurement.

## 14.13    SENSITIVITY AND STRAIGHT LINE FITTING

In Section 12.5 we have dealt with the problem of fitting a straight line to data when both variables are subject to error. The procedure discussed in that section is somewhat more complicated than that pertaining to the "classical" case (Section 12.2). For many sets of data the results obtained by use of the simpler method constitute such a close approximation to those obtained by the correct, but more elaborate method that the extra work and additional complexity of the latter is hardly justified. We now examine under what conditions this is the case.

We consider once more the model

$$Y = \alpha + \beta X \tag{14.34}$$

with

$$x = X + \varepsilon \qquad y = Y + \delta \tag{14.35}$$

where $\varepsilon$ and $\delta$ are random errors affecting the measurements $x$ and $y$ respectively.   By definition we have

$$\lambda = \frac{V(\varepsilon)}{V(\delta)} \tag{14.36}$$

The correct way of fitting a straight line to the data is to apply Eq. 12.45, i.e.,

$$\beta = \frac{\lambda w - u + \sqrt{(u - \lambda w)^2 + 4\lambda p^2}}{2\lambda p} \tag{14.37}$$

We have:

$$u = N \sum (x - \bar{x})^2$$
$$= N[\sum (X - \bar{X})^2 + 2 \sum (X - \bar{X})(\varepsilon - \bar{\varepsilon}) + \sum (\varepsilon - \bar{\varepsilon})^2]$$

Taking the expected value of $u$, we obtain:

$$E(u) = N[\sum (X - \bar{X})^2 + E\{\sum (\varepsilon - \bar{\varepsilon})^2\}]$$

since the expected value of the cross-product term is zero, as a result of the statistical independence of $X$ and $\varepsilon$.   Now, by definition of the variance, we have

$$E\left\{\frac{\sum (\varepsilon - \bar{\varepsilon})^2}{N - 1}\right\} = \sigma_\varepsilon^2$$

Hence

$$E(u) = N[\sum (X - \bar{X})^2 + (N - 1)\sigma_\varepsilon^2] \tag{14.38}$$

Similarly, we obtain

$$E(w) = N[\sum (Y - \bar{Y})^2 + (N - 1)\sigma_\delta^2] \tag{14.39}$$

and

$$E(p) = N[\sum (X - \bar{X})(Y - \bar{Y})] \tag{14.40}$$

Using Eq. 14.34 we obtain readily:

$$\sum (Y - \bar{Y})^2 = \beta^2 \sum (X - \bar{X})^2 \tag{14.41}$$

and

$$\sum (X - \bar{X})(Y - \bar{Y}) = \beta \sum (X - \bar{X})^2 \tag{14.42}$$

Let us represent the quantity $\sum (X - \bar{X})^2$ by $T$:

$$T = \sum (X - \bar{X})^2 \tag{14.43}$$

Then we may write

$$
\left.\begin{array}{l}
(1/N)E(u) = T + (N-1)\sigma_\varepsilon^2 = T + (N-1)\lambda\sigma_\delta^2 \\[6pt]
(1/N)E(w) = \beta^2 T + (N-1)\sigma_\delta^2 \\[6pt]
(1/N)E(p) = \beta T
\end{array}\right\} \qquad (14.44)
$$

We are now in a position to compare Eq. 14.37 with the formula for the classical case:

$$
\hat\beta = \frac{p}{u} \qquad (14.45)
$$

Substituting expected values for the quantities $u$ and $p$, we have:

$$
\frac{p}{u} \approx \frac{\beta T}{T + (N-1)\sigma_\varepsilon^2} = \beta\,\frac{1}{1 + \sigma_\varepsilon^2/[T/(N-1)]} \qquad (14.46)
$$

Thus, the quantity $p/u$, as an estimate of $\beta$, is subject to a bias which depends on the relative magnitude of $\sigma_\varepsilon^2/[T/(N-1)]$ with respect to unity. If we consider the regression of $x$ on $y$, rather than that of $y$ on $x$, the slope will be estimated by $p/w$. The quantity estimated by this expression is the reciprocal of the slope $\beta$ of $y$ with respect to $x$. Hence, another estimate of $\beta$ is given by $w/p$. Using Eqs. 14.44, we have:

$$
\frac{w}{p} \approx \frac{\beta^2 T + (N-1)\sigma_\delta^2}{\beta T} = \beta\left[1 + \frac{\sigma_\delta^2}{\beta^2 T/(N-1)}\right] \qquad (14.47)
$$

This estimate, too, is biased, by a quantity depending on the relative magnitude of $\sigma_\delta^2/[\beta^2 T/(N-1)]$ with respect to unity.

If, on the other hand, we substitute the expected values given by Eqs. 14.44 for $u$, $w$, and $p$, in Eq. 14.37 *we will find no bias at all.** The question we now ask is whether either $p/u$ or $w/p$ is a satisfactory approximation to the unbiased value given by Eq. 14.37, and which of these two is the better one.

A closer examination of the biasing quantity $\sigma_\varepsilon^2/[T/(N-1)]$ shows that it is a measure for the relative magnitude of the error of $x$ with respect to the total range of the $X$ values in the experiment, since:

$$
\frac{\sigma_\varepsilon^2}{T/(N-1)} = \left[\frac{\sigma_\varepsilon}{[\sum(X-\bar X)^2/(N-1)]^{1/2}}\right]^2
$$

Similarly:

$$
\frac{\sigma_\delta^2}{\beta^2 T/(N-1)} = \frac{\sigma_\delta^2}{\sum(Y-\bar Y)^2/(N-1)} = \left[\frac{\sigma_\delta}{[\sum(Y-\bar Y)^2/(N-1)]^{1/2}}\right]^2
$$

* In fact, a rather simple way of deriving Eq. 14.37 is to "eliminate" the two quantities, $\sigma_\delta^2$, and $T$, between the three equations of Eq. 14.44.

The quantities $[\sum (X - \bar{X})^2/(N - 1)]^{1/2}$ and $[\sum (Y - \bar{Y})^2/(N - 1)]^{1/2}$ are of course measures of the spread, or range, of the $X$ and $Y$ values, respectively. We thus see that in order that $p/u$ be an acceptable value for the slope it is necessary that the error of $x$, $\sigma_\varepsilon$, be small with respect to the range of $X$-values; and in order that $w/p$ be acceptable, it is necessary that $\sigma_\delta$ be small with respect to the range of $Y$ values. Fortunately, properly designed experiments for the examination of a relation between two quantities $x$ and $y$ are generally carried out over a range of values, for either variable, that is appreciably larger than the corresponding error of measurement. Under those conditions, $\sigma_\varepsilon^2$ will be quite small in comparison with $T/(N - 1)$ and Eq. 14.46 can be written in the form*:

$$\frac{p}{u} \approx \beta\left[1 - \frac{\sigma_\varepsilon^2}{T/(N - 1)}\right] \tag{14.48}$$

Examination of Eqs. 14.48 and 14.47 shows that the statistic $p/u$ systematically *underestimates* the slope $\beta$, while the statistic $w/p$ systematically *overestimates* it. The bias can be roughly estimated from an approximate knowledge of $\sigma_\varepsilon$ and $\sigma_\delta$. It is, however, important to note that even when such knowledge is not available, it is still possible to decide which of the two regressions gives a less biased slope estimate, provided that $\lambda$, the ratio of the error-variances $(\sigma_\varepsilon^2/\sigma_\delta^2)$, is known. Indeed, by taking the ratio of the biases of Eqs. 14.48 and 14.47 we obtain:

$$\frac{\sigma_\varepsilon^2/[T/(N - 1)]}{\sigma_\delta^2/[\beta^2 T/(N - 1)]}$$

or

$$\frac{\beta^2}{\sigma_\delta^2/\sigma_\varepsilon^2}$$

But this is simply the square of the relative sensitivity of $y$ with respect to $x$. Hence, if the relative sensitivity of $y$ with respect to $x$ is *small* (as compared to unity), $p/u$ is the better estimate, i.e., the regression should be made of $y$ on $x$. If the relative sensitivity of $y$ with respect to $x$ is *large* (as compared to unity), $w/p$ is the better estimate; then the regression should be made in terms of $x$ on $y$ and the reciprocal taken of the slope.

The criterion is easily applied provided that $\lambda$ be known (a necessary condition for solving the problem under any circumstances): the slope $\beta$ is first roughly estimated (in most cases a visual estimate is satisfactory) and the sensitivity ratio of $y$ with respect to $x$ is calculated:

$$RS(y/x) = \frac{\beta}{\sigma_\delta/\sigma_\varepsilon} = \beta\frac{\sigma_\varepsilon}{\sigma_\delta} = \beta\sqrt{\lambda} \tag{14.49}$$

---

* When a quantity $a$ is small with respect to unity, we have by the binomial theorem $1/(1 + a) = (1 + a)^{-1} \approx 1 - a$.

If this quantity is appreciably smaller than unity, the regression should be calculated in terms of $y$ on $x$. If on the other hand, the quantity $\beta\sqrt{\lambda}$ is considerably larger than unity, it is better to make the regression analysis in terms of $x$ on $y$. As an illustration, consider the data in Table 14.4 for which it is known that $\lambda = 4$. The quantities $u$, $w$, and $p$, are also shown in the table. From these are obtained the three estimates $p/u$, $w/p$, and $\hat{\beta}$, the latter being calculated by Eq. 14.37, using $\lambda = 4$.

Even a cursory examination of the data shows that the slope is of the order of 5. Thus:

$$RS(y/x) \approx 5\sqrt{\lambda} = 5\sqrt{4} = 10$$

Thus we would have been able to tell, *before* making the indicated computations, that it is preferable to regress $x$ on $y$, rather than $y$ on $x$. This is borne out by a comparison of the three estimates of the slope.

TABLE 14.4   An Example of Linear Regression.[a]   Comparison of Three Slope Estimates

| $x$ | $y$ | |
|---|---|---|
| 1.179 | 5.094 | Given: |
| 1.971 | 9.982 | $\lambda = 4$ |
| 2.922 | 14.935 | |
| 3.900 | 19.914 | Calculations: |
| 4.904 | 24.912 | $u = 335.009399$ |
| 6.265 | 30.014 | $w = 8328.818479$ |
| 7.270 | 35.089 | $p = 1666.375347$ |
| 7.764 | 39.727 | |
| | | Estimates for slope of $y$ versus $x$: |
| Sum 36.175 | 179.667 | $p/u = 4.9741$ |
| | | $w/p = 4.9982$ |
| | | $\hat{\beta} = 4.9979$ |

[a] The data in this table were artificially constructed by adding random normal deviates, of standard deviations 0.2 and 0.1 respectively, to a set of $(X, Y)$ data for which $Y = 5X$.

## 14.14   AN APPLICATION OF SENSITIVITY IN SPECTROPHOTOMETRY

It is customary, in spectrophotometric work in which Beer's law (2) applies, to select a working concentration in the neighborhood of 37 per cent transmittance. The reason usually advanced for this choice is the fact that the quantity

$$\frac{1}{c}\left|\frac{dc}{dI}\right|$$

where $c$ is concentration and $I$ is the transmitted intensity, is smallest when $I/I_0 = e^{-1} = 0.37$. Here $I_0$ is the incident intensity and it is assumed that Beer's law,

$$I/I_0 = \exp(-kc) \tag{14.50}$$

holds.

This line of reasoning appears plausible, since $(1/c)(dc/dI)$ is the relative change of concentration per unit change of intensity. However, when viewed in the light of the sensitivity concept, as derived from Eq. 14.3, the argument is found to take account of only one of the two pertinent factors that determine the optimum condition. This can be seen as follows.

The real unknown is the concentration, say $c_0$, of a given chemical solution. The sample subjected to the measurement is a dilution of this solution. Let $D$ be the dilution factor; thus

$$c = \frac{c_0}{D} \tag{14.51}$$

We wish to determine the value of $D$ that will lead to the best experimental value for $c_0$.

In terms of Eq. 14.3, the property of interest, $Q$, is $c_0$; the measurement $M$ is $I/I_0$. Hence, Beer's law becomes:

$$M = \frac{I}{I_0} = \exp(-kc) = \exp\left(\frac{-kc_0}{D}\right) = \exp\left(\frac{-k}{D}Q\right) \tag{14.52}$$

According to Eq. 14.3, we are to find the smallest value of $\sigma_{\hat{Q}}$, where $\hat{Q}$ is, by virtue of Eq. 14.52, equal to:

$$\hat{Q} = -\frac{D}{k}\ln M \tag{14.53}$$

The derivative of this function with respect to $M$, in absolute value, is:

$$\left|\frac{d\hat{Q}}{dM}\right| = \frac{D}{k}\exp\left(\frac{k}{D}Q\right)$$

Therefore:

$$\sigma_{\hat{Q}} = \left|\frac{d\hat{Q}}{dM}\right|\sigma_M = \frac{D}{k}\exp\left(\frac{k}{D}Q\right)\sigma_M \tag{14.54}$$

Now, *if $\sigma_M$ is constant*, this quantity will be a minimum when

$$\frac{D}{k} \exp \left(\frac{kQ}{D}\right)$$

is minimum.  By differentiating this expression with respect to $D$ and equating the derivative to zero, it is easily found that its minimum occurs for $D = kQ$.  But then we have, according to Eq. 14.52:

$$M = \frac{I}{I_0} = e^{-1} = 0.37$$

Thus, *for $\sigma_M = constant$*, an argument based on the sensitivity concept leads to the "classical" result.  But it is quite possible that the standard deviation of $M$, i.e., of the transmittance $I/I_0$, varies with the transmittance value.  In that case the optimum will no longer necessarily occur at 37 per cent transmittance.  If $\sigma_M$ is known as a function of $M$, say

$$\sigma_M = f(M) \tag{14.55}$$

the optimum will be obtained by finding that value of $D$ for which the quantity given by Eq. 14.54,

$$\frac{D}{k} \exp \left(\frac{kQ}{D}\right) f(M) \tag{14.56}$$

is a minimum.

As an application, suppose that it has been found experimentally that

$$\sigma_{I/I_0} = K \left(\frac{I}{I_0}\right)^{-1/2}$$

Then $\sigma_M = KM^{-1/2} = K \exp [(k/2D)Q]$, and we are to find the value of $D$ that minimizes the expression $[(D/k) \exp (kQ/D)] \cdot [K \exp (kQ/2D)]$.

Equating to zero the derivative of this expression with respect to $D$, we obtain: $D = (3/2)kQ$, which leads to the transmittance

$$I/I_0 = e^{-2/3} = 0.513 = 51.3 \text{ per cent}$$

It is not difficult to see that the application we have discussed in this section is but one instance of an extensive class of situations of frequent occurrence in the laboratory.  The common problem in all these situations is to determine an *optimum* range for the measured value provided by a given method of test.  The correct analysis of problems of this type must involve both aspects of the sensitivity measure: the amount by which the measurement varies when the property varies, and the uncertainty of the measurement as a result of experimental error.

## 14.15 SUMMARY

Of great practical importance is the comparison of different methods for the measurement of the same property of a material. The first consideration that enters into such a comparison is the intrinsic merit of a method of test, regardless of its ease of operation and its cost. But the latter two, as well as other factors of an economic nature cannot be ignored. In this chapter a concept, called sensitivity, is introduced for the characterization of the technical merit of a test method. Using this concept, the comparison of two different methods can be made without relating either one directly to the property they represent; all that is required is the theoretical or empirical relation between the values given by the two methods for the same series of samples (materials or physical systems), and measures for their respective reproducibilities. This information can be completely obtained in the course of a single, well-planned experiment. This is illustrated by the detailed discussion of a numerical example. The ratio of the sensitivities of two methods is shown to be unaffected by any transformation of scale. It is also shown to be a more generally reliable criterion than the ratio of the coefficients of variation of the two methods. Its geometric meaning is explained and it is shown to provide a rational basis for the calculation of the number of determinations required by either method in order to achieve a desired degree of precision. This calculation provides the link between technical and economic merit for the over-all evaluation of a method of test.

The sensitivity ratio is also useful in an entirely different context, namely for deciding in what way straight-line data should be fitted under certain circumstances.

Finally, the sensitivity concept provides a rational way for determining the optimum operating conditions for certain types of measurement. This is illustrated for the problem of determining the optimum transmittance value in spectrophotometric applications involving Beer's law.

## REFERENCES

1. Mandel, J., and R. D. Stiehler, "Sensitivity—A Criterion for the Comparison of Methods of Test," *J. Research Natl. Bur. Standards*, **53**, No. 3, 155–159 (Sept. 1954).
2. Meites, Louis, *Handbook of Analytical Chemistry*, McGraw-Hill, New York, 1963.
3. Stiehler, R. D., Private communication.

# APPENDIX

**TABLE I** Cumulative Probabilities of the Standard Normal Deviate

| x | 0.00 | 0.01 | 0.02 | 0.03 | 0.04 | 0.05 | 0.06 | 0.07 | 0.08 | 0.09 |
|---|------|------|------|------|------|------|------|------|------|------|
| 0.0 | .5000 | .5040 | .5080 | .5120 | .5160 | .5199 | .5239 | .5279 | .5319 | .5359 |
| 0.1 | .5398 | .5438 | .5478 | .5517 | .5557 | .5596 | .5636 | .5675 | .5714 | .5753 |
| 0.2 | .5793 | .5832 | .5871 | .5910 | .5948 | .5987 | .6026 | .6064 | .6103 | .6141 |
| 0.3 | .6179 | .6217 | .6255 | .6293 | .6331 | .6368 | .6406 | .6443 | .6480 | .6517 |
| 0.4 | .6554 | .6591 | .6628 | .6664 | .6700 | .6736 | .6772 | .6808 | .6844 | .6879 |
| 0.5 | .6915 | .6950 | .6985 | .7019 | .7054 | .7088 | .7123 | .7157 | .7190 | .7224 |
| 0.6 | .7257 | .7291 | .7324 | .7357 | .7389 | .7422 | .7454 | .7486 | .7517 | .7549 |
| 0.7 | .7580 | .7611 | .7642 | .7673 | .7704 | .7734 | .7764 | .7794 | .7823 | .7852 |
| 0.8 | .7881 | .7910 | .7939 | .7967 | .7995 | .8023 | .8051 | .8078 | .8106 | .8133 |
| 0.9 | .8159 | .8186 | .8212 | .8238 | .8264 | .8289 | .8315 | .8340 | .8365 | .8389 |
| 1.0 | .8413 | .8438 | .8461 | .8485 | .8508 | .8531 | .8554 | .8577 | .8599 | .8621 |
| 1.1 | .8643 | .8665 | .8686 | .8708 | .8729 | .8749 | .8770 | .8790 | .8810 | .8830 |
| 1.2 | .8849 | .8869 | .8888 | .8907 | .8925 | .8944 | .8962 | .8980 | .8997 | .9015 |
| 1.3 | .9032 | .9049 | .9066 | .9082 | .9099 | .9115 | .9131 | .9147 | .9162 | .9177 |
| 1.4 | .9192 | .9207 | .9222 | .9236 | .9251 | .9265 | .9279 | .9292 | .9306 | .9319 |
| 1.5 | .9332 | .9345 | .9357 | .9370 | .9382 | .9394 | .9406 | .9418 | .9429 | .9441 |
| 1.6 | .9452 | .9463 | .9474 | .9484 | .9495 | .9505 | .9515 | .9525 | .9535 | .9545 |
| 1.7 | .9554 | .9564 | .9573 | .9582 | .9591 | .9599 | .9608 | .9616 | .9625 | .9633 |
| 1.8 | .9641 | .9649 | .9656 | .9664 | .9671 | .9678 | .9686 | .9693 | .9699 | .9706 |
| 1.9 | .9713 | .9719 | .9726 | .9732 | .9738 | .9744 | .9750 | .9756 | .9761 | .9767 |
| 2.0 | .9772 | .9778 | .9783 | .9788 | .9793 | .9798 | .9803 | .9808 | .9812 | .9817 |
| 2.1 | .9821 | .9826 | .9830 | .9834 | .9838 | .9842 | .9846 | .9850 | .9854 | .9857 |
| 2.2 | .9861 | .9864 | .9868 | .9871 | .9875 | .9878 | .9881 | .9884 | .9887 | .9890 |
| 2.3 | .9893 | .9896 | .9898 | .9901 | .9904 | .9906 | .9909 | .9911 | .9913 | .9916 |
| 2.4 | .9918 | .9920 | .9922 | .9925 | .9927 | .9929 | .9931 | .9932 | .9934 | .9936 |
| 2.5 | .9938 | .9940 | .9941 | .9943 | .9945 | .9946 | .9948 | .9949 | .9951 | .9952 |
| 2.6 | .9953 | .9955 | .9956 | .9957 | .9959 | .9960 | .9961 | .9962 | .9963 | .9964 |
| 2.7 | .9965 | .9966 | .9967 | .9968 | .9969 | .9970 | .9971 | .9972 | .9973 | .9974 |
| 2.8 | .9974 | .9975 | .9976 | .9977 | .9977 | .9978 | .9979 | .9979 | .9980 | .9981 |
| 2.9 | .9981 | .9982 | .9982 | .9983 | .9984 | .9984 | .9985 | .9985 | .9986 | .9986 |
| 3.0 | .9987 | .9987 | .9987 | .9988 | .9988 | .9989 | .9989 | .9989 | .9990 | .9990 |
| 3.1 | .9990 | .9991 | .9991 | .9991 | .9992 | .9992 | .9992 | .9992 | .9993 | .9993 |
| 3.2 | .9993 | .9993 | .9994 | .9994 | .9994 | .9994 | .9994 | .9995 | .9995 | .9995 |
| 3.3 | .9995 | .9995 | .9995 | .9996 | .9996 | .9996 | .9996 | .9996 | .9996 | .9997 |
| 3.4 | .9997 | .9997 | .9997 | .9997 | .9997 | .9997 | .9997 | .9997 | .9997 | .9998 |
| 3.6 | .9998 | .9998 | .9999 | .9999 | .9999 | .9999 | .9999 | .9999 | .9999 | .9999 |

Source: E. Parzen, *Modern Probability Theory and Its Applications*, Wiley, 1960. Table I, p. 441.

**TABLE II** Percentage Points of Student's *t*

| P<br>DF | 0.750 | 0.900 | 0.950 | 0.975 | 0.990 | 0.995 | 0.999 |
|---|---|---|---|---|---|---|---|
| 1 | 1.000 | 3.078 | 6.314 | 12.706 | 31.821 | 63.657 | 318 |
| 2 | 0.816 | 1.886 | 2.920 | 4.303 | 6.965 | 9.925 | 22.3 |
| 3 | 0.765 | 1.638 | 2.353 | 3.182 | 4.541 | 5.841 | 10.2 |
| 4 | 0.741 | 1.533 | 2.132 | 2.776 | 3.747 | 4.604 | 7.173 |
| 5 | 0.727 | 1.476 | 2.015 | 2.571 | 3.365 | 4.032 | 5.893 |
| 6 | 0.718 | 1.440 | 1.943 | 2.447 | 3.143 | 3.707 | 5.208 |
| 7 | 0.711 | 1.415 | 1.895 | 2.365 | 2.998 | 3.499 | 4.785 |
| 8 | 0.706 | 1.397 | 1.860 | 2.306 | 2.896 | 3.355 | 4.501 |
| 9 | 0.703 | 1.383 | 1.833 | 2.262 | 2.821 | 3.250 | 4.297 |
| 10 | 0.700 | 1.372 | 1.812 | 2.228 | 2.764 | 3.169 | 4.144 |
| 11 | 0.697 | 1.363 | 1.796 | 2.201 | 2.718 | 3.106 | 4.025 |
| 12 | 0.695 | 1.356 | 1.782 | 2.179 | 2.681 | 3.055 | 3.930 |
| 13 | 0.694 | 1.350 | 1.771 | 2.160 | 2.650 | 3.012 | 3.852 |
| 14 | 0.692 | 1.345 | 1.761 | 2.145 | 2.624 | 2.977 | 3.787 |
| 15 | 0.691 | 1.341 | 1.753 | 2.131 | 2.602 | 2.947 | 3.733 |
| 16 | 0.690 | 1.337 | 1.746 | 2.120 | 2.583 | 2.921 | 3.686 |
| 17 | 0.689 | 1.333 | 1.740 | 2.110 | 2.567 | 2.898 | 3.646 |
| 18 | 0.688 | 1.330 | 1.734 | 2.101 | 2.552 | 2.878 | 3.610 |
| 19 | 0.688 | 1.328 | 1.729 | 2.093 | 2.539 | 2.861 | 3.579 |
| 20 | 0.687 | 1.325 | 1.725 | 2.086 | 2.528 | 2.845 | 3.552 |
| 21 | 0.686 | 1.323 | 1.721 | 2.080 | 2.518 | 2.831 | 3.527 |
| 22 | 0.686 | 1.321 | 1.717 | 2.074 | 2.508 | 2.819 | 3.505 |
| 23 | 0.685 | 1.319 | 1.714 | 2.069 | 2.500 | 2.807 | 3.485 |
| 24 | 0.685 | 1.318 | 1.711 | 2.064 | 2.492 | 2.797 | 3.467 |
| 25 | 0.684 | 1.316 | 1.708 | 2.060 | 2.485 | 2.787 | 3.450 |
| 26 | 0.684 | 1.315 | 1.706 | 2.056 | 2.479 | 2.779 | 3.435 |
| 27 | 0.684 | 1.314 | 1.703 | 2.052 | 2.473 | 2.771 | 3.421 |
| 28 | 0.683 | 1.313 | 1.701 | 2.048 | 2.467 | 2.763 | 3.408 |
| 29 | 0.683 | 1.311 | 1.699 | 2.045 | 2.462 | 2.756 | 3.396 |
| 30 | 0.683 | 1.310 | 1.697 | 2.042 | 2.457 | 2.750 | 3.385 |
| 40 | 0.681 | 1.303 | 1.684 | 2.021 | 2.423 | 2.704 | 3.307 |
| 60 | 0.679 | 1.296 | 1.671 | 2.000 | 2.390 | 2.660 | 3.232 |
| 120 | 0.677 | 1.289 | 1.658 | 1.980 | 2.358 | 2.617 | 3.160 |
| ∞ | 0.674 | 1.282 | 1.645 | 1.960 | 2.326 | 2.576 | 3.090 |

Source: K. O. Brownlee, *Statistical Theory and Methodology in Science and Engineering*, Wiley, 1960. Table II, p. 548.

**TABLE III**  Percentage Points of the $F$ Distribution, Level of Significance: 5%

| $v_1$ / $v_2$ | 1 | 2 | 3 | 4 | 5 | 6 | 7 | 8 | 9 |
|---|---|---|---|---|---|---|---|---|---|
| 1 | 161.45 | 199.50 | 215.71 | 224.58 | 230.16 | 233.99 | 236.77 | 238.88 | 240.54 |
| 2 | 18.513 | 19.000 | 19.164 | 19.247 | 19.296 | 19.330 | 19.353 | 19.371 | 19.385 |
| 3 | 10.128 | 9.5521 | 9.2766 | 9.1172 | 9.0135 | 8.9406 | 8.8868 | 8.8452 | 8.8123 |
| 4 | 7.7086 | 6.9443 | 6.5914 | 6.3883 | 6.2560 | 6.1631 | 6.0942 | 6.0410 | 5.9988 |
| 5 | 6.6079 | 5.7861 | 5.4095 | 5.1922 | 5.0503 | 4.9503 | 4.8759 | 4.8183 | 4.7725 |
| 6 | 5.9874 | 5.1433 | 4.7571 | 4.5337 | 4.3874 | 4.2839 | 4.2066 | 4.1468 | 4.0990 |
| 7 | 5.5914 | 4.7374 | 4.3468 | 4.1203 | 3.9715 | 3.8660 | 3.7870 | 3.7257 | 3.6767 |
| 8 | 5.3177 | 4.4590 | 4.0662 | 3.8378 | 3.6875 | 3.5806 | 3.5005 | 3.4381 | 3.3881 |
| 9 | 5.1174 | 4.2565 | 3.8626 | 3.6331 | 3.4817 | 3.3738 | 3.2927 | 3.2296 | 3.1789 |
| 10 | 4.9646 | 4.1028 | 3.7083 | 3.4780 | 3.3258 | 3.2172 | 3.1355 | 3.0717 | 3.0204 |
| 11 | 4.8443 | 3.9823 | 3.5874 | 3.3567 | 3.2039 | 3.0946 | 3.0123 | 2.9480 | 2.8962 |
| 12 | 4.7472 | 3.8853 | 3.4903 | 3.2592 | 3.1059 | 2.9961 | 2.9134 | 2.8486 | 2.7964 |
| 13 | 4.6672 | 3.8056 | 3.4105 | 3.1791 | 3.0254 | 2.9153 | 2.8321 | 2.7669 | 2.7144 |
| 14 | 4.6001 | 3.7389 | 3.3439 | 3.1122 | 2.9582 | 2.8477 | 2.7642 | 2.6987 | 2.6458 |
| 15 | 4.5431 | 3.6823 | 3.2874 | 3.0556 | 2.9013 | 2.7905 | 2.7066 | 2.6408 | 2.5876 |
| 16 | 4.4940 | 3.6337 | 3.2389 | 3.0069 | 2.8524 | 2.7413 | 2.6572 | 2.5911 | 2.5377 |
| 17 | 4.4513 | 3.5915 | 3.1968 | 2.9647 | 2.8100 | 2.6987 | 2.6143 | 2.5480 | 2.4943 |
| 18 | 4.4139 | 3.5546 | 3.1599 | 2.9277 | 2.7729 | 2.6613 | 2.5767 | 2.5102 | 2.4563 |
| 19 | 4.3808 | 3.5219 | 3.1274 | 2.8951 | 2.7401 | 2.6283 | 2.5435 | 2.4768 | 2.4227 |
| 20 | 4.3513 | 3.4928 | 3.0984 | 2.8661 | 2.7109 | 2.5990 | 2.5140 | 2.4471 | 2.3928 |
| 21 | 4.3248 | 3.4668 | 3.0725 | 2.8401 | 2.6848 | 2.5727 | 2.4876 | 2.4205 | 2.3661 |
| 22 | 4.3009 | 3.4434 | 3.0491 | 2.8167 | 2.6613 | 2.5491 | 2.4638 | 2.3965 | 2.3419 |
| 23 | 4.2793 | 3.4221 | 3.0280 | 2.7955 | 2.6400 | 2.5277 | 2.4422 | 2.3748 | 2.3201 |
| 24 | 4.2597 | 3.4028 | 3.0088 | 2.7763 | 2.6207 | 2.5082 | 2.4226 | 2.3551 | 2.3002 |
| 25 | 4.2417 | 3.3852 | 2.9912 | 2.7587 | 2.6030 | 2.4904 | 2.4047 | 2.3371 | 2.2821 |
| 26 | 4.2252 | 3.3690 | 2.9751 | 2.7426 | 2.5868 | 2.4741 | 2.3883 | 2.3205 | 2.2655 |
| 27 | 4.2100 | 3.3541 | 2.9604 | 2.7278 | 2.5719 | 2.4591 | 2.3732 | 2.3053 | 2.2501 |
| 28 | 4.1960 | 3.3404 | 2.9467 | 2.7141 | 2.5581 | 2.4453 | 2.3593 | 2.2913 | 2.2360 |
| 29 | 4.1830 | 3.3277 | 2.9340 | 2.7014 | 2.5454 | 2.4324 | 2.3463 | 2.2782 | 2.2229 |
| 30 | 4.1709 | 3.3158 | 2.9223 | 2.6896 | 2.5336 | 2.4205 | 2.3343 | 2.2662 | 2.2107 |
| 40 | 4.0848 | 3.2317 | 2.8387 | 2.6060 | 2.4495 | 2.3359 | 2.2490 | 2.1802 | 2.1240 |
| 60 | 4.0012 | 3.1504 | 2.7581 | 2.5252 | 2.3683 | 2.2540 | 2.1665 | 2.0970 | 2.0401 |
| 120 | 3.9201 | 3.0718 | 2.6802 | 2.4472 | 2.2900 | 2.1750 | 2.0867 | 2.0164 | 1.9588 |
| $\infty$ | 3.8415 | 2.9957 | 2.6049 | 2.3719 | 2.2141 | 2.0986 | 2.0096 | 1.9384 | 1.8799 |

**TABLE III** (*continued*)

| $\nu_2$ \ $\nu_1$ | 10 | 12 | 15 | 20 | 24 | 30 | 40 | 60 | 120 | $\infty$ |
|---|---|---|---|---|---|---|---|---|---|---|
| 1 | 241.88 | 243.91 | 245.95 | 248.01 | 249.05 | 250.09 | 251.14 | 252.20 | 253.25 | 254.32 |
| 2 | 19.396 | 19.413 | 19.429 | 19.446 | 19.454 | 19.462 | 19.471 | 19.479 | 19.487 | 19.496 |
| 3 | 8.7855 | 8.7446 | 8.7029 | 8.6602 | 8.6385 | 8.6166 | 8.5944 | 8.5720 | 8.5494 | 8.5265 |
| 4 | 5.9644 | 5.9117 | 5.8578 | 5.8025 | 5.7744 | 5.7459 | 5.7170 | 5.6878 | 5.6581 | 5.6281 |
| 5 | 4.7351 | 4.6777 | 4.6188 | 4.5581 | 4.5272 | 4.4957 | 4.4638 | 4.4314 | 4.3984 | 4.3650 |
| 6 | 4.0600 | 3.9999 | 3.9381 | 3.8742 | 3.8415 | 3.8082 | 3.7743 | 3.7398 | 3.7047 | 3.6688 |
| 7 | 3.6365 | 3.5747 | 3.5108 | 3.4445 | 3.4105 | 3.3758 | 3.3404 | 3.3043 | 3.2674 | 3.2298 |
| 8 | 3.3472 | 3.2840 | 3.2184 | 3.1503 | 3.1152 | 3.0794 | 3.0428 | 3.0053 | 2.9669 | 2.9276 |
| 9 | 3.1373 | 3.0729 | 3.0061 | 2.9365 | 2.9005 | 2.8637 | 2.8259 | 2.7872 | 2.7475 | 2.7067 |
| 10 | 2.9782 | 2.9130 | 2.8450 | 2.7740 | 2.7372 | 2.6996 | 2.6609 | 2.6211 | 2.5801 | 2.5379 |
| 11 | 2.8536 | 2.7876 | 2.7186 | 2.6464 | 2.6090 | 2.5705 | 2.5309 | 2.4901 | 2.4480 | 2.4045 |
| 12 | 2.7534 | 2.6866 | 2.6169 | 2.5436 | 2.5055 | 2.4663 | 2.4259 | 2.3842 | 2.3410 | 2.2962 |
| 13 | 2.6710 | 2.6037 | 2.5331 | 2.4589 | 2.4202 | 2.3803 | 2.3392 | 2.2966 | 2.2524 | 2.2064 |
| 14 | 2.6021 | 2.5342 | 2.4630 | 2.3879 | 2.3487 | 2.3082 | 2.2664 | 2.2230 | 2.1778 | 2.1307 |
| 15 | 2.5437 | 2.4753 | 2.4035 | 2.3275 | 2.2878 | 2.2468 | 2.2043 | 2.1601 | 2.1141 | 2.0658 |
| 16 | 2.4935 | 2.4247 | 2.3522 | 2.2756 | 2.2354 | 2.1938 | 2.1507 | 2.1058 | 2.0589 | 2.0096 |
| 17 | 2.4499 | 2.3807 | 2.3077 | 2.2304 | 2.1898 | 2.1477 | 2.1040 | 2.0584 | 2.0107 | 1.9604 |
| 18 | 2.4117 | 2.3421 | 2.2686 | 2.1906 | 2.1497 | 2.1071 | 2.0629 | 2.0166 | 1.9681 | 1.9168 |
| 19 | 2.3779 | 2.3080 | 2.2341 | 2.1555 | 2.1141 | 2.0712 | 2.0264 | 1.9796 | 1.9302 | 1.8780 |
| 20 | 2.3479 | 2.2776 | 2.2033 | 2.1242 | 2.0825 | 2.0391 | 1.9938 | 1.9464 | 1.8963 | 1.8432 |
| 21 | 2.3210 | 2.2504 | 2.1757 | 2.0960 | 2.0540 | 2.0102 | 1.9645 | 1.9165 | 1.8657 | 1.8117 |
| 22 | 2.2967 | 2.2258 | 2.1508 | 2.0707 | 2.0283 | 1.9842 | 1.9380 | 1.8895 | 1.8380 | 1.7831 |
| 23 | 2.2747 | 2.2036 | 2.1282 | 2.0476 | 2.0050 | 1.9605 | 1.9139 | 1.8649 | 1.8128 | 1.7570 |
| 24 | 2.2547 | 2.1834 | 2.1077 | 2.0267 | 1.9838 | 1.9390 | 1.8920 | 1.8424 | 1.7897 | 1.7331 |
| 25 | 2.2365 | 2.1649 | 2.0889 | 2.0075 | 1.9643 | 1.9192 | 1.8718 | 1.8217 | 1.7684 | 1.7110 |
| 26 | 2.2197 | 2.1479 | 2.0716 | 1.9898 | 1.9464 | 1.9010 | 1.8533 | 1.8027 | 1.7488 | 1.6906 |
| 27 | 2.2043 | 2.1323 | 2.0558 | 1.9736 | 1.9299 | 1.8842 | 1.8361 | 1.7851 | 1.7307 | 1.6717 |
| 28 | 2.1900 | 2.1179 | 2.0411 | 1.9586 | 1.9147 | 1.8687 | 1.8203 | 1.7689 | 1.7138 | 1.6541 |
| 29 | 2.1768 | 2.1045 | 2.0275 | 1.9446 | 1.9005 | 1.8543 | 1.8055 | 1.7537 | 1.6981 | 1.6377 |
| 30 | 2.1646 | 2.0921 | 2.0148 | 1.9317 | 1.8874 | 1.8409 | 1.7918 | 1.7396 | 1.6835 | 1.6223 |
| 40 | 2.0772 | 2.0035 | 1.9245 | 1.8389 | 1.7929 | 1.7444 | 1.6928 | 1.6373 | 1.5766 | 1.5089 |
| 60 | 1.9926 | 1.9174 | 1.8364 | 1.7480 | 1.7001 | 1.6491 | 1.5943 | 1.5343 | 1.4673 | 1.3893 |
| 120 | 1.9105 | 1.8337 | 1.7505 | 1.6587 | 1.6084 | 1.5543 | 1.4952 | 1.4290 | 1.3519 | 1.2539 |
| $\infty$ | 1.8307 | 1.7522 | 1.6664 | 1.5705 | 1.5173 | 1.4591 | 1.3940 | 1.3180 | 1.2214 | 1.0000 |

Source: Reproduced by permission of Professor E. S. Pearson from "Tables of Percentage Points of the Inverted Beta ($F$) Distribution," *Biometrika*, **33** (1943), pp. 73–88, by Maxine Merrington and Catherine M. Thompson.

**TABLE IV**  Percentage Points of the $F$ Distribution, Level of Significance: 1 %

| $v_2$ \ $v_1$ | 1 | 2 | 3 | 4 | 5 | 6 | 7 | 8 | 9 |
|---|---|---|---|---|---|---|---|---|---|
| 1 | 4052.2 | 4999.5 | 5403.3 | 5624.6 | 5763.7 | 5859.0 | 5928.3 | 5981.6 | 6022.5 |
| 2 | 98.503 | 99.000 | 99.166 | 99.249 | 99.299 | 99.332 | 99.356 | 99.374 | 99.388 |
| 3 | 34.116 | 30.817 | 29.457 | 28.710 | 28.237 | 27.911 | 27.672 | 27.489 | 27.345 |
| 4 | 21.198 | 18.000 | 16.694 | 15.977 | 15.522 | 15.207 | 14.976 | 14.799 | 14.659 |
| 5 | 16.258 | 13.274 | 12.060 | 11.392 | 10.967 | 10.672 | 10.456 | 10.289 | 10.158 |
| 6 | 13.745 | 10.925 | 9.7795 | 9.1483 | 8.7459 | 8.4661 | 8.2600 | 8.1016 | 7.9761 |
| 7 | 12.246 | 9.5466 | 8.4513 | 7.8467 | 7.4604 | 7.1914 | 6.9928 | 6.8401 | 6.7188 |
| 8 | 11.259 | 8.6491 | 7.5910 | 7.0060 | 6.6318 | 6.3707 | 6.1776 | 6.0289 | 5.9106 |
| 9 | 10.561 | 8.0215 | 6.9919 | 6.4221 | 6.0569 | 5.8018 | 5.6129 | 5.4671 | 5.3511 |
| 10 | 10.044 | 7.5594 | 6.5523 | 5.9943 | 5.6363 | 5.3858 | 5.2001 | 5.0567 | 4.9424 |
| 11 | 9.6460 | 7.2057 | 6.2167 | 5.6683 | 5.3160 | 5.0692 | 4.8861 | 4.7445 | 4.6315 |
| 12 | 9.3302 | 6.9266 | 5.9526 | 5.4119 | 5.0643 | 4.8206 | 4.6395 | 4.4994 | 4.3875 |
| 13 | 9.0738 | 6.7010 | 5.7394 | 5.2053 | 4.8616 | 4.6204 | 4.4410 | 4.3021 | 4.1911 |
| 14 | 8.8616 | 6.5149 | 5.5639 | 5.0354 | 4.6950 | 4.4558 | 4.2779 | 4.1399 | 4.0297 |
| 15 | 8.6831 | 6.3589 | 5.4170 | 4.8932 | 4.5556 | 4.3183 | 4.1415 | 4.0045 | 3.8948 |
| 16 | 8.5310 | 6.2262 | 5.2922 | 4.7726 | 4.4374 | 4.2016 | 4.0259 | 3.8896 | 3.7804 |
| 17 | 8.3997 | 6.1121 | 5.1850 | 4.6690 | 4.3359 | 4.1015 | 3.9267 | 3.7910 | 3.6822 |
| 18 | 8.2854 | 6.0129 | 5.0919 | 4.5790 | 4.2479 | 4.0146 | 3.8406 | 3.7054 | 3.5971 |
| 19 | 8.1850 | 5.9259 | 5.0103 | 4.5003 | 4.1708 | 3.9386 | 3.7653 | 3.6305 | 3.5225 |
| 20 | 8.0960 | 5.8489 | 4.9382 | 4.4307 | 4.1027 | 3.8714 | 3.6987 | 3.5644 | 3.4567 |
| 21 | 8.0166 | 5.7804 | 4.8740 | 4.3688 | 4.0421 | 3.8117 | 3.6396 | 3.5056 | 3.3981 |
| 22 | 7.9454 | 5.7190 | 4.8166 | 4.3134 | 3.9880 | 3.7583 | 3.5867 | 3.4530 | 3.3458 |
| 23 | 7.8811 | 5.6637 | 4.7649 | 4.2635 | 3.9392 | 3.7102 | 3.5390 | 3.4057 | 3.2986 |
| 24 | 7.8229 | 5.6136 | 4.7181 | 4.2184 | 3.8951 | 3.6667 | 3.4959 | 3.3629 | 3.2560 |
| 25 | 7.7698 | 5.5680 | 4.6755 | 4.1774 | 3.8550 | 3.6272 | 3.4568 | 3.3239 | 3.2172 |
| 26 | 7.7213 | 5.5263 | 4.6366 | 4.1400 | 3.8183 | 3.5911 | 3.4210 | 3.2884 | 3.1818 |
| 27 | 7.6767 | 5.4881 | 4.6009 | 4.1056 | 3.7848 | 3.5580 | 3.3882 | 3.2558 | 3.1494 |
| 28 | 7.6356 | 5.4529 | 4.5681 | 4.0740 | 3.7539 | 3.5276 | 3.3581 | 3.2259 | 3.1195 |
| 29 | 7.5976 | 5.4205 | 4.5378 | 4.0449 | 3.7254 | 3.4995 | 3.3302 | 3.1982 | 3.0920 |
| 30 | 7.5625 | 5.3904 | 4.5097 | 4.0179 | 3.6990 | 3.4735 | 3.3045 | 3.1726 | 3.0665 |
| 40 | 7.3141 | 5.1785 | 4.3126 | 3.8283 | 3.5138 | 3.2910 | 3.1238 | 2.9930 | 2.8876 |
| 60 | 7.0771 | 4.9774 | 4.1259 | 3.6491 | 3.3389 | 3.1187 | 2.9530 | 2.8233 | 2.7185 |
| 120 | 6.8510 | 4.7865 | 3.9493 | 3.4796 | 3.1735 | 2.9559 | 2.7918 | 2.6629 | 2.5586 |
| $\infty$ | 6.6349 | 4.6052 | 3.7816 | 3.3192 | 3.0173 | 2.8020 | 2.6393 | 2.5113 | 2.4073 |

**TABLE IV** (*continued*)

| $\nu_1$ / $\nu_2$ | 10 | 12 | 15 | 20 | 24 | 30 | 40 | 60 | 120 | $\infty$ |
|---|---|---|---|---|---|---|---|---|---|---|
| 1 | 6055.8 | 6106.3 | 6157.3 | 6208.7 | 6234.6 | 6260.7 | 6286.8 | 6313.0 | 6339.4 | 6366.0 |
| 2 | 99.399 | 99.416 | 99.432 | 99.449 | 99.458 | 99.466 | 99.474 | 99.483 | 99.491 | 99.501 |
| 3 | 27.229 | 27.052 | 26.872 | 26.690 | 26.598 | 26.505 | 26.411 | 26.316 | 26.221 | 26.125 |
| 4 | 14.546 | 14.374 | 14.198 | 14.020 | 13.929 | 13.838 | 13.745 | 13.652 | 13.558 | 13.463 |
| 5 | 10.051 | 9.8883 | 9.7222 | 9.5527 | 9.4665 | 9.3793 | 9.2912 | 9.2020 | 9.1118 | 9.0204 |
| 6 | 7.8741 | 7.7183 | 7.5590 | 7.3958 | 7.3127 | 7.2285 | 7.1432 | 7.0568 | 6.9690 | 6.8801 |
| 7 | 6.6201 | 6.4691 | 6.3143 | 6.1554 | 6.0743 | 5.9921 | 5.9084 | 5.8236 | 5.7372 | 5.6495 |
| 8 | 5.8143 | 5.6668 | 5.5151 | 5.3591 | 5.2793 | 5.1981 | 5.1156 | 5.0316 | 4.9460 | 4.8588 |
| 9 | 5.2565 | 5.1114 | 4.9621 | 4.8080 | 4.7290 | 4.6486 | 4.5667 | 4.4831 | 4.3978 | 4.3105 |
| 10 | 4.8492 | 4.7059 | 4.5582 | 4.4054 | 4.3269 | 4.2469 | 4.1653 | 4.0819 | 3.9965 | 3.9090 |
| 11 | 4.5393 | 4.3974 | 4.2509 | 4.0990 | 4.0209 | 3.9411 | 3.8596 | 3.7761 | 3.6904 | 3.6025 |
| 12 | 4.2961 | 4.1553 | 4.0096 | 3.8584 | 3.7805 | 3.7008 | 3.6192 | 3.5355 | 3.4494 | 3.3608 |
| 13 | 4.1003 | 3.9603 | 3.8154 | 3.6646 | 3.5868 | 3.5070 | 3.4253 | 3.3413 | 3.2548 | 3.1654 |
| 14 | 3.9394 | 3.8001 | 3.6557 | 3.5052 | 3.4274 | 3.3476 | 3.2656 | 3.1813 | 3.0942 | 3.0040 |
| 15 | 3.8049 | 3.6662 | 3.5222 | 3.3719 | 3.2940 | 3.2141 | 3.1319 | 3.0471 | 2.9595 | 2.8684 |
| 16 | 3.6909 | 3.5527 | 3.4089 | 3.2588 | 3.1808 | 3.1007 | 3.0182 | 2.9330 | 2.8447 | 2.7528 |
| 17 | 3.5931 | 3.4552 | 3.3117 | 3.1615 | 3.0835 | 3.0032 | 2.9205 | 2.8348 | 2.7459 | 2.6530 |
| 18 | 3.5082 | 3.3706 | 3.2273 | 3.0771 | 2.9990 | 2.9185 | 2.8354 | 2.7493 | 2.6597 | 2.5660 |
| 19 | 3.4338 | 3.2965 | 3.1533 | 3.0031 | 2.9249 | 2.8442 | 2.7608 | 2.6742 | 2.5839 | 2.4893 |
| 20 | 3.3682 | 3.2311 | 3.0880 | 2.9377 | 2.8594 | 2.7785 | 2.6947 | 2.6077 | 2.5168 | 2.4212 |
| 21 | 3.3098 | 3.1729 | 3.0299 | 2.8796 | 2.8011 | 2.7200 | 2.6359 | 2.5484 | 2.4568 | 2.3603 |
| 22 | 3.2576 | 3.1209 | 2.9780 | 2.8274 | 2.7488 | 2.6675 | 2.5831 | 2.4951 | 2.4029 | 2.3055 |
| 23 | 3.2106 | 3.0740 | 2.9311 | 2.7805 | 2.7017 | 2.6202 | 2.5355 | 2.4471 | 2.3542 | 2.2559 |
| 24 | 3.1681 | 3.0316 | 2.8887 | 2.7380 | 2.6591 | 2.5773 | 2.4923 | 2.4035 | 2.3099 | 2.2107 |
| 25 | 3.1294 | 2.9931 | 2.8502 | 2.6993 | 2.6203 | 2.5383 | 2.4530 | 2.3637 | 2.2695 | 2.1694 |
| 26 | 3.0941 | 2.9579 | 2.8150 | 2.6640 | 2.5848 | 2.5026 | 2.4170 | 2.3273 | 2.2325 | 2.1315 |
| 27 | 3.0618 | 2.9256 | 2.7827 | 2.6316 | 2.5522 | 2.4699 | 2.3840 | 2.2938 | 2.1984 | 2.0965 |
| 28 | 3.0320 | 2.8959 | 2.7530 | 2.6017 | 2.5223 | 2.4397 | 2.3535 | 2.2629 | 2.1670 | 2.0642 |
| 29 | 3.0045 | 2.8685 | 2.7256 | 2.5742 | 2.4946 | 2.4118 | 2.3253 | 2.2344 | 2.1378 | 2.0342 |
| 30 | 2.9791 | 2.8431 | 2.7002 | 2.5487 | 2.4689 | 2.3860 | 2.2992 | 2.2079 | 2.1107 | 2.0062 |
| 40 | 2.8005 | 2.6648 | 2.5216 | 2.3689 | 2.2880 | 2.2034 | 2.1142 | 2.0194 | 1.9172 | 1.8047 |
| 60 | 2.6318 | 2.4961 | 2.3523 | 2.1978 | 2.1154 | 2.0285 | 1.9360 | 1.8363 | 1.7263 | 1.6006 |
| 120 | 2.4721 | 2.3363 | 2.1915 | 2.0346 | 1.9500 | 1.8600 | 1.7628 | 1.6557 | 1.5330 | 1.3805 |
| $\infty$ | 2.3209 | 2.1848 | 2.0385 | 1.8783 | 1.7908 | 1.6964 | 1.5923 | 1.4730 | 1.3246 | 1.0000 |

Source: Reproduced by permission of Professor E. S. Pearson from "Tables of Percentage Points of the Inverted Beta ($F$) Distribution," *Biometrika*, 33 (1943), pp. 73–88, by Maxine Merrington and Catherine M. Thompson.

**TABLE V** Percentage Points of the $\chi^2$ Distribution.

| DF \ P | 0.995 | 0.990 | 0.975 | 0.950 | 0.900 | 0.750 |
|---|---|---|---|---|---|---|
| 1 | $392{,}704 \times 10^{-10}$ | $157{,}088 \times 10^{-9}$ | $982{,}069 \times 10^{-9}$ | $393{,}214 \times 10^{-8}$ | 0.0157908 | 0.1015308 |
| 2 | 0.0100251 | 0.0201007 | 0.0506356 | 0.102587 | 0.210720 | 0.575364 |
| 3 | 0.0717212 | 0.114832 | 0.215795 | 0.351846 | 0.584375 | 1.212534 |
| 4 | 0.206990 | 0.297110 | 0.484419 | 0.710721 | 1.063623 | 1.92255 |
| 5 | 0.411740 | 0.554300 | 0.831211 | 1.145476 | 1.61031 | 2.67460 |
| 6 | 0.675727 | 0.872085 | 1.237347 | 1.63539 | 2.20413 | 3.45460 |
| 7 | 0.989265 | 1.239043 | 1.68987 | 2.16735 | 2.83311 | 4.25485 |
| 8 | 1.344419 | 1.646482 | 2.17973 | 2.73264 | 3.48954 | 5.07064 |
| 9 | 1.734926 | 2.087912 | 2.70039 | 3.32511 | 4.16816 | 5.89883 |
| 10 | 2.15585 | 2.55821 | 3.24697 | 3.94030 | 4.86518 | 6.73720 |
| 11 | 2.60321 | 3.05347 | 3.81575 | 4.57481 | 5.57779 | 7.58412 |
| 12 | 3.07382 | 3.57056 | 4.40379 | 5.22603 | 6.30380 | 8.43842 |
| 13 | 3.56503 | 4.10691 | 5.00874 | 5.89186 | 7.04150 | 9.29906 |
| 14 | 4.07468 | 4.66043 | 5.62872 | 6.57063 | 7.78953 | 10.1653 |
| 15 | 4.60094 | 5.22935 | 6.26214 | 7.26094 | 8.54675 | 11.0365 |
| 16 | 5.14224 | 5.81221 | 6.90766 | 7.96164 | 9.31223 | 11.9122 |
| 17 | 5.69724 | 6.40776 | 7.56418 | 8.67176 | 10.0852 | 12.7919 |
| 18 | 6.26481 | 7.01491 | 8.23075 | 9.39046 | 10.8649 | 13.6753 |
| 19 | 6.84398 | 7.63273 | 8.90655 | 10.1170 | 11.6509 | 14.5620 |
| 20 | 7.43386 | 8.26040 | 9.59083 | 10.8508 | 12.4426 | 15.4518 |
| 21 | 8.03366 | 8.89720 | 10.28293 | 11.5913 | 13.2396 | 16.3444 |
| 22 | 8.64272 | 9.54249 | 10.9823 | 12.3380 | 14.0415 | 17.2396 |
| 23 | 9.26042 | 10.19567 | 11.6885 | 13.0905 | 14.8479 | 18.1373 |
| 24 | 9.88623 | 10.8564 | 12.4001 | 13.8484 | 15.6587 | 19.0372 |
| 25 | 10.5197 | 11.5240 | 13.1197 | 14.6114 | 16.4734 | 19.9393 |
| 26 | 11.1603 | 12.1981 | 13.8439 | 15.3791 | 17.2919 | 20.8434 |
| 27 | 11.8076 | 12.8786 | 14.5733 | 16.1513 | 18.1138 | 21.7494 |
| 28 | 12.4613 | 13.5648 | 15.3079 | 16.9279 | 18.9392 | 22.6572 |
| 29 | 13.1211 | 14.2565 | 16.0471 | 17.7083 | 19.7677 | 23.5666 |
| 30 | 13.7867 | 14.9535 | 16.7908 | 18.4926 | 20.5992 | 24.4776 |
| 40 | 20.7065 | 22.1643 | 24.4331 | 26.5093 | 29.0505 | 33.6603 |
| 50 | 27.9907 | 29.7067 | 32.3574 | 34.7642 | 37.6886 | 42.9421 |
| 60 | 35.5346 | 37.4848 | 40.4817 | 43.1879 | 46.4589 | 52.2938 |
| 70 | 43.2752 | 45.4418 | 48.7576 | 51.7393 | 55.3290 | 61.6983 |
| 80 | 51.1720 | 53.5400 | 57.1532 | 60.3915 | 64.2778 | 71.1445 |
| 90 | 59.1963 | 61.7541 | 65.6466 | 69.1260 | 73.2912 | 80.6247 |
| 100 | 67.3276 | 70.0648 | 74.2219 | 77.9295 | 82.3581 | 90.1332 |
| $t_\alpha$ | $-2.5758$ | $-2.3263$ | $-1.9600$ | $-1.6449$ | $-1.2816$ | $-0.6745$ |

**TABLE V** (*continued*)

| P<br>DF | 0.500 | 0.250 | 0.100 | 0.050 | 0.025 | 0.010 | 0.005 |
|---|---|---|---|---|---|---|---|
| 1 | 0.454937 | 1.32330 | 2.70554 | 3.84146 | 5.02389 | 6.63490 | 7.87944 |
| 2 | 1.38629 | 2.77259 | 4.60517 | 5.99147 | 7.37776 | 9.21034 | 10.5966 |
| 3 | 2.36597 | 4.10835 | 6.25139 | 7.81473 | 9.34840 | 11.3449 | 12.8381 |
| 4 | 3.35670 | 5.38527 | 7.77944 | 9.48773 | 11.1433 | 13.2767 | 14.8602 |
| 5 | 4.35146 | 6.62568 | 9.23635 | 11.0705 | 12.8325 | 15.0863 | 16.7496 |
| 6 | 5.34812 | 7.84080 | 10.6446 | 12.5916 | 14.4494 | 16.8119 | 18.5476 |
| 7 | 6.34581 | 9.03715 | 12.0170 | 14.0671 | 16.0128 | 18.4753 | 20.2777 |
| 8 | 7.34412 | 10.2188 | 13.3616 | 15.5073 | 17.5346 | 20.0902 | 21.9550 |
| 9 | 8.34283 | 11.3887 | 14.6837 | 16.9190 | 19.0228 | 21.6660 | 23.5893 |
| 10 | 9.34182 | 12.5489 | 15.9871 | 18.3070 | 20.4831 | 23.2093 | 25.1882 |
| 11 | 10.3410 | 13.7007 | 17.2750 | 19.6751 | 21.9200 | 24.7250 | 26.7569 |
| 12 | 11.3403 | 14.8454 | 18.5494 | 21.0261 | 23.3367 | 26.2170 | 28.2995 |
| 13 | 12.3398 | 15.9839 | 19.8119 | 22.3621 | 24.7356 | 27.6883 | 29.8194 |
| 14 | 13.3393 | 17.1170 | 21.0642 | 23.6848 | 26.1190 | 29.1413 | 31.3193 |
| 15 | 14.3389 | 18.2451 | 22.3072 | 24.9958 | 27.4884 | 30.5779 | 32.8013 |
| 16 | 15.3385 | 19.3688 | 23.5418 | 26.2962 | 28.8454 | 31.9999 | 34.2672 |
| 17 | 16.3381 | 20.4887 | 24.7690 | 27.5871 | 30.1910 | 33.4087 | 35.7185 |
| 18 | 17.3379 | 21.6049 | 25.9894 | 28.8693 | 31.5264 | 34.8053 | 37.1564 |
| 19 | 18.3376 | 22.7178 | 27.2036 | 30.1435 | 32.8523 | 36.1908 | 38.5822 |
| 20 | 19.3374 | 23.8277 | 28.4120 | 31.4104 | 34.1696 | 37.5662 | 39.9968 |
| 21 | 20.3372 | 24.9348 | 29.6151 | 32.6705 | 35.4789 | 38.9321 | 41.4010 |
| 22 | 21.3370 | 26.0393 | 30.8133 | 33.9244 | 36.7807 | 40.2894 | 42.7956 |
| 23 | 22.3369 | 27.1413 | 32.0069 | 35.1725 | 38.0757 | 41.6384 | 44.1813 |
| 24 | 23.3367 | 28.2412 | 33.1963 | 36.4151 | 39.3641 | 42.9798 | 45.5585 |
| 25 | 24.3366 | 29.3389 | 34.3816 | 37.6525 | 40.6465 | 44.3141 | 46.9278 |
| 26 | 25.3364 | 30.4345 | 35.5631 | 38.8852 | 41.9232 | 45.6417 | 48.2899 |
| 27 | 26.3363 | 31.5284 | 36.7412 | 40.1133 | 43.1944 | 46.9630 | 49.6449 |
| 28 | 27.3363 | 32.6205 | 37.9159 | 41.3372 | 44.4607 | 48.2782 | 50.9933 |
| 29 | 28.3362 | 33.7109 | 39.0875 | 42.5569 | 45.7222 | 49.5879 | 52.3356 |
| 30 | 29.3360 | 34.7998 | 40.2560 | 43.7729 | 46.9792 | 50.8922 | 53.6720 |
| 40 | 39.3354 | 45.6160 | 51.8050 | 55.7585 | 59.3417 | 63.6907 | 66.7659 |
| 50 | 49.3349 | 56.3336 | 63.1671 | 67.5048 | 71.4202 | 76.1539 | 79.4900 |
| 60 | 59.3347 | 66.9814 | 74.3970 | 79.0819 | 83.2976 | 88.3794 | 91.9517 |
| 70 | 69.3344 | 77.5766 | 85.5271 | 90.5312 | 95.0231 | 100.425 | 104.215 |
| 80 | 79.3343 | 88.1303 | 96.5782 | 101.879 | 106.629 | 112.329 | 116.321 |
| 90 | 89.3342 | 98.6499 | 107.565 | 113.145 | 118.136 | 124.116 | 128.299 |
| 100 | 99.3341 | 109.141 | 118.498 | 124.342 | 129.561 | 135.807 | 140.169 |
| $t_\alpha$ | 0.0000 | +0.6745 | +1.2816 | +1.6449 | +1.9600 | +2.3263 | +2.5758 |

Source: Reproduced by permission of Professor E. S. Pearson from "Tables of the Percentage Points of the $\chi^2$-Distribution," *Biometrika*, **32** (1941), pp. 188–189, by Catherine M. Thompson.

**TABLE VI**  Tolerance Limits for the Normal Distribution

$P$ denotes the probability that at least a fraction $\gamma$ of the normal distribution will be included between the tolerance limits $\bar{x} \pm sl$, where $\bar{x}$ and $s$ are the mean and the standard deviation of the $N$ observations and $DF = N - 1$.

| Degrees of freedom DF | Confidence coefficient $P = 0.90$ Proportion of distribution $\gamma$ | | | | Confidence coefficient $P = 0.95$ Proportion of distribution $\gamma$ | | | | Confidence coefficient $P = 0.99$ Proportion of distribution $\gamma$ | | | |
|---|---|---|---|---|---|---|---|---|---|---|---|---|
| | 0.90 | 0.95 | 0.99 | 0.999 | 0.90 | 0.95 | 0.99 | 0.999 | 0.90 | 0.95 | 0.99 | 0.999 |
| 4 | 3·51 | 4·18 | 5·49 | 7·02 | 4·29 | 5·11 | 6·72 | 8·58 | 6·64 | 7·92 | 10·40 | 13·29 |
| 5 | 3·14 | 3·74 | 4·92 | 6·28 | 3·72 | 4·44 | 5·83 | 7·45 | 5·35 | 6·38 | 8·38 | 10·71 |
| 6 | 2·91 | 3·47 | 4·55 | 5·82 | 3·38 | 4·02 | 5·29 | 6·75 | 4·62 | 5·51 | 7·24 | 9·25 |
| 7 | 2·75 | 3·27 | 4·30 | 5·50 | 3·14 | 3·74 | 4·92 | 6·28 | 4·15 | 4·95 | 6·51 | 8·31 |
| 8 | 2·63 | 3·13 | 4·12 | 5·26 | 2·97 | 3·54 | 4·65 | 5·94 | 3·83 | 4·56 | 5·98 | 7·66 |
| 9 | 2·54 | 3·02 | 3·97 | 5·08 | 2·84 | 3·39 | 4·45 | 5·69 | 3·59 | 4·27 | 5·62 | 7·17 |
| 10 | 2·47 | 2·94 | 3·86 | 4·93 | 2·74 | 3·26 | 4·29 | 5·48 | 3·40 | 4·05 | 5·32 | 6·80 |
| 12 | 2·36 | 2·81 | 3·69 | 4·72 | 2·59 | 3·08 | 4·05 | 5·18 | 3·13 | 3·73 | 4·90 | 6·26 |
| 14 | 2·28 | 2·72 | 3·57 | 4·56 | 2·49 | 2·96 | 3·89 | 4·96 | 2·95 | 3·52 | 4·61 | 5·89 |
| 16 | 2·22 | 2·65 | 3·48 | 4·44 | 2·40 | 2·86 | 3·76 | 4·80 | 2·81 | 3·35 | 4·40 | 5·62 |
| 18 | 2·17 | 2·59 | 3·40 | 4·35 | 2·34 | 2·79 | 3·66 | 4·68 | 2·70 | 3·22 | 4·23 | 5·41 |
| 20 | 2·14 | 2·54 | 3·34 | 4·27 | 2·29 | 2·72 | 3·58 | 4·57 | 2·62 | 3·12 | 4·10 | 5·24 |
| 25 | 2·07 | 2·46 | 3·23 | 4·13 | 2·19 | 2·61 | 3·43 | 4·39 | 2·47 | 2·94 | 3·87 | 4·94 |
| 30 | 2·02 | 2·40 | 3·16 | 4·04 | 2·13 | 2·54 | 3·33 | 4·26 | 2·37 | 2·82 | 3·71 | 4·74 |
| 40 | 1·95 | 2·33 | 3·06 | 3·91 | 2·05 | 2·44 | 3·20 | 4·09 | 2·24 | 2·67 | 3·50 | 4·47 |
| 50 | 1·91 | 2·28 | 3·00 | 3·83 | 1·99 | 2·37 | 3·12 | 3·99 | 2·16 | 2·57 | 3·38 | 4·31 |
| 60 | 1·89 | 2·25 | 2·95 | 3·77 | 1·96 | 2·33 | 3·06 | 3·91 | 2·10 | 2·50 | 3·29 | 4·20 |
| 70 | 1·86 | 2·22 | 2·92 | 3·73 | 1·93 | 2·30 | 3·02 | 3·85 | 2·06 | 2·45 | 3·22 | 4·11 |
| 80 | 1·85 | 2·20 | 2·89 | 3·69 | 1·91 | 2·27 | 2·98 | 3·81 | 2·02 | 2·41 | 3·17 | 4·05 |
| 90 | 1·83 | 2·18 | 2·87 | 3·67 | 1·89 | 2·25 | 2·96 | 3·78 | 2·00 | 2·38 | 3·13 | 3·99 |
| 100 | 1·82 | 2·17 | 2·85 | 3·64 | 1·87 | 2·23 | 2·93 | 3·75 | 1·98 | 2·35 | 3·09 | 3·95 |
| 200 | 1·76 | 2·10 | 2·76 | 3·53 | 1·80 | 2·14 | 2·82 | 3·60 | 1·87 | 2·22 | 2·92 | 3·73 |
| 300 | 1·74 | 2·07 | 2·73 | 3·48 | 1·77 | 2·11 | 2·77 | 3·54 | 1·82 | 2·17 | 2·85 | 3·64 |
| 400 | 1·73 | 2·06 | 2·70 | 3·45 | 1·75 | 2·08 | 2·74 | 3·50 | 1·79 | 2·14 | 2·81 | 3·59 |
| 500 | 1·72 | 2·05 | 2·69 | 3·43 | 1·74 | 2·07 | 2·72 | 3·48 | 1·78 | 2·12 | 2·78 | 3·55 |
| 600 | 1·71 | 2·04 | 2·68 | 3·42 | 1·73 | 2·06 | 2·71 | 3·46 | 1·76 | 2·10 | 2·76 | 3·53 |
| 800 | 1·70 | 2·03 | 2·66 | 3·40 | 1·72 | 2·05 | 2·69 | 3·43 | 1·75 | 2·08 | 2·74 | 3·50 |
| 1000 | 1·70 | 2·02 | 2·65 | 3·39 | 1·71 | 2·04 | 2·68 | 3·42 | 1·74 | 2·07 | 2·72 | 3·47 |
| ∞ | 1·64 | 1·96 | 2·58 | 3·29 | 1·64 | 1·96 | 2·58 | 3·29 | 1·64 | 1·96 | 2·58 | 3·29 |

Source: A. Hald, *Statistical Theory with Engineering Applications*, Wiley, 1952.  Table 11.10, p. 315.

```
95 97 03 44 96   49 87 13 01 69   56 19 27 30 31   16 95 62 74 55   62 27 80 55 00
78 07 89 48 22   97 68 14 69 88   21 52 73 88 28   08 27 09 48 86   34 88 92 17 36
24 08 18 32 54   17 10 47 62 05   13 45 48 09 65   88 39 09 21 48   12 46 88 81 22
02 62 56 96 61   58 40 50 70 94   07 38 72 82 85   68 21 50 32 70   50 99 54 26 39
55 85 71 69 31   00 03 29 96 17   55 28 81 32 56   58 65 37 17 49   62 61 84 05 49

96 13 18 60 70   62 09 04 94 05   03 75 77 32 37   13 74 76 59 41   60 49 08 17 32
33 73 96 86 33   43 15 91 12 62   49 52 62 12 27   85 64 63 67 13   72 58 85 85 79
37 10 24 50 84   63 32 25 04 92   56 86 34 21 10   46 28 41 11 27   65 01 80 44 11
83 63 98 19 87   48 93 24 60 58   14 20 24 31 90   06 21 43 33 17   01 06 38 29 32
31 28 23 11 97   92 99 45 16 23   88 32 80 13 98   26 36 83 38 31   40 86 82 06 47

01 55 37 09 08   81 17 93 34 68   48 50 08 41 90   60 26 43 21 86   50 58 48 36 73
38 62 21 77 17   19 83 36 01 24   85 14 14 31 77   37 05 31 81 51   18 19 62 50 12
42 94 20 19 46   97 74 57 42 36   41 15 88 09 02   71 08 54 31 04   44 33 74 61 78
02 94 74 35 47   44 49 19 83 14   64 35 71 96 78   39 82 22 51 72   72 82 37 27 37
98 93 56 53 31   43 87 88 26 07   51 33 57 32 67   94 12 59 72 70   40 60 24 24 88

04 69 96 20 56   34 82 13 40 44   35 73 12 04 47   89 03 15 13 38   57 23 90 10 64
46 17 32 80 88   50 94 76 71 00   66 80 63 10 19   43 71 14 39 67   18 68 83 99 86
03 03 53 25 33   48 56 38 29 97   72 26 86 66 59   45 72 22 50 57   85 61 96 34 51
51 93 74 83 62   36 47 66 48 33   77 88 04 09 60   16 52 91 29 36   82 02 81 08 41
12 57 43 34 86   58 28 03 45 04   17 56 69 15 72   92 11 29 73 60   32 39 72 92 29

53 84 34 35 10   80 58 23 33 53   41 31 36 65 19   36 95 29 39 82   36 44 83 29 54
55 38 24 04 52   34 62 02 75 71   31 79 80 57 90   83 04 81 87 06   39 94 97 18 98
44 18 54 27 28   95 03 69 23 00   22 27 93 81 82   27 62 41 14 88   15 55 08 23 49
70 40 80 38 20   65 19 35 17 50   53 98 92 85 83   17 17 27 36 58   28 81 79 50 86
04 73 99 66 15   80 49 90 75 98   81 37 70 45 42   88 08 71 02 61   36 66 22 05 45

35 42 90 60 81   67 17 29 99 11   27 47 34 75 24   58 34 16 99 76   47 96 74 27 60
33 09 18 41 03   44 38 21 48 83   47 02 60 64 06   02 79 41 48 98   70 99 85 86 05
70 73 27 11 03   32 39 41 95 31   48 92 07 52 74   04 60 86 62 54   74 82 96 12 31
52 54 07 60 09   11 28 74 68 44   65 54 62 25 13   90 38 80 51 09   63 66 08 69 63
76 73 21 52 45   47 21 92 11 66   09 43 96 59 57   34 35 27 05 22   59 22 31 16 52

82 51 57 91 88   66 47 82 53 13   27 84 57 65 23   27 64 73 11 55   49 12 71 94 95
35 81 86 44 22   67 86 41 16 79   76 11 42 25 48   69 34 95 67 54   37 70 36 35 96
71 15 85 43 23   14 87 31 12 31   46 97 62 75 56   98 94 05 25 97   77 35 25 77 47
68 51 40 99 65   59 34 38 61 88   74 60 22 09 94   69 36 04 99 84   57 92 33 78 45
55 33 10 60 67   39 62 16 35 78   84 93 94 14 10   56 11 50 64 11   13 78 75 52 94

43 56 50 63 55   60 09 73 81 27   36 83 55 12 43   83 05 88 57 87   96 61 34 01 50
49 60 92 82 67   31 06 48 13 20   59 96 66 97 84   72 73 16 83 42   31 61 35 11 32
00 01 71 82 97   15 55 79 08 53   56 52 49 15 44   16 86 30 72 18   54 09 89 88 36
39 18 47 54 33   55 91 02 90 29   81 74 34 98 16   95 03 65 94 61   87 12 02 44 75
50 74 52 41 39   70 85 23 26 19   13 27 95 69 73   54 97 86 65 16   23 66 41 51 77

08 33 98 06 13   75 58 12 37 63   00 43 89 06 97   92 67 83 52 14   40 24 35 63 04
90 78 83 08 33   13 86 55 41 35   26 91 94 70 72   10 46 68 93 19   63 53 12 78 48
55 77 40 85 45   71 63 09 79 12   53 67 21 77 12   01 16 30 72 76   47 79 27 39 59
32 37 88 40 16   08 22 32 48 18   85 65 10 25 54   18 45 24 66 79   89 75 88 70 59
36 14 97 98 50   80 62 48 58 87   03 56 33 46 00   23 79 76 36 62   77 78 78 29 17

82 01 52 21 90   27 95 27 62 34   99 79 54 37 02   69 99 83 70 68   07 39 06 50 24
89 87 92 23 49   06 96 93 21 97   59 73 66 29 74   39 30 89 05 10   02 57 59 89 34
37 67 24 02 62   68 25 66 66 24   48 12 13 94 93   47 54 64 03 40   66 52 45 98 42
44 24 63 02 09   81 52 96 73 00   20 67 70 19 65   80 69 01 80 47   36 75 39 66 96
60 61 42 61 47   50 35 25 28 26   66 25 62 99 76   45 73 32 96 07   64 46 87 73 66
```

401

**TABLE VII** (*continued*)

```
97 58 55 23 12   87 39 84 32 23   26 91 01 11 26   01 24 06 58 20   33 46 38 86 23
84 95 87 34 95   31 23 12 64 75   89 28 38 15 91   81 89 08 86 08   88 20 02 11 67
11 52 38 09 94   32 47 35 42 67   39 33 89 97 16   28 94 86 93 86   96 13 43 85 99
38 69 94 97 10   44 42 85 46 88   56 56 63 58 22   89 19 26 82 25   94 15 54 65 62
23 99 36 33 41   99 76 22 29 19   92 53 92 15 71   47 57 74 69 03   65 57 90 53 17

09 15 95 74 87   09 63 82 63 29   84 57 45 80 07   13 57 40 58 34   21 93 90 39 21
55 75 91 36 57   38 30 89 64 42   01 84 83 12 79   32 09 56 03 81   90 88 00 71 02
84 62 29 92 42   03 92 37 46 19   90 75 68 84 49   53 80 62 19 20   31 14 42 11 17
79 25 70 07 80   85 32 53 87 11   33 79 14 20 04   12 40 31 74 39   80 21 37 65 20
40 10 91 52 27   21 18 64 61 04   85 55 16 90 71   31 95 15 86 74   87 80 75 71 27

93 18 86 63 72   22 53 44 23 89   38 06 46 04 79   67 77 33 21 75   40 51 74 60 53
63 71 69 30 23   12 85 90 05 07   67 33 56 52 60   21 50 72 26 28   48 67 31 87 61
05 29 95 78 06   10 41 62 18 37   42 91 98 43 33   20 58 62 80 65   19 90 07 84 49
30 04 29 90 89   64 25 66 36 41   99 59 15 43 86   34 10 05 99 83   08 02 18 01 22
75 50 83 42 46   80 76 77 34 16   04 05 06 28 86   60 70 04 13 28   98 76 78 43 69

68 82 44 11 33   11 20 42 00 22   40 03 06 12 45   06 32 34 44 18   01 26 36 78 42
51 38 78 69 65   25 98 73 40 31   12 04 99 51 09   49 04 32 68 68   54 64 15 25 68
98 41 81 63 70   58 43 39 93 18   54 46 98 33 01   47 85 39 81 11   48 84 07 64 76
08 44 37 01 53   59 67 11 11 53   16 98 16 52 52   39 32 22 18 22   04 03 06 77 17
17 30 92 82 09   42 37 88 43 35   11 54 89 05 61   10 46 27 43 33   88 92 72 62 01

74 87 89 10 02   19 45 29 65 70   77 81 98 78 67   05 62 57 08 79   30 32 62 91 87
61 81 52 99 80   11 55 21 98 02   08 26 01 20 16   07 42 88 56 51   31 96 14 85 49
55 08 43 08 22   50 28 03 18 00   80 79 60 18 33   92 36 13 50 41   43 59 82 16 65
44 38 47 15 16   96 03 51 42 15   35 96 40 87 91   56 91 13 58 85   40 06 36 04 30
12 45 97 68 57   62 36 61 03 29   46 60 79 85 99   91 13 99 95 58   75 14 74 88 12

19 95 23 05 45   01 87 81 18 92   36 94 07 14 08   90 32 51 29 61   50 60 34 92 25
71 55 86 72 94   77 08 55 65 50   33 53 94 81 52   36 31 53 12 74   88 59 99 35 95
07 32 94 03 20   66 29 98 75 65   70 30 56 59 08   24 51 75 48 73   11 29 77 08 36
10 35 58 59 25   89 62 60 77 71   24 13 38 20 83   02 48 11 67 95   38 97 15 58 18
62 99 34 08 06   81 46 09 16 82   95 17 13 46 36   51 36 87 56 10   80 79 40 48 82

19 44 35 31 20   16 05 25 26 38   98 94 18 38 88   10 90 29 01 12   48 85 52 97 22
77 76 94 64 49   45 39 58 07 88   32 11 43 09 51   32 69 31 63 02   33 47 08 94 85
97 43 81 59 46   59 26 04 63 86   87 31 55 50 66   11 37 04 68 14   57 17 08 82 48
09 77 93 46 95   36 98 08 77 39   71 44 48 10 19   54 80 24 83 47   06 79 01 78 43
71 09 43 23 16   33 93 21 87 89   16 53 05 53 16   98 96 30 89 49   83 32 23 13 32

25 19 47 70 48   16 91 39 59 80   66 77 96 02 08   59 58 48 91 81   04 31 64 65 15
43 23 23 81 42   61 42 37 17 76   75 40 18 81 33   51 68 04 41 00   72 82 28 68 03
50 57 81 53 79   98 04 75 77 30   49 18 17 01 70   06 01 53 04 76   49 93 39 68 00
81 04 78 50 20   33 21 64 10 00   49 43 08 86 53   25 50 24 70 63   01 08 52 66 67
19 62 59 60 23   26 11 30 12 63   26 60 61 15 83   27 41 02 61 80   72 19 91 56 53

32 52 48 94 61   60 43 08 29 67   86 20 90 03 18   48 22 42 82 59   84 31 00 92 15
79 73 88 64 27   89 92 95 64 78   40 06 16 28 66   54 93 14 19 00   39 11 13 27 55
05 12 93 24 38   18 25 64 65 51   81 15 80 43 36   94 49 89 58 80   80 76 25 65 69
59 72 45 18 64   49 67 78 83 66   72 92 63 42 78   21 14 35 00 16   05 92 74 20 31
22 75 30 52 34   00 43 50 50 91   10 64 18 60 30   48 99 84 23 37   20 03 50 50 05

86 21 48 23 45   01 80 49 33 99   57 92 46 06 55   60 98 81 40 20   72 45 67 83 67
47 02 27 40 96   41 44 06 54 76   83 52 32 56 15   09 45 22 54 07   49 70 54 48 84
36 76 21 72 44   85 55 63 87 29   62 84 18 48 29   23 75 29 90 68   02 56 04 32 34
43 84 04 45 20   18 42 25 25 95   70 15 92 80 82   47 10 21 18 57   83 54 02 09 53
88 82 00 84 16   82 67 66 77 89   78 31 98 11 56   27 07 76 59 71   87 56 99 27 28
```

Source: A. Hald, *Statistical Tables and Formulas*, Wiley, 1952. Table XIX, pp. 9⁴ and 97.

# Index

A CATALOG OF SELECTED
# DOVER BOOKS
## IN SCIENCE AND MATHEMATICS

# A CATALOG OF SELECTED
# DOVER BOOKS
## IN SCIENCE AND MATHEMATICS

QUALITATIVE THEORY OF DIFFERENTIAL EQUATIONS, V.V. Nemytskii and V.V. Stepanov. Classic graduate-level text by two prominent Soviet mathematicians covers classical differential equations as well as topological dynamics and ergodic theory. Bibliographies. 523pp. 5⅜ × 8½. 65954-2 Pa. $10.95

MATRICES AND LINEAR ALGEBRA, Hans Schneider and George Phillip Barker. Basic textbook covers theory of matrices and its applications to systems of linear equations and related topics such as determinants, eigenvalues and differential equations. Numerous exercises. 432pp. 5⅜ × 8½. 66014-1 Pa. $9.95

QUANTUM THEORY, David Bohm. This advanced undergraduate-level text presents the quantum theory in terms of qualitative and imaginative concepts, followed by specific applications worked out in mathematical detail. Preface. Index. 655pp. 5⅜ × 8½. 65969-0 Pa. $13.95

ATOMIC PHYSICS (8th edition), Max Born. Nobel laureate's lucid treatment of kinetic theory of gases, elementary particles, nuclear atom, wave-corpuscles, atomic structure and spectral lines, much more. Over 40 appendices, bibliography. 495pp. 5⅜ × 8½. 65984-4 Pa. $11.95

ELECTRONIC STRUCTURE AND THE PROPERTIES OF SOLIDS: The Physics of the Chemical Bond, Walter A. Harrison. Innovative text offers basic understanding of the electronic structure of covalent and ionic solids, simple metals, transition metals and their compounds. Problems. 1980 edition. 582pp. 6⅛ × 9¼. 66021-4 Pa. $14.95

BOUNDARY VALUE PROBLEMS OF HEAT CONDUCTION, M. Necati Özisik. Systematic, comprehensive treatment of modern mathematical methods of solving problems in heat conduction and diffusion. Numerous examples and problems. Selected references. Appendices. 505pp. 5⅜ × 8½. 65990-9 Pa. $11.95

A SHORT HISTORY OF CHEMISTRY (3rd edition), J.R. Partington. Classic exposition explores origins of chemistry, alchemy, early medical chemistry, nature of atmosphere, theory of valency, laws and structure of atomic theory, much more. 428pp. 5⅜ × 8½. (Available in U.S. only) 65977-1 Pa. $10.95

A HISTORY OF ASTRONOMY, A. Pannekoek. Well-balanced, carefully reasoned study covers such topics as Ptolemaic theory, work of Copernicus, Kepler, Newton, Eddington's work on stars, much more. Illustrated. References. 521pp. 5⅜ × 8½. 65994-1 Pa. $11.95

PRINCIPLES OF METEOROLOGICAL ANALYSIS, Walter J. Saucier. Highly respected, abundantly illustrated classic reviews atmospheric variables, hydrostatics, static stability, various analyses (scalar, cross-section, isobaric, isentropic, more). For intermediate meteorology students. 454pp. 6⅛ × 9¼. 65979-8 Pa. $12.95

RELATIVITY, THERMODYNAMICS AND COSMOLOGY, Richard C. Tolman. Landmark study extends thermodynamics to special, general relativity; also applications of relativistic mechanics, thermodynamics to cosmological models. 501pp. 5⅜ × 8½. 65383-8 Pa. $12.95

APPLIED ANALYSIS, Cornelius Lanczos. Classic work on analysis and design of finite processes for approximating solution of analytical problems. Algebraic equations, matrices, harmonic analysis, quadrature methods, much more. 559pp. 5⅜ × 8½. 65656-X Pa. $12.95

SPECIAL RELATIVITY FOR PHYSICISTS, G. Stephenson and C.W. Kilmister. Concise elegant account for nonspecialists. Lorentz transformation, optical and dynamical applications, more. Bibliography. 108pp. 5⅜ × 8½. 65519-9 Pa. $4.95

INTRODUCTION TO ANALYSIS, Maxwell Rosenlicht. Unusually clear, accessible coverage of set theory, real number system, metric spaces, continuous functions, Riemann integration, multiple integrals, more. Wide range of problems. Undergraduate level. Bibliography. 254pp. 5⅜ × 8½. 65038-3 Pa. $7.95

INTRODUCTION TO QUANTUM MECHANICS With Applications to Chemistry, Linus Pauling & E. Bright Wilson, Jr. Classic undergraduate text by Nobel Prize winner applies quantum mechanics to chemical and physical problems. Numerous tables and figures enhance the text. Chapter bibliographies. Appendices. Index. 468pp. 5⅜ × 8½. 64871-0 Pa. $11.95

ASYMPTOTIC EXPANSIONS OF INTEGRALS, Norman Bleistein & Richard A. Handelsman. Best introduction to important field with applications in a variety of scientific disciplines. New preface. Problems. Diagrams. Tables. Bibliography. Index. 448pp. 5⅜ × 8½. 65082-0 Pa. $11.95

MATHEMATICS APPLIED TO CONTINUUM MECHANICS, Lee A. Segel. Analyzes models of fluid flow and solid deformation. For upper-level math, science and engineering students. 608pp. 5⅜ × 8½. 65369-2 Pa. $13.95

ELEMENTS OF REAL ANALYSIS, David A. Sprecher. Classic text covers fundamental concepts, real number system, point sets, functions of a real variable, Fourier series, much more. Over 500 exercises. 352pp. 5⅜ × 8½. 65385-4 Pa. $9.95

PHYSICAL PRINCIPLES OF THE QUANTUM THEORY, Werner Heisenberg. Nobel Laureate discusses quantum theory, uncertainty, wave mechanics, work of Dirac, Schroedinger, Compton, Wilson, Einstein, etc. 184pp. 5⅜ × 8½. 60113-7 Pa. $4.95

INTRODUCTORY REAL ANALYSIS, A.N. Kolmogorov, S.V. Fomin. Translated by Richard A. Silverman. Self-contained, evenly paced introduction to real and functional analysis. Some 350 problems. 403pp. 5⅜ × 8½. 61226-0 Pa. $9.95

PROBLEMS AND SOLUTIONS IN QUANTUM CHEMISTRY AND PHYSICS, Charles S. Johnson, Jr. and Lee G. Pedersen. Unusually varied problems, detailed solutions in coverage of quantum mechanics, wave mechanics, angular momentum, molecular spectroscopy, scattering theory, more. 280 problems plus 139 supplementary exercises. 430pp. 6½ × 9¼. 65236-X Pa. $11.95

ASYMPTOTIC METHODS IN ANALYSIS, N.G. de Bruijn. An inexpensive, comprehensive guide to asymptotic methods—the pioneering work that teaches by explaining worked examples in detail. Index. 224pp. 5⅜ × 8½.    64221-6 Pa. $6.95

OPTICAL RESONANCE AND TWO-LEVEL ATOMS, L. Allen and J.H. Eberly. Clear, comprehensive introduction to basic principles behind all quantum optical resonance phenomena. 53 illustrations. Preface. Index. 256pp. 5⅜ × 8½.
65533-4 Pa. $7.95

COMPLEX VARIABLES, Francis J. Flanigan. Unusual approach, delaying complex algebra till harmonic functions have been analyzed from real variable viewpoint. Includes problems with answers. 364pp. 5⅜ × 8½.    61388-7 Pa. $7.95

ATOMIC SPECTRA AND ATOMIC STRUCTURE, Gerhard Herzberg. One of best introductions; especially for specialist in other fields. Treatment is physical rather than mathematical. 80 illustrations. 257pp. 5⅜ × 8½.    60115-3 Pa. $5.95

APPLIED COMPLEX VARIABLES, John W. Dettman. Step-by-step coverage of fundamentals of analytic function theory—plus lucid exposition of five important applications: Potential Theory; Ordinary Differential Equations; Fourier Transforms; Laplace Transforms; Asymptotic Expansions. 66 figures. Exercises at chapter ends. 512pp. 5⅜ × 8½.    64670-X Pa. $10.95

ULTRASONIC ABSORPTION: An Introduction to the Theory of Sound Absorption and Dispersion in Gases, Liquids and Solids, A.B. Bhatia. Standard reference in the field provides a clear, systematically organized introductory review of fundamental concepts for advanced graduate students, research workers. Numerous diagrams. Bibliography. 440pp. 5⅜ × 8½.    64917-2 Pa. $11.95

UNBOUNDED LINEAR OPERATORS: Theory and Applications, Seymour Goldberg. Classic presents systematic treatment of the theory of unbounded linear operators in normed linear spaces with applications to differential equations. Bibliography. 199pp. 5⅜ × 8½.    64830-3 Pa. $7.95

LIGHT SCATTERING BY SMALL PARTICLES, H.C. van de Hulst. Comprehensive treatment including full range of useful approximation methods for researchers in chemistry, meteorology and astronomy. 44 illustrations. 470pp. 5⅜ × 8½.    64228-3 Pa. $10.95

CONFORMAL MAPPING ON RIEMANN SURFACES, Harvey Cohn. Lucid, insightful book presents ideal coverage of subject. 334 exercises make book perfect for self-study. 55 figures. 352pp. 5⅜ × 8¼.    64025-6 Pa. $8.95

OPTICKS, Sir Isaac Newton. Newton's own experiments with spectroscopy, colors, lenses, reflection, refraction, etc., in language the layman can follow. Foreword by Albert Einstein. 532pp. 5⅜ × 8½.    60205-2 Pa. $9.95

GENERALIZED INTEGRAL TRANSFORMATIONS, A.H. Zemanian. Graduate-level study of recent generalizations of the Laplace, Mellin, Hankel, K. Weierstrass, convolution and other simple transformations. Bibliography. 320pp. 5⅜ × 8½.    65375-7 Pa. $7.95

THE ELECTROMAGNETIC FIELD, Albert Shadowitz. Comprehensive undergraduate text covers basics of electric and magnetic fields, builds up to electromagnetic theory. Also related topics, including relativity. Over 900 problems. 768pp. 5⅜ × 8¼. 65660-8 Pa. $17.95

FOURIER SERIES, Georgi P. Tolstov. Translated by Richard A. Silverman. A valuable addition to the literature on the subject, moving clearly from subject to subject and theorem to theorem. 107 problems, answers. 336pp. 5⅜ × 8½. 63317-9 Pa. $7.95

THEORY OF ELECTROMAGNETIC WAVE PROPAGATION, Charles Herach Papas. Graduate-level study discusses the Maxwell field equations, radiation from wire antennas, the Doppler effect and more. xiii + 244pp. 5⅜ × 8½. 65678-0 Pa. $6.95

DISTRIBUTION THEORY AND TRANSFORM ANALYSIS: An Introduction to Generalized Functions, with Applications, A.H. Zemanian. Provides basics of distribution theory, describes generalized Fourier and Laplace transformations. Numerous problems. 384pp. 5⅜ × 8½. 65479-6 Pa. $9.95

THE PHYSICS OF WAVES, William C. Elmore and Mark A. Heald. Unique overview of classical wave theory. Acoustics, optics, electromagnetic radiation, more. Ideal as classroom text or for self-study. Problems. 477pp. 5⅜ × 8½. 64926-1 Pa. $11.95

CALCULUS OF VARIATIONS WITH APPLICATIONS, George M. Ewing. Applications-oriented introduction to variational theory develops insight and promotes understanding of specialized books, research papers. Suitable for advanced undergraduate/graduate students as primary, supplementary text. 352pp. 5⅜ × 8½. 64856-7 Pa. $8.95

A TREATISE ON ELECTRICITY AND MAGNETISM, James Clerk Maxwell. Important foundation work of modern physics. Brings to final form Maxwell's theory of electromagnetism and rigorously derives his general equations of field theory. 1,084pp. 5⅜ × 8½. 60636-8, 60637-6 Pa., Two-vol. set $19.90

AN INTRODUCTION TO THE CALCULUS OF VARIATIONS, Charles Fox. Graduate-level text covers variations of an integral, isoperimetrical problems, least action, special relativity, approximations, more. References. 279pp. 5⅜ × 8½. 65499-0 Pa. $7.95

HYDRODYNAMIC AND HYDROMAGNETIC STABILITY, S. Chandrasekhar. Lucid examination of the Rayleigh-Benard problem; clear coverage of the theory of instabilities causing convection. 704pp. 5⅜ × 8¼. 64071-X Pa. $14.95

CALCULUS OF VARIATIONS, Robert Weinstock. Basic introduction covering isoperimetric problems, theory of elasticity, quantum mechanics, electrostatics, etc. Exercises throughout. 326pp. 5⅜ × 8½. 63069-2 Pa. $7.95

DYNAMICS OF FLUIDS IN POROUS MEDIA, Jacob Bear. For advanced students of ground water hydrology, soil mechanics and physics, drainage and irrigation engineering and more. 335 illustrations. Exercises, with answers. 784pp. 6⅛ × 9¼. 65675-6 Pa. $19.95

NUMERICAL METHODS FOR SCIENTISTS AND ENGINEERS, Richard Hamming. Classic text stresses frequency approach in coverage of algorithms, polynomial approximation, Fourier approximation, exponential approximation, other topics. Revised and enlarged 2nd edition. 721pp. 5⅜ × 8½.

65241-6 Pa. $14.95

THEORETICAL SOLID STATE PHYSICS, Vol. I: Perfect Lattices in Equilibrium; Vol. II: Non-Equilibrium and Disorder, William Jones and Norman H. March. Monumental reference work covers fundamental theory of equilibrium properties of perfect crystalline solids, non-equilibrium properties, defects and disordered systems. Appendices. Problems. Preface. Diagrams. Index. Bibliography. Total of 1,301pp. 5⅜ × 8½. Two volumes. Vol. I 65015-4 Pa. $12.95

Vol. II 65016-2 Pa. $12.95

OPTIMIZATION THEORY WITH APPLICATIONS, Donald A. Pierre. Broad-spectrum approach to important topic. Classical theory of minima and maxima, calculus of variations, simplex technique and linear programming, more. Many problems, examples. 640pp. 5⅜ × 8½. 65205-X Pa. $13.95

THE MODERN THEORY OF SOLIDS, Frederick Seitz. First inexpensive edition of classic work on theory of ionic crystals, free-electron theory of metals and semiconductors, molecular binding, much more. 736pp. 5⅜ × 8½.

65482-6 Pa. $15.95

ESSAYS ON THE THEORY OF NUMBERS, Richard Dedekind. Two classic essays by great German mathematician: on the theory of irrational numbers; and on transfinite numbers and properties of natural numbers. 115pp. 5⅜ × 8½.

21010-3 Pa. $4.95

THE FUNCTIONS OF MATHEMATICAL PHYSICS, Harry Hochstadt. Comprehensive treatment of orthogonal polynomials, hypergeometric functions, Hill's equation, much more. Bibliography. Index. 322pp. 5⅜ × 8½. 65214-9 Pa. $9.95

NUMBER THEORY AND ITS HISTORY, Oystein Ore. Unusually clear, accessible introduction covers counting, properties of numbers, prime numbers, much more. Bibliography. 380pp. 5⅜ × 8½. 65620-9 Pa. $8.95

THE VARIATIONAL PRINCIPLES OF MECHANICS, Cornelius Lanczos. Graduate level coverage of calculus of variations, equations of motion, relativistic mechanics, more. First inexpensive paperbound edition of classic treatise. Index. Bibliography. 418pp. 5⅜ × 8½. 65067-7 Pa. $10.95

MATHEMATICAL TABLES AND FORMULAS, Robert D. Carmichael and Edwin R. Smith. Logarithms, sines, tangents, trig functions, powers, roots, reciprocals, exponential and hyperbolic functions, formulas and theorems. 269pp. 5⅜ × 8½. 60111-0 Pa. $5.95

THEORETICAL PHYSICS, Georg Joos, with Ira M. Freeman. Classic overview covers essential math, mechanics, electromagnetic theory, thermodynamics, quantum mechanics, nuclear physics, other topics. First paperback edition. xxiii + 885pp. 5⅜ × 8½. 65227-0 Pa. $18.95

HANDBOOK OF MATHEMATICAL FUNCTIONS WITH FORMULAS, GRAPHS, AND MATHEMATICAL TABLES, edited by Milton Abramowitz and Irene A. Stegun. Vast compendium: 29 sets of tables, some to as high as 20 places. 1,046pp. 8 × 10½. 61272-4 Pa. $22.95

MATHEMATICAL METHODS IN PHYSICS AND ENGINEERING, John W. Dettman. Algebraically based approach to vectors, mapping, diffraction, other topics in applied math. Also generalized functions, analytic function theory, more. Exercises. 448pp. 5⅜ × 8¼. 65649-7 Pa. $8.95

A SURVEY OF NUMERICAL MATHEMATICS, David M. Young and Robert Todd Gregory. Broad self-contained coverage of computer-oriented numerical algorithms for solving various types of mathematical problems in linear algebra, ordinary and partial, differential equations, much more. Exercises. Total of 1,248pp. 5⅜ × 8½. Two volumes. Vol. I 65691-8 Pa. $14.95
Vol. II 65692-6 Pa. $14.95

TENSOR ANALYSIS FOR PHYSICISTS, J.A. Schouten. Concise exposition of the mathematical basis of tensor analysis, integrated with well-chosen physical examples of the theory. Exercises. Index. Bibliography. 289pp. 5⅜ × 8½. 65582-2 Pa. $7.95

INTRODUCTION TO NUMERICAL ANALYSIS (2nd Edition), F.B. Hildebrand. Classic, fundamental treatment covers computation, approximation, interpolation, numerical differentiation and integration, other topics. 150 new problems. 669pp. 5⅜ × 8½. 65363-3 Pa. $14.95

INVESTIGATIONS ON THE THEORY OF THE BROWNIAN MOVEMENT, Albert Einstein. Five papers (1905–8) investigating dynamics of Brownian motion and evolving elementary theory. Notes by R. Fürth. 122pp. 5⅜ × 8½. 60304-0 Pa. $4.95

NUMERICAL METHODS FOR SCIENTISTS AND ENGINEERS, Richard Hamming. Classic text stresses frequency approach in coverage of algorithms, polynomial approximation, Fourier approximation, exponential approximation, other topics. Revised and enlarged 2nd edition. 721pp. 5⅜ × 8½. 65241-6 Pa. $14.95

AN INTRODUCTION TO STATISTICAL THERMODYNAMICS, Terrell L. Hill. Excellent basic text offers wide-ranging coverage of quantum statistical mechanics, systems of interacting molecules, quantum statistics, more. 523pp. 5⅜ × 8½. 65242-4 Pa. $11.95

ELEMENTARY DIFFERENTIAL EQUATIONS, William Ted Martin and Eric Reissner. Exceptionally clear, comprehensive introduction at undergraduate level. Nature and origin of differential equations, differential equations of first, second and higher orders. Picard's Theorem, much more. Problems with solutions. 331pp. 5⅜ × 8½. 65024-3 Pa. $8.95

STATISTICAL PHYSICS, Gregory H. Wannier. Classic text combines thermodynamics, statistical mechanics and kinetic theory in one unified presentation of thermal physics. Problems with solutions. Bibliography. 532pp. 5⅜ × 8½. 65401-X Pa. $11.95

ORDINARY DIFFERENTIAL EQUATIONS, Morris Tenenbaum and Harry Pollard. Exhaustive survey of ordinary differential equations for undergraduates in mathematics, engineering, science. Thorough analysis of theorems. Diagrams. Bibliography. Index. 818pp. 5⅜ × 8½. 64940-7 Pa. $16.95

STATISTICAL MECHANICS: Principles and Applications, Terrell L. Hill. Standard text covers fundamentals of statistical mechanics, applications to fluctuation theory, imperfect gases, distribution functions, more. 448pp. 5⅜ × 8½. 65390-0 Pa. $9.95

ORDINARY DIFFERENTIAL EQUATIONS AND STABILITY THEORY: An Introduction, David A. Sánchez. Brief, modern treatment. Linear equation, stability theory for autonomous and nonautonomous systems, etc. 164pp. 5⅜ × 8¼. 63828-6 Pa. $5.95

THIRTY YEARS THAT SHOOK PHYSICS: The Story of Quantum Theory, George Gamow. Lucid, accessible introduction to influential theory of energy and matter. Careful explanations of Dirac's anti-particles, Bohr's model of the atom, much more. 12 plates. Numerous drawings. 240pp. 5⅜ × 8½. 24895-X Pa. $5.95

THEORY OF MATRICES, Sam Perlis. Outstanding text covering rank, non-singularity and inverses in connection with the development of canonical matrices under the relation of equivalence, and without the intervention of determinants. Includes exercises. 237pp. 5⅜ × 8½. 66810-X Pa. $7.95

GREAT EXPERIMENTS IN PHYSICS: Firsthand Accounts from Galileo to Einstein, edited by Morris H. Shamos. 25 crucial discoveries: Newton's laws of motion, Chadwick's study of the neutron, Hertz on electromagnetic waves, more. Original accounts clearly annotated. 370pp. 5⅜ × 8½. 25346-5 Pa. $9.95

INTRODUCTION TO PARTIAL DIFFERENTIAL EQUATIONS WITH AP-PLICATIONS, E.C. Zachmanoglou and Dale W. Thoe. Essentials of partial differential equations applied to common problems in engineering and the physical sciences. Problems and answers. 416pp. 5⅜ × 8½. 65251-3 Pa. $10.95

BURNHAM'S CELESTIAL HANDBOOK, Robert Burnham, Jr. Thorough guide to the stars beyond our solar system. Exhaustive treatment. Alphabetical by constellation: Andromeda to Cetus in Vol. 1; Chamaeleon to Orion in Vol. 2; and Pavo to Vulpecula in Vol. 3. Hundreds of illustrations. Index in Vol. 3. 2,000pp. 6⅛ × 9¼. 23567-X, 23568-8, 23673-0 Pa., Three-vol. set $41.85

ASYMPTOTIC EXPANSIONS FOR ORDINARY DIFFERENTIAL EQUA-TIONS, Wolfgang Wasow. Outstanding text covers asymptotic power series, Jordan's canonical form, turning point problems, singular perturbations, much more. Problems. 384pp. 5⅜ × 8½. 65456-7 Pa. $9.95

AMATEUR ASTRONOMER'S HANDBOOK, J.B. Sidgwick. Timeless, comprehensive coverage of telescopes, mirrors, lenses, mountings, telescope drives, micrometers, spectroscopes, more. 189 illustrations. 576pp. 5⅜ × 8¼. (USO) 24034-7 Pa. $9.95

SPECIAL FUNCTIONS, N.N. Lebedev. Translated by Richard Silverman. Famous Russian work treating more important special functions, with applications to specific problems of physics and engineering. 38 figures. 308pp. 5⅜ × 8½.
60624-4 Pa. $7.95

OBSERVATIONAL ASTRONOMY FOR AMATEURS, J.B. Sidgwick. Mine of useful data for observation of sun, moon, planets, asteroids, aurorae, meteors, comets, variables, binaries, etc. 39 illustrations. 384pp. 5⅜ × 8¼. (Available in U.S. only)
24033-9 Pa. $8.95

INTEGRAL EQUATIONS, F.G. Tricomi. Authoritative, well-written treatment of extremely useful mathematical tool with wide applications. Volterra Equations, Fredholm Equations, much more. Advanced undergraduate to graduate level. Exercises. Bibliography. 238pp. 5⅜ × 8½.
64828-1 Pa. $6.95

CELESTIAL OBJECTS FOR COMMON TELESCOPES, T.W. Webb. Inestimable aid for locating and identifying nearly 4,000 celestial objects. 77 illustrations. 645pp. 5⅜ × 8½.
20917-2, 20918-0 Pa., Two-vol. set $12.00

MODERN NONLINEAR EQUATIONS, Thomas L. Saaty. Emphasizes practical solution of problems; covers seven types of equations. ". . . a welcome contribution to the existing literature. . . ."—*Math Reviews.* 490pp. 5⅜ × 8½. 64232-1 Pa. $9.95

FUNDAMENTALS OF ASTRODYNAMICS, Roger Bate et al. Modern approach developed by U.S. Air Force Academy. Designed as a first course. Problems, exercises. Numerous illustrations. 455pp. 5⅜ × 8½.
60061-0 Pa. $8.95

INTRODUCTION TO LINEAR ALGEBRA AND DIFFERENTIAL EQUATIONS, John W. Dettman. Excellent text covers complex numbers, determinants, orthonormal bases, Laplace transforms, much more. Exercises with solutions. Undergraduate level. 416pp. 5⅜ × 8½.
65191-6 Pa. $9.95

INCOMPRESSIBLE AERODYNAMICS, edited by Bryan Thwaites. Covers theoretical and experimental treatment of the uniform flow of air and viscous fluids past two-dimensional aerofoils and three-dimensional wings; many other topics. 654pp. 5⅜ × 8½.
65465-6 Pa. $16.95

INTRODUCTION TO DIFFERENCE EQUATIONS, Samuel Goldberg. Exceptionally clear exposition of important discipline with applications to sociology, psychology, economics. Many illustrative examples; over 250 problems. 260pp. 5⅜ × 8½.
65084-7 Pa. $7.95

LAMINAR BOUNDARY LAYERS, edited by L. Rosenhead. Engineering classic covers steady boundary layers in two- and three-dimensional flow, unsteady boundary layers, stability, observational techniques, much more. 708pp. 5⅜ × 8½.
65646-2 Pa. $15.95

LECTURES ON CLASSICAL DIFFERENTIAL GEOMETRY, Second Edition, Dirk J. Struik. Excellent brief introduction covers curves, theory of surfaces, fundamental equations, geometry on a surface, conformal mapping, other topics. Problems. 240pp. 5⅜ × 8½.
65609-8 Pa. $6.95

ROTARY-WING AERODYNAMICS, W.Z. Stepniewski. Clear, concise text covers aerodynamic phenomena of the rotor and offers guidelines for helicopter performance evaluation. Originally prepared for NASA. 537 figures. 640pp. 6⅛ × 9¼.
64647-5 Pa. $14.95

DIFFERENTIAL GEOMETRY, Heinrich W. Guggenheimer. Local differential geometry as an application of advanced calculus and linear algebra. Curvature, transformation groups, surfaces, more. Exercises. 62 figures. 378pp. 5⅜ × 8½.
63433-7 Pa. $7.95

INTRODUCTION TO SPACE DYNAMICS, William Tyrrell Thomson. Comprehensive, classic introduction to space-flight engineering for advanced undergraduate and graduate students. Includes vector algebra, kinematics, transformation of coordinates. Bibliography. Index. 352pp. 5⅜ × 8½.     65113-4 Pa. $8.95

A SURVEY OF MINIMAL SURFACES, Robert Osserman. Up-to-date, in-depth discussion of the field for advanced students. Corrected and enlarged edition covers new developments. Includes numerous problems. 192pp. 5⅜ × 8½.
64998-9 Pa. $8.95

ANALYTICAL MECHANICS OF GEARS, Earle Buckingham. Indispensable reference for modern gear manufacture covers conjugate gear-tooth action, gear-tooth profiles of various gears, many other topics. 263 figures. 102 tables. 546pp. 5⅜ × 8½.                                                    65712-4 Pa. $11.95

SET THEORY AND LOGIC, Robert R. Stoll. Lucid introduction to unified theory of mathematical concepts. Set theory and logic seen as tools for conceptual understanding of real number system. 496pp. 5⅜ × 8¼.     63829-4 Pa. $10.95

A HISTORY OF MECHANICS, René Dugas. Monumental study of mechanical principles from antiquity to quantum mechanics. Contributions of ancient Greeks, Galileo, Leonardo, Kepler, Lagrange, many others. 671pp. 5⅜ × 8½.
65632-2 Pa. $14.95

FAMOUS PROBLEMS OF GEOMETRY AND HOW TO SOLVE THEM, Benjamin Bold. Squaring the circle, trisecting the angle, duplicating the cube: learn their history, why they are impossible to solve, then solve them yourself. 128pp. 5⅜ × 8½.                                                    24297-8 Pa. $3.95

MECHANICAL VIBRATIONS, J.P. Den Hartog. Classic textbook offers lucid explanations and illustrative models, applying theories of vibrations to a variety of practical industrial engineering problems. Numerous figures. 233 problems, solutions. Appendix. Index. Preface. 436pp. 5⅜ × 8½.     64785-4 Pa. $9.95

CURVATURE AND HOMOLOGY, Samuel I. Goldberg. Thorough treatment of specialized branch of differential geometry. Covers Riemannian manifolds, topology of differentiable manifolds, compact Lie groups, other topics. Exercises. 315pp. 5⅜ × 8½.                                                    64314-X Pa. $8.95

HISTORY OF STRENGTH OF MATERIALS, Stephen P. Timoshenko. Excellent historical survey of the strength of materials with many references to the theories of elasticity and structure. 245 figures. 452pp. 5⅜ × 8½. 61187-6 Pa. $10.95

GEOMETRY OF COMPLEX NUMBERS, Hans Schwerdtfeger. Illuminating, widely praised book on analytic geometry of circles, the Moebius transformation, and two-dimensional non-Euclidean geometries. 200pp. 5⅜ × 8¼.
63830-8 Pa. $6.95

MECHANICS, J.P. Den Hartog. A classic introductory text or refresher. Hundreds of applications and design problems illuminate fundamentals of trusses, loaded beams and cables, etc. 334 answered problems. 462pp. 5⅜ × 8½. 60754-2 Pa. $8.95

TOPOLOGY, John G. Hocking and Gail S. Young. Superb one-year course in classical topology. Topological spaces and functions, point-set topology, much more. Examples and problems. Bibliography. Index. 384pp. 5⅜ × 8¼.
65676-4 Pa. $8.95

STRENGTH OF MATERIALS, J.P. Den Hartog. Full, clear treatment of basic material (tension, torsion, bending, etc.) plus advanced material on engineering methods, applications. 350 answered problems. 323pp. 5⅜ × 8½. 60755-0 Pa. $7.50

ELEMENTARY CONCEPTS OF TOPOLOGY, Paul Alexandroff. Elegant, intuitive approach to topology from set-theoretic topology to Betti groups; how concepts of topology are useful in math and physics. 25 figures. 57pp. 5⅜ × 8¼.
60747-X Pa. $2.95

ADVANCED STRENGTH OF MATERIALS, J.P. Den Hartog. Superbly written advanced text covers torsion, rotating disks, membrane stresses in shells, much more. Many problems and answers. 388pp. 5⅜ × 8½. 65407-9 Pa. $9.95

COMPUTABILITY AND UNSOLVABILITY, Martin Davis. Classic graduate-level introduction to theory of computability, usually referred to as theory of recurrent functions. New preface and appendix. 288pp. 5⅜ × 8½. 61471-9 Pa. $6.95

GENERAL CHEMISTRY, Linus Pauling. Revised 3rd edition of classic first-year text by Nobel laureate. Atomic and molecular structure, quantum mechanics, statistical mechanics, thermodynamics correlated with descriptive chemistry. Problems. 992pp. 5⅜ × 8½. 65622-5 Pa. $19.95

AN INTRODUCTION TO MATRICES, SETS AND GROUPS FOR SCIENCE STUDENTS, G. Stephenson. Concise, readable text introduces sets, groups, and most importantly, matrices to undergraduate students of physics, chemistry, and engineering. Problems. 164pp. 5⅜ × 8½. 65077-4 Pa. $6.95

THE HISTORICAL BACKGROUND OF CHEMISTRY, Henry M. Leicester. Evolution of ideas, not individual biography. Concentrates on formulation of a coherent set of chemical laws. 260pp. 5⅜ × 8½. 61053-5 Pa. $6.95

THE PHILOSOPHY OF MATHEMATICS: An Introductory Essay, Stephan Körner. Surveys the views of Plato, Aristotle, Leibniz & Kant concerning propositions and theories of applied and pure mathematics. Introduction. Two appendices. Index. 198pp. 5⅜ × 8½. 25048-2 Pa. $6.95

THE DEVELOPMENT OF MODERN CHEMISTRY, Aaron J. Ihde. Authoritative history of chemistry from ancient Greek theory to 20th-century innovation. Covers major chemists and their discoveries. 209 illustrations. 14 tables. Bibliographies. Indices. Appendices. 851pp. 5⅜ × 8½. 64235-6 Pa. $17.95

DE RE METALLICA, Georgius Agricola. The famous Hoover translation of greatest treatise on technological chemistry, engineering, geology, mining of early modern times (1556). All 289 original woodcuts. 638pp. 6¾ × 11.
60006-8 Pa. $17.95

SOME THEORY OF SAMPLING, William Edwards Deming. Analysis of the problems, theory and design of sampling techniques for social scientists, industrial managers and others who find statistics increasingly important in their work. 61 tables. 90 figures. xvii + 602pp. 5⅜ × 8½. 64684-X Pa. $15.95

THE VARIOUS AND INGENIOUS MACHINES OF AGOSTINO RAMELLI: A Classic Sixteenth-Century Illustrated Treatise on Technology, Agostino Ramelli. One of the most widely known and copied works on machinery in the 16th century. 194 detailed plates of water pumps, grain mills, cranes, more. 608pp. 9 × 12. (EBE)
25497-6 Clothbd. $34.95

LINEAR PROGRAMMING AND ECONOMIC ANALYSIS, Robert Dorfman, Paul A. Samuelson and Robert M. Solow. First comprehensive treatment of linear programming in standard economic analysis. Game theory, modern welfare economics, Leontief input-output, more. 525pp. 5⅜ × 8½. 65491-5 Pa. $13.95

ELEMENTARY DECISION THEORY, Herman Chernoff and Lincoln E. Moses. Clear introduction to statistics and statistical theory covers data processing, probability and random variables, testing hypotheses, much more. Exercises. 364pp. 5⅜ × 8½. 65218-1 Pa. $9.95

THE COMPLEAT STRATEGYST: Being a Primer on the Theory of Games of Strategy, J.D. Williams. Highly entertaining classic describes, with many illustrated examples, how to select best strategies in conflict situations. Prefaces. Appendices. 268pp. 5⅜ × 8½. 25101-2 Pa. $6.95

MATHEMATICAL METHODS OF OPERATIONS RESEARCH, Thomas L. Saaty. Classic graduate-level text covers historical background, classical methods of forming models, optimization, game theory, probability, queueing theory, much more. Exercises. Bibliography. 448pp. 5⅜ × 8¼. 65703-5 Pa. $12.95

CONSTRUCTIONS AND COMBINATORIAL PROBLEMS IN DESIGN OF EXPERIMENTS, Damaraju Raghavarao. In-depth reference work examines orthogonal Latin squares, incomplete block designs, tactical configuration, partial geometry, much more. Abundant explanations, examples. 416pp. 5⅜ × 8¼.
65685-3 Pa. $10.95

THE ABSOLUTE DIFFERENTIAL CALCULUS (CALCULUS OF TENSORS), Tullio Levi-Civita. Great 20th-century mathematician's classic work on material necessary for mathematical grasp of theory of relativity. 452pp. 5⅜ × 8½.
63401-9 Pa. $9.95

VECTOR AND TENSOR ANALYSIS WITH APPLICATIONS, A.I. Borisenko and I.E. Tarapov. Concise introduction. Worked-out problems, solutions, exercises. 257pp. 5⅜ × 8¼. 63833-2 Pa. $6.95

THE FOUR-COLOR PROBLEM: Assaults and Conquest, Thomas L. Saaty and Paul G. Kainen. Engrossing, comprehensive account of the century-old combinatorial topological problem, its history and solution. Bibliographies. Index. 110 figures. 228pp. 5⅜ × 8½. 65092-8 Pa. $6.95

CATALYSIS IN CHEMISTRY AND ENZYMOLOGY, William P. Jencks. Exceptionally clear coverage of mechanisms for catalysis, forces in aqueous solution, carbonyl- and acyl-group reactions, practical kinetics, more. 864pp. 5⅜ × 8½. 65460-5 Pa. $19.95

PROBABILITY: An Introduction, Samuel Goldberg. Excellent basic text covers set theory, probability theory for finite sample spaces, binomial theorem, much more. 360 problems. Bibliographies. 322pp. 5⅜ × 8½. 65252-1 Pa. $8.95

LIGHTNING, Martin A. Uman. Revised, updated edition of classic work on the physics of lightning. Phenomena, terminology, measurement, photography, spectroscopy, thunder, more. Reviews recent research. Bibliography. Indices. 320pp. 5⅜ × 8¼. 64575-4 Pa. $8.95

PROBABILITY THEORY: A Concise Course, Y.A. Rozanov. Highly readable, self-contained introduction covers combination of events, dependent events, Bernoulli trials, etc. Translation by Richard Silverman. 148pp. 5⅜ × 8¼. 63544-9 Pa. $5.95

THE CEASELESS WIND: An Introduction to the Theory of Atmospheric Motion, John A. Dutton. Acclaimed text integrates disciplines of mathematics and physics for full understanding of dynamics of atmospheric motion. Over 400 problems. Index. 97 illustrations. 640pp. 6 × 9. 65096-0 Pa. $17.95

STATISTICS MANUAL, Edwin L. Crow, et al. Comprehensive, practical collection of classical and modern methods prepared by U.S. Naval Ordnance Test Station. Stress on use. Basics of statistics assumed. 288pp. 5⅜ × 8½. 60599-X Pa. $6.95

DICTIONARY/OUTLINE OF BASIC STATISTICS, John E. Freund and Frank J. Williams. A clear concise dictionary of over 1,000 statistical terms and an outline of statistical formulas covering probability, nonparametric tests, much more. 208pp. 5⅜ × 8½. 66796-0 Pa. $6.95

STATISTICAL METHOD FROM THE VIEWPOINT OF QUALITY CONTROL, Walter A. Shewhart. Important text explains regulation of variables, uses of statistical control to achieve quality control in industry, agriculture, other areas. 192pp. 5⅜ × 8½. 65232-7 Pa. $6.95

THE INTERPRETATION OF GEOLOGICAL PHASE DIAGRAMS, Ernest G. Ehlers. Clear, concise text emphasizes diagrams of systems under fluid or containing pressure; also coverage of complex binary systems, hydrothermal melting, more. 288pp. 6½ × 9¼. 65389-7 Pa. $10.95

STATISTICAL ADJUSTMENT OF DATA, W. Edwards Deming. Introduction to basic concepts of statistics, curve fitting, least squares solution, conditions without parameter, conditions containing parameters. 26 exercises worked out. 271pp. 5⅜ × 8½. 64685-8 Pa. $7.95

TENSOR CALCULUS, J.L. Synge and A. Schild. Widely used introductory text covers spaces and tensors, basic operations in Riemannian space, non-Riemannian spaces, etc. 324pp. 5⅜ × 8¼. 63612-7 Pa. $7.95

A CONCISE HISTORY OF MATHEMATICS, Dirk J. Struik. The best brief history of mathematics. Stresses origins and covers every major figure from ancient Near East to 19th century. 41 illustrations. 195pp. 5⅜ × 8½. 60255-9 Pa. $7.95

A SHORT ACCOUNT OF THE HISTORY OF MATHEMATICS, W.W. Rouse Ball. One of clearest, most authoritative surveys from the Egyptians and Phoenicians through 19th-century figures such as Grassman, Galois, Riemann. Fourth edition. 522pp. 5⅜ × 8½. 20630-0 Pa. $10.95

HISTORY OF MATHEMATICS, David E. Smith. Nontechnical survey from ancient Greece and Orient to late 19th century; evolution of arithmetic, geometry, trigonometry, calculating devices, algebra, the calculus. 362 illustrations. 1,355pp. 5⅜ × 8½. 20429-4, 20430-8 Pa., Two-vol. set $23.90

THE GEOMETRY OF RENÉ DESCARTES, René Descartes. The great work founded analytical geometry. Original French text, Descartes' own diagrams, together with definitive Smith-Latham translation. 244pp. 5⅜ × 8½.
60068-8 Pa. $6.95

THE ORIGINS OF THE INFINITESIMAL CALCULUS, Margaret E. Baron. Only fully detailed and documented account of crucial discipline: origins; development by Galileo, Kepler, Cavalieri; contributions of Newton, Leibniz, more. 304pp. 5⅜ × 8½. (Available in U.S. and Canada only) 65371-4 Pa. $9.95

THE HISTORY OF THE CALCULUS AND ITS CONCEPTUAL DEVELOPMENT, Carl B. Boyer. Origins in antiquity, medieval contributions, work of Newton, Leibniz, rigorous formulation. Treatment is verbal. 346pp. 5⅜ × 8½.
60509-4 Pa. $7.95

THE THIRTEEN BOOKS OF EUCLID'S ELEMENTS, translated with introduction and commentary by Sir Thomas L. Heath. Definitive edition. Textual and linguistic notes, mathematical analysis. 2,500 years of critical commentary. Not abridged. 1,414pp. 5⅜ × 8½. 60088-2, 60089-0, 60090-4 Pa., Three-vol. set $29.85

GAMES AND DECISIONS: Introduction and Critical Survey, R. Duncan Luce and Howard Raiffa. Superb nontechnical introduction to game theory, primarily applied to social sciences. Utility theory, zero-sum games, n-person games, decision-making, much more. Bibliography. 509pp. 5⅜ × 8½. 65943-7 Pa. $11.95

THE HISTORICAL ROOTS OF ELEMENTARY MATHEMATICS, Lucas N.H. Bunt, Phillip S. Jones, and Jack D. Bedient. Fundamental underpinnings of modern arithmetic, algebra, geometry and number systems derived from ancient civilizations. 320pp. 5⅜ × 8½. 25563-8 Pa. $8.95

CALCULUS REFRESHER FOR TECHNICAL PEOPLE, A. Albert Klaf. Covers important aspects of integral and differential calculus via 756 questions. 566 problems, most answered. 431pp. 5⅜ × 8½. 20370-0 Pa. $8.95

**CHALLENGING MATHEMATICAL PROBLEMS WITH ELEMENTARY SOLUTIONS, A.M. Yaglom and I.M. Yaglom.** Over 170 challenging problems on probability theory, combinatorial analysis, points and lines, topology, convex polygons, many other topics. Solutions. Total of 445pp. 5⅜ × 8½. Two-vol. set.

Vol. I 65536-9 Pa. $6.95
Vol. II 65537-7 Pa. $6.95

**FIFTY CHALLENGING PROBLEMS IN PROBABILITY WITH SOLUTIONS, Frederick Mosteller.** Remarkable puzzlers, graded in difficulty, illustrate elementary and advanced aspects of probability. Detailed solutions. 88pp. 5⅜ × 8½.

65355-2 Pa. $3.95

**EXPERIMENTS IN TOPOLOGY, Stephen Barr.** Classic, lively explanation of one of the byways of mathematics. Klein bottles, Moebius strips, projective planes, map coloring, problem of the Koenigsberg bridges, much more, described with clarity and wit. 43 figures. 210pp. 5⅜ × 8½. 25933-1 Pa. $5.95

**RELATIVITY IN ILLUSTRATIONS, Jacob T. Schwartz.** Clear nontechnical treatment makes relativity more accessible than ever before. Over 60 drawings illustrate concepts more clearly than text alone. Only high school geometry needed. Bibliography. 128pp. 6⅛ × 9¼. 25965-X Pa. $5.95

**AN INTRODUCTION TO ORDINARY DIFFERENTIAL EQUATIONS, Earl A. Coddington.** A thorough and systematic first course in elementary differential equations for undergraduates in mathematics and science, with many exercises and problems (with answers). Index. 304pp. 5⅜ × 8½. 65942-9 Pa. $7.95

**FOURIER SERIES AND ORTHOGONAL FUNCTIONS, Harry F. Davis.** An incisive text combining theory and practical example to introduce Fourier series, orthogonal functions and applications of the Fourier method to boundary-value problems. 570 exercises. Answers and notes. 416pp. 5⅜ × 8½. 65973-9 Pa. $9.95

**THE THEORY OF BRANCHING PROCESSES, Theodore E. Harris.** First systematic, comprehensive treatment of branching (i.e. multiplicative) processes and their applications. Galton-Watson model, Markov branching processes, electron-photon cascade, many other topics. Rigorous proofs. Bibliography. 240pp. 5⅜ × 8½. 65952-6 Pa. $6.95

**AN INTRODUCTION TO ALGEBRAIC STRUCTURES, Joseph Landin.** Superb self-contained text covers "abstract algebra": sets and numbers, theory of groups, theory of rings, much more. Numerous well-chosen examples, exercises. 247pp. 5⅜ × 8½. 65940-2 Pa. $6.95

---

*Prices subject to change without notice.*
Available at your book dealer or write for free Mathematics and Science Catalog to Dept. GI, Dover Publications, Inc., 31 East 2nd St., Mineola, N.Y. 11501. Dover publishes more than 175 books each year on science, elementary and advanced mathematics, biology, music, art, literature, history, social sciences and other areas.